Mineral and Electrolyte Metabolism

KARGER

Printed in Switzerland on acid-free paper by Boehm Hutter AG, Reinach

Appears bimonthly: 1 volume per year (6 issues)

Guidelines for Authors

Mineral and Electrolyte Metabolism

Submission

'Mineral and Electrolyte Metabolism' publishes original articles on both clinical and experimental research covering the physiology and disorders of mineral and electrolyte metabolism. Only papers written in English are considered and should be sent to:

Shaul G. Massry, MD
University of Southern California
School of Medicine
2025 Zonal Avenue
Los Angeles, CA 90033 (USA)

Manuscripts should be submitted in triplicate (with three sets of illustrations of which one is an original), typewritten double-spaced on one side of the paper, with a wide margin.

Conditions

All manuscripts are subject to editorial review. Manuscripts are received with the explicit understanding that they are not under simultaneous consideration by any other publication. Submission of an article for publication implies the transfer of the copyright from the author to the publisher upon acceptance. Accepted papers become the permanent property of 'Mineral and Electrolyte Metabolism' and may not be reproduced by any means, in whole or in part, without the written consent of the publisher. It is the author's responsibility to obtain permission to reproduce illustrations, tables, etc. from other publications.

Arrangement

Title page: The first page of each paper should indicate the title (main title underlined), the authors' names, and the institute where the work was conducted. A short title for use as running head as well as the full address of the author to whom correspondence should be sent are also required.

Full address: The exact postal address complete with postal code must be given at the bottom of the title page. Please also supply phone and fax numbers, as well as your e-mail address.

Key words: For indexing purposes, a list of 3–10 key words in English is essential.

Abstract: Each paper needs an abstract of up to 200 words. It should be organized as follows:

Background/Aims:	What is the major problem that prompted the study?
Methods:	How was the study done?
Results:	Most important findings?
Conclusion:	Most important conclusion?

Small type: Paragraphs which can or must be set in smaller type (case histories, test methods, etc.) should be indicated with a 'p' (petit) in the margin on the left-hand side.

Footnotes: Avoid footnotes. When essential, they are numbered consecutively and typed at the foot of the appropriate page.

Tables and illustrations: Tables and illustrations (both numbered in Arabic numerals) should be prepared on separate sheets. Tables require a heading and figures a legend, also prepared on a separate sheet. For the reproduction of illustrations, only good drawings and original photographs can be accepted; negatives or photocopies cannot be used. When possible, group several illustrations on one block for reproduction (max. size 181 × 223 mm) or provide crop marks. On the back of each illustration, indicate its number, the author's name, and 'top'.

Color illustrations: Up to 6 color illustrations per page can be integrated with the text at the reduced price of CHF 600.– / USD 545.00 per page. Color illustrations are reproduced at the author's expense.

References: In the text identify references by Arabic numerals [in square brackets]. Material submitted for publication but not yet accepted should be noted as 'unpublished data' and not be included in the reference list. The list of references should include only those publications which are cited in the text. Do not alphabetize; number references in the order in which they are first mentioned in the text. The surnames of the authors followed by initials should be given. There should be no punctuation other than a comma to separate the authors. Cite all authors, 'et al' is not sufficient. Abbreviate journal names according to the Index Medicus system. (Also see International Committee of Medical Journal Editors: Uniform requirements for manuscripts submitted to biomedical journals. Br Med J 1988;296:401–405.)

Examples
(a) Papers published in periodicals: Sun J, Koto H, Chung KF: Interaction of ozone and allergen challenges on bronchial responsiveness and inflammation in sensitised guinea pigs. Int Arch Allergy Immunol 1997;112:191–195.
(b) Monographs: Matthews DE, Farewell VT: Using and Understanding Medical Statistics, ed 3, revised. Basel, Karger, 1996.
(c) Edited books: Parren PWHI, Burton DR: Antibodies against HIV-1 from phage display libraries: Mapping of an immune response and progress towards antiviral immunotherapy; in Capra JD (ed): Antibody Engineering. Chem Immunol. Basel, Karger, 1997, vol 65, pp 18–56.

Further details: Authors may consult our leaflet 'Rules for the Preparation of Manuscripts', available on request from the Publisher, S. Karger AG, P.O. Box, CH–4009 Basel (Switzerland).

Page charges

A charge of USD 50.00 / CHF 95.– per page will be levied for the first 4 printed pages of an article. Any additional page will be charged to the author at USD 196.00 / CHF 245.–. The allotted size of a paper is equal to approx. 11 manuscript pages (including tables, illustrations and references).

Galley proofs

Unless indicated otherwise, galley proofs are sent to the first-named author and should be returned with the least possible delay. Alterations made in galley proofs, other than the correction of printer's errors, are charged to the author. No page proofs are supplied.

Reprints

Order forms and a price list are sent with the galley proofs. Orders submitted after the issue is printed are subject to considerably higher prices. Allow five weeks from date of publication for delivery of reprints.

KARGER

E-Mail karger@karger.ch
Fax + 41 61 306 12 34
http://www.karger.ch

The Guidelines for Authors are available at:
http://www.karger.ch/journals/mem/meminst.htm

Mineral and Electrolyte Metabolism

General Information

Publication data: 'Mineral and Electrolyte Metabolism' is published 6 times annually. Volume 23 with 6 issues appears in 1997.

Subscription rates: Subscriptions run for a full calendar year. Prices are given per volume.
Personal subscription:
Subscription + postage and handling
CHF 235.60 + CHF 23.50 Europe
CHF 235.60 + CHF 35.50 Overseas
USD 196.40 + USD 23.50
(Must be in the name of, billed to, and paid by an individual. Order must be marked 'personal subscription'.)
Institutional subscription:
Subscription + postage and handling
CHF 673.– + CHF 29.40 Europe
CHF 673.– + CHF 44.40 Overseas
USD 561.00 + USD 29.40
Airmail postage: CHF 36.–, USD 30.00.
(Extra per volume.)

Microforms: Please direct all enquiries and orders to: University Microfilms Inc.
300 North Zeeb Road, P.O. Box 1346
Ann Arbor, MI 48106-1346 (USA)

Subscription orders: Orders can be placed at agencies, bookstores, directly with the Publisher
S. Karger AG, P.O. Box,
CH–4009 Basel (Switzerland)
or with any of the following distributors:
Deutschland: S. Karger GmbH, Postfach
D–79095 Freiburg
France: Librairie Luginbühl
36, bd de Latour-Maubourg, F–75007 Paris
Great Britain, Ireland: S. Karger Publishers
11 Ravenscroft Road
Beckenham, Kent BR3 4TP
USA: S. Karger Publishers, Inc.
26 West Avon Road, P.O. Box 529
Farmington, CT 06085
India, Bangladesh, Sri Lanka:
S. Karger India Subscription Agency
B–5/132 Safdarjung Enclave
New Delhi-110029
Japan: Katakura Libri, Inc.
Take 1 Bldg, 2nd Floor, 4-5-7 Yushima
Bunkyo-ku, Tokyo 113

Change of address: Both old and new address should be sent to the subscription source.

Single issues and back volumes: Information can be obtained through the Publisher.

Advertising: Correspondence and rate requests should be addressed to the Publisher.

Bibliographic indices: This journal is regularly listed in bibliographic services, including *Current Contents®*.

Book Reviews: Books are accepted for review by special agreement.

Copying: This journal has been registered with the Copyright Clearance Center (CCC), as indicated by the code appearing on the first page of each article.
For readers in the US, this code signals consent for copying of articles for personal or internal use, or for the personal or internal use of specific clients, provided that the stated fee is paid per copy directly to
Copyright Clearance Center Inc.
222 Rosewood Drive
Danvers, MA 01923 (USA)
A copy of the first page of the article must accompany payment. Consent does not extend to copying for general distribution, for promotion, for creating new works, or for resale. In these cases, specific written permission must be obtained from the copyright owner,
S. Karger AG,
P.O. Box, CH–4009 Basel (Switzerland).

KARGER

E-Mail karger@karger.ch
Fax + 41 61 306 12 34
http://www.karger.ch

The Journal Home Page is available at:
http://www.karger.ch/journals/mem/memdes.htm

PD SOLUTIONS

PORTFOLIO

Baxter continues to lead the way in PD therapy improvements by providing the largest range of quality solutions that allow physicians to further individualize the patient prescription.

Nutrineal® is a part of the Baxter PD solutions portfolio:

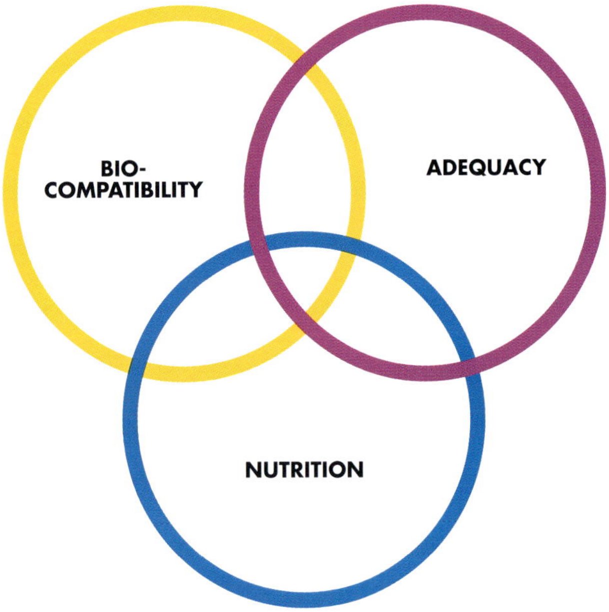

- *The near physiologic pH of Nutrineal® enhances solution biocompatibility, as measured by in vitro function and viablilty of polymorpholeukocytes derived from the drained effluent.*[1]
- *With 1.1% aminoacids replacing glucose as an osmotic agent, Nutrineal® can provide nutritional supplementation for malnourished PD patients.*[2]
- *Nutrineal® addresses the issues of adequacy and nutrition by offering effective dialysis treatment and nutrition in one bag.*

NUTRINEAL® PD4 SOLUTION

ABBREVIATED PRESCRIBING INFORMATION

COMPOSITION

L-Tirosine	0,30g/l	L-Leucine	1,02g/l
L-Tryptophan	0,27g/l	L-Lysine	0,76g/l
L-Phenylalanine	0,57g/l	L-Histidine	0,71g/l
L-Threonine	0,65g/l	L-Arginine	1,07g/l
L-Serine	0,51g/l	Calcium Chloride	
L-Proline	0,60g/l	($2H_2O$)	0,184g/l
Glycine	0,51g/l	Magnesium Chloride	
L-Alanine	0,95g/l	($6H_2O$)	0,051g/l
L-Valine	0,39g/l	Sodium Lactate	4,480g/l
L-Methionine	0,85g/l	Sodium Chloride	5,380g/l
L-Isoleucine	0,85g/l		

USE

Nutritional supplement for malnourished renal patients (albumin concentration lower than 35 g/l) being maintained on peritoneal dialysis.

DOSAGE AND ADMINISTRATION

For intraperitoneal administration only. Dosage schedule, frequency of treatment and duration of dwell should be selected by the prescribing physician according to the patient's clinical status. Treatment should be re-evaluated after 3 months if there is no clinical or biochemical improvement in the status of the patient. Recommended dietary allowance of protein is approximately 1,2-1,3 g/kg for dialysed adults and 2-3 g/kg for infants. Higher amounts may be necessary in several catabolic states.
For information, a 2 litre bag of Nutrineal® with 1,1% amino acids provides 22 g of amino acids which is equivalent to 0,30g/kg/day for a 70 kg adult.

CONTRAINDICATIONS

Some clinical conditions such as recent abdominal surgery and gastrointestinal disturbances have to be considered as contraindications.
In each case the benefits of treatment must be weighted against the possible complications.
Uraemia higher than 38 mmol/litre (2,4 g/l).

PRECAUTIONS AND WARNING

The safety and efficacy of this solution have not been assessed in children. Caution is recommended in cases of uncorrected acidosis, severe hepatic failure, hyperammonaemia or hypersensitivity to one or more amino acids. Fluid and electrolyte balance, acid base balance and changes in nitrogen balance should be monitored. Home patients on continuous ambulatory peritoneal dialysis should be trained in specialised centres. Medication doses should be adapted. Kaliaemia should be monitored in patients treated with digitalis. Diabetic patients require careful monitoring of insulin dosage or hypoglycaemic treatment. There is no evidence on the drug's safety in human pregnancy and animal studies have not been conducted. Avoid in pregnancy unless use is considered essential.

ADVERSE REACTIONS

Adverse reactions are those of peritoneal dialysis, in particular: peritoneal infections; electrolyte and fluid imbalances; metabolic acidosis; hyperazotaemia.

PHARMACEUTICAL PARTICULARS

Check compatibility before mixing additives. There is no incompatibility between Nutrineal® and the addition of Heparin or Insulin. Currently, no information is available on the stability of Nutrineal® with aminoglycosides. Nutrineal® has a 2 year shelf-life. Store below 30°C and protect from light.

OTHER INFORMATION

Nutrineal® is available as:
2,5 litre bag - 2,0 litre bag

FURTHER INFORMATION IS AVAILABLE ON REQUEST.

F 97106

Dialysis and Nutrition in one bag

- Easy to administer nutritional supplement
- Improves nutrition in malnourished patients[3,4]
- Improved physiological profile
- Glucose free
- No phosphorus

NUTRINEAL®

	NOT DIFFERENT	DIFFERENT
IMPORTANT		*Nutrition* *No phosphorus* *Easy to use* *Glucose free* *Higher pH*
NOT IMPORTANT		

REFERENCES

1. Brulez H.F., Heezius D.C., de Fijter C.W., Oe L.P., Verhoef J., Verbrugh H.A.: In vitro compatibility of a 1,1% amino acid containing peritoneal dialysis fluid with Phagocyte function. Adv. Perit. Dial. 1994; 10: 241-244
2. Jones M.R. et al "Amino Acid Solutions for CAPD: Rationale and Clinical Experience". Mineral and Electrolyte Metabolism, 1992; 18: 309-315
3. Young G.A., Kopple J.D., Lindholm B., et al., "Nutritional assessment of continuous ambulatory peritoneal dialysis patient : An international Study", Am. J. Kidney Dis. 1991; 38: 487-494
4. Jones M.R. et al. "Amino Acid Solutions for CAPD: Rationale and Clinical Experience". Mineral and Electrolyte Metabolism, 1992; 18: 309-315

Peritoneal Dialysis

Baxter

F 97107

Mineral and Electrolyte Metabolism

Nutrition and Metabolism in Renal Disease

Proceedings of the 8th International Congress on Renal Nutrition and Metabolism

Naples, October 9–12, 1996

Editors
Joel D. Kopple, Torrance, Calif., USA
Natale G. De Santo, Naples, Italy
Giovambattista Capasso, Naples, Italy
Shaul G. Massry, Los Angeles, Calif., USA

63 figures, 1 in color, and 44 tables, 1997

KARGER

Basel · Freiburg · Paris · London · New York ·
New Delhi · Bangkok · Singapore · Tokyo · Sydney

We gratefully acknowledge the generous financial support from

Baxter Healthcare Corporation
Sigma Tau Pharmaceuticals, Inc.
Second University of Naples

KARGER

S. Karger
Medical and Scientific Publishers
Basel · Freiburg · Paris · London
New York · New Delhi · Bangkok
Singapore · Tokyo · Sydney

Drug Dosage
The authors and the publisher have exerted every effort to ensure that drug selection and dosage set forth in this text are in accord with current recommendations and practice at the time of publication. However, in view of ongoing research, changes in government regulations, and the constant flow of information relating to drug therapy and drug reactions, the reader is urged to check the package insert for each drug for any change in indications and dosage and for added warnings and precautions. This is particularly important when the recommended agent is a new and/or infrequently employed drug.

Printed in Switzerland on acid-free paper by
Boehm-Hutter AG, Reinach, BL
ISBN 3–8055–6586–0

Contents Vol. 23, No. 3–6, 1997

Mineral and Electrolyte Metabolism

KARGER

E-Mail karger@karger.ch
Fax+ 41 61 306 12 34
http://www.karger.ch

The list of contents is available at:
http://www.karger.ch/journals/mem/memcont.htm

Mineral and Electrolyte Metabolism

Miner Electrolyte Metab 1997;23:129

Introduction

We are pleased to present the Proceedings of the Eighth International Congress on Renal Nutrition and Metabolism. This meeting, which was held in Naples, Italy, 9–12 October. 1996, was the eighth in a series of conferences on nutritional and metabolic aspects of renal disease and renal failure. Professor Natale G. De Santo chaired the local organizing committee.

This was the fourth such meeting since the establishment of the International Society for Renal Nutrition and Metabolism in 1988. As with past conferences, this meeting was attended by several hundred physicians, scientists, renal nutritionists, and other health care workers from many countries and several continents. It was indeed an international congress. The topics addressed ranged from cell and molecular biology, physiology, biochemistry, and epidemiology to clinical interventions and large-scale clinical trials.

These meetings as well as the International Society for Renal Nutrition and Metabolism have continued to enjoy support from individuals worldwide. Such support reflects the perception among researchers and health care workers that nutritional and metabolic disorders are common in patients with renal disease and acute or chronic renal failure and that treatment may reduce their morbidity and mortality.

The current officers of the Society are as follows: Walter H. Hörl, MD, PhD, President; Shaul G. Massry, MD, Immediate Past President; William Mitch, MD, Vice President; George Kaysen, MD, Secretary-Treasurer; Giuseppe d'Amico, MD, Natale G. De Santo, MD, Timothy Goodship, MD, Ralph Rabkin, MD, Marianne Vennegor, RD, John Walls, MD, and Marsha Wolfson, MD, Councilors, and Joel D. Kopple, MD, Councilor ex officio.

The manuscripts for the dietitians' sessions of this past congress are published in the *Journal of Renal Nutrition.* As the papers in the present proceedings indicate, many challenges still confront the health care worker and scientist, but the potential for improving the clinical status of the patient with renal disease and renal failure remain bright. We invite you to join the Society, to attend its biannual conferences, and to participate in the efforts of the Society to improve the knowledge and methods of treating the nutritional and metabolic disorders of patients with acute or chronic renal disease.

We are pleased to announce that the Ninth International Congress on Renal Nutrition and Metabolism will be held from August 31–September 3, 1998, in Vienna, Austria, under the chairmanship of Professor Walter H. Hörl. We encourage you to contact Professor Hörl for information concerning the next congress.

Joel D. Kopple, MD, Los Angeles, Calif., USA
Natale G. De Santo, MD, Naples, Italy
Giovambattista Capasso, MD, Naples, Italy
Shaul G. Massry, MD, Los Angeles, Calif., USA

KARGER

E-Mail karger@karger.ch
Fax +41 61 306 12 34
http://www.karger.ch

Mineral and Electrolyte Metabolism

Miner Electrolyte Metab 1997;23:130–134

Steven D. Clarke

Department of Human Ecology,
The University of Texas,
Austin, Tex., USA

Nutrient Regulation of Gene Expression: A Methodological Strategy

Key Words
Fatty acid synthase
γ-Glutamylcysteine synthase
Nutrients
Transcription

Abstract
Nutrients should be viewed as the earliest of the hormone signals which allowed an organism to respond to changes in its nutrient environment. Defining nutrient function at the nuclear level will permit us to understand how our dietary environment is related to the development of nutritionally related pathophysiologies such as diabetes, heart disease, and cancer. Moreover, through the application of molecular techniques, we will begin to understand how specific nutrients govern the developmental pattern of specific organs such as the kidney. In this review, the fatty acid synthase gene is employed as a model to demonstrate how one sequentially approaches questions pertaining to the regulation of gene expression by a nutrient, and the article presents a nuclear explanation for how dietary polyunsaturated fats decrease blood triglycerides. More importantly, studies of this nature provide a basis for screening genetically susceptible populations and information that will allow the nutritionists of the 21st century to customize a diet for patients at risk.

Introduction

During their primordial development, unicellular organisms had to respond to numerous environmental challenges. A particular need for any life form was and continues to be its ability to respond to the nutrient supply and increase/decrease the expression of genes essential to nutrient synthesis and assimilation. Thus, nutrient availability became an important and essential modulator of cellular gene expression. More important, nutrients should be viewed as the earliest hormone signals which allowed an organism to respond to changes in its nutrient environment. The regulation of gene expression by specific nutrients is not restricted to the unicellular world: nutrient control of genes involved in cell differentiation, organ development, and cellular metabolism continues to be a vital component of mammalian biology. Examples of such essential biological activities include retinoid modulation of cell differentiation, vitamin D control of gene expression, iron regulation of iron transport and storage, and fatty acid control of enzymes in lipid synthesis and degradation [1–4]. It is these actions of nutrients which may well explain the beneficial as well as detrimental role that nutrients play in nutritionally related pathophysiologies such as diabetes, kidney disease, cancer, and heart disease.

Nutrient function at the genetic level and the role that nutrient control of gene expression plays in health and disease have only recently begun to be elucidated with the application of molecular techniques. The purpose of this review is to describe the application of these techniques as I explore how our understanding of the fatty acid regulation of lipogenic gene expression evolved over the past 20 years.

Suppression of Lipogenesis by Dietary Fat

Studies in the early 1960s revealed that feeding a high-fat diet suppressed the conversion of glucose to fatty acids [5]. Moreover, such suppression could be observed within a short period of time after ingestion of the fat [5]. Initial mechanistic explanations focused on the metabolite (i.e., allosteric) control of key regulatory enzymes such as phosphofructokinase and acetyl-CoA carboxylase [6–8]. However, as monospecific antibodies to the various lipogenic enzymes were developed, data began to appear which indicated that the amount of enzyme protein could be reduced by high-fat diets, and that these proteins increased in response to feeding carbohydrate [9–14].

In 1965, Allmann and Gibson [9] asked a very significant but simple question: does the inhibition of lipogenesis by dietary fat vary with the type of fatty acid in the fat. Somewhat surprisingly they found that when palmitate and linoleate were compared, only linoleate could block the rise in hepatic lipogenesis associated with carbohydrate feeding. Moreover, they demonstrated that linoleate appeared to inhibit lipogenesis by decreasing the amount of lipogenic enzyme protein [9]. During the early 1970s, several different

KARGER
E-Mail karger@karger.ch
Fax +41 61 306 12 34
http://www.karger.ch

0378–0392/97/0236–0130 $12.00/0

Steven D. Clarke
Department of Human Ecology, 115 GEA, The University of Texas
Austin, TX 78712 (USA)
Tel. +1 512 232 1537, fax +1 512 471 5630
E-Mail stevedclarke@mail.utexas.edu

fatty acids were evaluated for their ability to reduce the activity of enzymes such as fatty acid synthase, malic enzyme, glucose-6-phosphate dehydrogenase, and glucokinase [10–14]. The consistent finding was that only polyunsaturated fatty acids (PUFAs) of the n–6 and n–3 families were able to block the induction of the lipogenic enzyme activity that resulted from feeding a high-carbohydrate diet.

While these findings potentially explained how polyunsaturated fats could reduce the level of circulating triglycerides in people fed high-carbohydrate diets, they provided little in the way of a mechanistic understanding of the action of the fatty acids. The next mechanistic step was to ascertain if PUFAs inhibited the synthesis of lipogenic enzymes or accelerated the degradation of these proteins. Using the technique of immunoprecipitation, Flick et al. [11] demonstrated that dietary safflower oil, which is rich in linoleic acid, suppressed the incorporation of radiolabelled leucine into hepatic fatty acid synthase. Similarly, Schwartz and Abraham [12, 13] found that dietary PUFAs inhibited leucine incorporation into malic enzyme as well as fatty acid synthase. Through the application of immunotechniques, these investigators were able to unequivocally demonstrate that PUFAs inhibited the synthesis of hepatic lipogenic enzymes. Additional studies with isolated hepatocytes in culture revealed that the suppression of lipogenic and glycolytic enzymes by PUFAs did not require peripheral tissue metabolism of the fatty acid, nor was it related to altered release of insulin, glucagon, or gut hormones. However, PUFAs were found to suppress the induction of lipogenic enzymes by insulin and/or glucocorticoids [14].

Is the Suppression of Enzyme Synthesis the Result of Decreased mRNA Abundance or Suppressed mRNA Translation?

In the early 1980s, methodologies became available to produce complementary DNAs (cDNAs) to specific transcripts. At this time Porter's group [15] identified the cDNA for rat liver fatty acid synthase. They accomplished this by preparing cDNAs to hepatic mRNA extracted from rats fed a high-carbohydrate diet and inserted these cDNAs into vectors which were used to transform *Escherichia coli* [15]. These constructs were prepared in a way which allowed the bacteria to transcribe and translate the cDNAs into peptides. Each bacteria contained one transcript. *E. coli* colonies which produced peptides specific to fatty acid synthase were identified by immunoreactivity with a monospecific fatty acid synthase antibody.

Using this cDNA in a Northern analysis, Clarke and associates [16, 17] established that dietary PUFAs reduced the abundance of hepatic fatty acid synthase mRNA by 90%. This was accomplished by feeding rats or mice a high-glucose diet that contained either 5% 18:1 n–9 or 5% 18:2 n–6. After 7 days of adaptation to the diet, total hepatic RNA was extracted and size fractionated by agarose gel electrophoresis. The RNA was subsequently transferred to a high-affinity paper (e.g., nitrocellulose), the blot was blocked as in Western analysis, and the mRNA transcript for fatty acid synthase was identified by hybridization with the radiolabelled cDNA for fatty acid. Additional studies have revealed that the decrease in mRNA abundance elicited by dietary PUFAs occurs very quickly (i.e., <3 h), and that PUFAs decrease the amount of mRNA for several lipogenic and glycolytic enzymes including glucose-6-phosphate dehydrogenase, stearoyl-CoA desaturase, malic enzyme, pyruvate kinase, and glucokinase [16–20].

Nutrient regulation of gene expression involves more than simply the macronutrients or the steroid-like vitamins (e.g., vitamin A), but also extends to several micronutrients including copper [21] and vitamin B_6 [22]. As an example, we have discovered that the activity of renal γ-glutamylcysteine synthetase, the rate-limiting enzyme in glutathione synthesis, is decreased by a dietary copper deficiency [21]. Moreover, Northern analysis has revealed that the 50% decrease in enzyme activity was paralleled by a 50% reduction in γ-glutamylcysteine synthetase mRNA abundance [21]. Thus, it would appear that the rate of glutathione synthesis in the liver may be dictated by the level of γ-glutamylcysteine synthetase mRNA in the kidney.

Transcriptional Regulation of a Gene

Changes in the tissue content of a transcript such as those observed for hepatic fatty acid synthase or renal γ-glutamylcysteine synthetase are the consequence of either altered rates of mRNA degradation or synthesis. Nuclear run-on assays allow us to ascertain which of these occurs [23, 24]. In essence, nuclei are isolated from the tissue of interest after a specific dietary or hormonal treatment. These nuclei are subsequently incubated with radiolabelled nucleotides, and the specific nascent transcript is 'retrieved' by complementing the radiolabelled transcript with its specific cDNA. In utilizing this technique, one should keep in mind that if the nuclear run-on assay fails to detect a treatment effect on gene transcription, the conclusion may not always be that gene expression is governed by mechanisms of mRNA stability. It is very feasible that the clone is hybridizing to a sequence in a second gene that is not the gene of interest. Thus nuclear run-on assays should be conducted with multiple cDNAs that represent various regions throughout the gene of interest [23].

Again, using hepatic fatty acid synthase expression as the example, we have demonstrated that the rise in fatty acid synthase mRNA associated with high-carbohydrate diets, and the fall in this mRNA following feeding of polyunsaturated fats, results entirely from the induction and suppression of fatty acid synthase gene transcription [24]. Similarly, we recently reported that the 50% decrease in kidney γ-glutamylcysteine synthetase activity and mRNA that occurs during copper deficiency results from an inhibition of gene transcription [21]. One cannot always assume that a change in the rate of transcription for a particular gene in one tissue means that all tissues regulate that same gene by altering rates of gene transcription. For example, the several-fold rise in fatty acid synthase mRNA that occurs during the differentiation of a fat cell is the product of a lower rate of transcript degradation rather than an induction of fatty acid synthase gene transcription [25]. However, fluctuations in fatty acid synthase mRNA due to changes in dietary or hormonal status in a mature fat cell are entirely due to altered rates of transcription [25]. This point also holds true when one is comparing data between cell models and intact animals because the mechanisms by which a nutrient modulates gene expression in a cell model (e.g., HepG2 liver cells) often are not the same mechanisms employed by the intact animal. As an example, the in vivo increase in hepatic fatty acid synthase mRNA associated with feeding a carbohydrate diet is entirely due to an induction of gene transcription [16], but in the HepG2 hepatoma model, glucose increases fatty acid synthase mRNA via changes in mRNA stability [24, 26]. Thus, it is extremely important to demonstrate that if a cell model is being used to study nutrient regulation of gene expression, the cell model must respond to nutrient manipulations in the same fashion as the intact animal. Moreover, conclusions regarding nutrient effects on mRNA synthesis versus degradation must be based upon data gathered for the tissue and cell type of interest, i.e., conclusions regarding gene expression from one tissue may not extend to all tissue types. Unfortunately this point is often overlooked and is a key flaw in many studies.

If the nuclear run-on assay indicates that a particular treatment does not modify the rate of transcript synthesis but rather modifies the rate of mRNA turnover, then the next step is to identify the sequences within the transcript that confer stability, and to identify the proteins involved in regulating transcript stability. A model for these studies is the classic work on iron metabolism and its role in regulating ferritin and transferrin receptor translation and mRNA stability [3, 27, 28]. On the other hand, if a nutrient or hormone modulates gene transcription, then the next logical steps become (a) mapping the affected DNA sequence in the gene, (b) identifying the proteins which interact with the sequence(s) to elicit the effect, and (c) elucidating the intracellular signaling mechanisms by which the affected proteins are controlled. There are numerous examples where this has been done in defining the regulation of gene transcription by a variety of hormones, but the mechanisms by which specific nutrients regulate gene transcription are less well defined. Certainly, the roles that steroid-like factors such as vitamins A and D play in gene expression have been extensively studied [29, 30], but the mechanisms by which glucose, fatty acids, and micronutrients (e.g., copper, vitamin B_6) govern gene expression are only beginning to be understood [4, 21, 22, 31, 32]. Moreover, the role that specific nutrients play in gene expression during select stages of development and aging, and the role that modifications in gene expression play in the development of nutritionally related pathophysiologies such as renal disease, diabetes, and cancer remain nebulous. However significant progress has been made in these areas through our studies with the glycolytic and lipogenic genes of the liver [4, 14, 16, 19].

In 1990, the entire fatty acid synthase gene, including all introns and exons, the 3′-flanking region, and 2.5 kb of the promoter and 5′-flanking sequence, was cloned. Since then, our goal has been to identify specific sequences in the fatty acid synthase gene which function to regulate the rate of gene transcription in response to hormonal and nutrient manipulations. Mapping these putative elements relies upon a combination of techniques that functionally test the suspected target sequences, and that examine changes in DNA/protein interactions within these sequences. One very useful method is the study of DNAse hypersensitivity sites [33]. In essence, this procedure relies upon the concept that regions within a gene which are intimately involved with the regulation of gene transcription will have an altered structure that renders the area more sensitive to being cleaved by DNAse. These sites can be evaluated by (a) isolating nuclei from an organ or cell model following a specific treatment (e.g., feeding saturated vs. polyunsaturated fats), (b) treating the nuclear DNA with DNAse, (c) size fractionating the DNA, and (d) identifying the specific gene fragments and the pattern of DNAse cutting by hybridizing the size-fractionated fragments with a cDNA specific for the gene of interest (e.g., fatty acid synthase). Once a DNA target region has been localized (e.g., < 500 bp), the impact of a dietary or hormonal treatment on the interactions of nuclear proteins with the DNA sequence may be examined by applying the techniques of footprinting and electromobility shift assays [34–36]. The assumptions of these assays are that the dietary or hormonal treatment has modified specific nuclear proteins, that these modifications can be preserved during the extraction of the proteins from the nucleus, and that the protein interaction with the putative regulatory DNA sequence will display itself as more or less protein binding. Unfortunately, it is not always feasible to preserve such modifications during protein extraction. Thus, techniques such as footprinting are helpful but protein/DNA studies are often not conclusive. Moreover, DNA/protein interaction studies in vitro do not always indicate that such interactions occur in vivo. All conclusions derived from DNA/protein interaction studies must be supported by functional studies (e.g. transfection analysis).

Using the techniques described above, we have found that the fatty acid synthase gene contains at least two hypersensitive sites, and that one of the sites was affected by dietary polyunsaturated fats (D.B. Jump and S.D. Clarke, unpublished data). These apparent regulatory sequences have been examined for functionality using transfection analysis in primary hepatocytes. Functional testing of a DNA sequence that may be involved in the regulation of gene transcription is a relatively simple concept. In essence, the suspected DNA sequence is linked to a basal promoter which in turn is connected to a gene that expresses a protein product that is not native to the cell type of interest, and is easily measured, e.g., luciferase, defective growth hormone, chloramphenicol acetyl transferase (CAT), β-galactosidase. A vector is constructed that carries the DNA construct of interest and the DNA is delivered to cells which respond to the nutrient or hormone in question. Again the key to the study of nutrient regulation of gene expression, and to the mapping of the specific DNA sequences of a gene affected by the nutrient is that the cell type used for transfection must display the same response as the intact animal. The failure to take this point into consideration is a fatal flaw for many nutrient/gene interaction studies. In the case of PUFA control of fatty acid synthase gene transcription, we have found that only hepatocytes in primary culture fulfill this requirement. However, studies mapping the insulin response element have revealed that several cell lines will mimic the in vivo responses, including HepG2 and H4-2E hepatoma cells and 3T3-L1 preadipocytes [36].

The first approach to mapping these elements was to transfect cells with a vector containing the region of −2195 to +65 (+1 is the first nucleotide in the mRNA) from the fatty acid synthase gene attached to the reporter protein (CAT), and to subsequently examine changes in CAT activity as affected by the hormones and nutrients of interest. Surprisingly, neither glucose nor PUFAs inhibited or altered CAT expression. On the other hand, insulin increased CAT expression 2- to 3-fold [36]. Moreover, recent data indicate that growth hormone suppresses insulin-induced CAT activity by 50–70% (T. Etherton and S.D. Clarke, unpublished data). The in vitro induction of CAT activity by insulin is only a small part of the in vivo response for fatty acid synthase [14, 36]. This observation suggested that while insulin may have a specific DNA response element in the region of −2195 to +65 bp, other enhancer response elements must be required to achieve in vivo levels of expression. Very recent cloning has shown that the element required for in vivo rates of transcription is a DNA sequence that binds the glucocorticoid receptor, and this response sequence is located in the region between −5500 bp and −10000 bp (M. Nakamura and S.D. Clarke, unpublished data).

Progressive deletions from the 5′-end of −2195 bp to −120 bp of the fatty acid synthase gene revealed that the insulin response sequence was located at −120 bp [36]. By mutating specific nucleotides in the putative insulin response region, Wang and Sul [37] were able to identify the specific response element and, more importantly, they identified the nuclear protein (USF) that mediated insulin control of the fatty acid synthase gene. Similar studies have also demonstrated that the sterol response binding protein (SREBP) modulates fatty acid synthase gene transcription by binding to the same area as the USF-binding sequence. Moreover, SREBP binding appears to enhance Sp1 binding [38]. Thus, it would appear that hormonally and nutritionally regulated transcription factors modulate gene transcription through a series of cooperative interactions.

Although the DNA region in the fatty acid synthase gene that is responsive to PUFAs has not yet been fully mapped, collaborative work with Jump et al. [39] has been successful in mapping the PUFA response element (PUFA-RE) in both the pyruvate kinase and S14 genes. As described previously for mapping the insulin element, a combination of deletion and mutation analyses was utilized to map the PUFA region. Interestingly, the PUFA-RE in the pyruvate kinase gene mapped to the same DNA sequence as the carbohydrate response element [19], while the PUFA-RE in the S14 gene was over 1000 bp away from the carbohydrate element [39]. Identification of the PUFA-RE and tentative identification of nuclear proteins that interact with this element have revealed that PUFAs do not govern gene transcription via activation of the much acclaimed peroxisome proliferator-activated receptors [34, 40]. A second important aspect of these studies is that they demonstrate that both hormonal and nutrient response sequences can and often are located hundreds of nucleotides apart, and that one does not have the entire gene sequence until it can be demonstrated that the constructed gene can yield in vivo rates of gene transcription. A second point worth noting is that not all nutritionally regulated genes are located in the sequence that is 5′ to the start site of transcription, i.e., DNA response elements can be located in the introns [41]. This point was recently reinforced when the apparent glucose response sequence for fatty acid synthase was found in the first intron of the fatty acid synthase gene [41]. Clearly, the nutritional and hormonal control of gene transcription represents a complex collection of DNA/protein interactions, and the site of these DNA sequences can never be assumed to be closely located to the start site of transcription.

Unequivocal conclusions regarding the in vivo function of a putative DNA response element require that the sequences be placed in an in vivo environment, and that the sequences mediate the expected in vivo outcome in response to hormonal or nutrient manipulation. Such transgenic studies have been conducted with a number of genes including fatty acid synthase [42]. The recent transgenic studies from Sul and associates [42] indicate that the putative insulin response element for fatty acid synthase does in fact respond in the transgenic mouse in a manner comparable to the native gene. However, these studies also verified that the region between −2195 and +67 does not contain the PUFA-RE and, more importantly, that the expression of the transgene is markedly below that of the native fatty acid synthase gene. Thus, even though over 2 kb of the fatty acid synthase promoter have been cloned and studied, many significant regulatory components of the gene have not yet been located.

Clearly from the previous discussions, we have been able to employ a wide array of molecular biological techniques to demonstrate that the changes in the flux of a pathway, e.g., the rate of hepatic fatty acid biosynthesis or the rate of renal glutathione synthesis, are regulated via control of gene transcription. Moreover, by employing cloning and DNA/protein interactions techniques, we have localized the specific DNA sequence affected by a nutrient (e.g., fatty acids), we have redefined the function of select nutrients (e.g., copper), and we have opened new doors to understanding the mechanisms by which nutrients modulate the development of nutritionally related pathophysiologies. These very powerful molecular techniques provide a tool for developing a sequential approach to answering questions of nutrient function at the nuclear level. However, a note of caution, the techniques are only as good as the biological hypotheses that an investigator develops.

References

1 Cooney AJ, Leng X, Tsai SY, O'Malley BWO, Tsai MJ: Multiple mechanisms of chicken ovalbumin upstream promoter transcription factor-dependent repression of transactivation by vitamin D, thyroid hormone, and retinoic acid receptors. J Biol Chem 1993;268:4152–4160.

2 Means AL, Gudas LJ: The roles of retinoids in vertebrate development. Annu Rev Biochem 1995; 64:201–233.

3 Klausner RD, Harford JB: Cis-trans models for post-transcriptional gene regulation. Science 1989; 246:870–873.

4 Clarke SD, Jump DB: Dietary polyunsaturated fatty acid regulation of gene transcription. Annu Rev Nutr 1994;14:83–98.

5 Hill R, Webster WW, Linazasoro JM, Chaikoff IL: Time occurrence of changes in the liver's capacity to utilize acetate for fatty acid and cholesterol synthesis after fat feeding. J Lipid Res 1960;1:150–153.

6 Shrago E, Ball M, Sul HS, Baquer NZ, McLean P: Interrelationship in the regulation of pyruvate dehydrogenase and adenine nucleotide translocase by palmitoyl-CoA in isolated mitochondria. Eur J Biochem 1977;75:83–89.

7 Clarke SD, Jump DB: Regulation of gene transcription by polyunsaturated fatty acids. Prog Lipid Res 1993;32:139–149.

8 Goodridge AG: Regulation of fatty acid synthesis in isolated hepatocytes: Evidence for a physiological role for long chain fatty acyl coenzyme A and citrate. J Biol Chem 1973;248:4318–4326.

9 Allmann DW, Gibson DM: Fatty acid synthesis during early linoleic acid deficiency in the mouse. J Lipid Res 1965;6:51–62.

10 Herzberg GR: The influence of dietary fatty acid composition on lipogenesis. Adv Nutr Res 1983;5: 221–239.

11 Flick PK, Chen JH, Vagelos PR: Effect of dietary linoleate on synthesis and degradation of fatty acid synthetase from rat liver. J Biol Chem 1977;252: 4242–4248.

12 Schwartz RS, Abraham S: Effect of dietary fat on the activity, content, rates of synthesis and degradation and translation of mRNA coding malic enzyme. Arch Biochem Biophys 1983;221:206–215.

13 Schwartz RS, Abraham S: Effect of dietary polyunsaturated fatty acids on the activity and content of fatty acid synthetase in mouse liver. Biochim Biophys Acta 1982;711:316–322.

14 Clarke SD, Jump D: Fatty acid regulation of gene expression: A unique role for polyunsaturated fats; in Berdanier CD, Hargrove JL (eds): Nutrition and Gene Expression. Boca Raton, CRC, 1992, pp 227–376.

15 Neprogroff CM, Adachi K, Yan C, Porter JW: Cloning of DNA complementary to rat liver fatty acid synthetase mRNA. Eur J Biochem 1984;140: 441–445.

16 Clarke SD, Armstrong MK, Jump DB: Dietary polyunsaturated fats uniquely suppress rat liver fatty acid synthase and S14 mRNA content. J Nutr 1990;120:225–231.

17 Jump DB, Clarke SD, Thelen A, Liimatta M: Coordinate regulation of glycolytic and lipogenic gene expression by polyunsaturated fatty acids. J Lipid Res 1994;35:1076–1084.

18 Tomlinson JE, Nakayama R, Holten D: Repression of pentose phosphate pathway dehydrogenase synthesis and mRNA by dietary fat in rats. J Nutr 1988;118:408–414.

19 Liimatta M, Towle HC, Clarke SD, Jump DB: Fatty acid regulation of hepatic pyruvate kinase gene transcription. Mol Endocrinol 1994;8:1147–1153.

20 Ntambi JM: Dietary regulation of stearoyl-CoA desaturase 1 gene expression in mouse liver. J Biol Chem 1992;267:10925–10930.

21 Kim S, Wilson JJ, Allen KGD, Clarke SD: Dietary copper deficiency suppresses renal gamma-glutamylcysteine synthetase. Biochim Biophys Acta 1996;1313:89–94.

22 Sampson DA, Harrison SC, Clarke SD, Yan X: Dietary protein quality alters ornithine decarboxylase activity but not vitamin B-6 nutritional status in rats. J Nutr 1995;125:2199–2287.

23 Goodridge AG, Crish JF, Fantozzi DA, Glynias MJ, Hillgartner BF, Klautky SA, Ma I, Mitchell DA, Salati LM: Regulation of avian genes involved in fatty acid synthesis. J Anim Sci 1990;66(suppl): 49–66.

24 Blake WL, Clarke SD: Suppression of hepatic fatty acid synthase and S14 gene transcription by dietary polyunsaturated fat. J Nutr 1990;120:1727–1729.

25 Moustaid N, Sul HS: Regulation of expression of the fatty acid synthase gene in 3T3-L1 cells by differentiation and triiodothyronine. J Biol Chem 1991;266:18550–18554.

26 Semenkovich CF, Coleman T, Goforth R: Physiologic concentrations of glucose regulate fatty acid synthase activity in Hep G2 cells by mediating fatty acid synthase mRNA stability. J Biol Chem 1993; 268:6961–6970.

27 Guo B, Phillips JD, Yu Y, Leibold EA: Iron regulates the intracellular degradation of iron regulatory protein 2 by the proteasome. J Biol Chem 1995; 270:21645–21651.

28 Leibold EA, Guo B: Iron-dependent regulation of ferritin and transferrin receptor expression by the iron-responsive element binding protein. Annu Rev Nutr 1992;12:345–368.

29 Liu SM, Koszewski N, Lupez M, Malluche HH, Olivera A, Russell J: Characterization of a response element in the 5′-flanking region of the avian (chicken) PTH gene that mediates negative regulation of gene transcription by 1,25-dihydroxyvitamin D3 and binds the vitamin D3 receptor. Mol Endocrinol 1996;10:206–215.

30 Minucci S, Ozato K: Retinoid receptors in transcriptional regulation. Curr Opin Genet Dev 1996; 6:567–574.

31 Wilson J, Kim S, Allen KGD, Baillie R, Clarke SD: Hepatic fatty acid synthase gene transcription is induced by dietary copper deficiency. Am J Physiol 1997;212:1174–1179.

32 Chin E, Bondy CA: Dietary protein-induced renal growth: Correlation between renal IGF-1 synthesis and hyperplasia. Am J Physiol 1994;266:1037–1045.

33 Jump DB, Bell A, Santiago V: Thyroid hormone and dietary carbohydrate interact to regulate rat liver S15 gene transcription and chromatin structure. J Biol Chem 1990;265:3474–3478.

34 Jump DB, Ren B, Clarke SD, Thelen A: Effects of fatty acids on hepatic gene expression. Leukotrienes Essen Fatty Acids 1995;52:107–111.

35 Rongnoparut P, Verdon CP, Gehnrich SC, Sul HS: Isolation and characterization of the transcriptionally regulated mouse liver (B-type) phosphofructokinase gene and its promoter. J Biol Chem 1991;266:8086–8091.

36 Moustaid N, Sakamoto K, Clarke S, Beyer RS, Sul HS: Regulation of fatty acid synthase transcription sequences that confer a positive insulin effect and differentiation dependent expression in 3T3-L1 preadipocytes are present in the 332 bp promoter. Biochem J 1993;292:767–772.

37 Wang D, Sul HS: Upstream stimulatory factors binding to insulin response sequence of the fatty acid synthase promoter. J Biol Chem 1995;270: 28716–28722.

38 Bennett MK, Lopez JM, Sanchez HG, Osborne TF: Sterol regulation of fatty acid synthase promoter. J Biol Chem 1995;270:25578–25583.

39 Jump DB, Clarke SD, MacDougald O, Thelen A: Polyunsaturated fatty acids inhibit S14 gene transcription in rat liver and cultured hepatocytes. Proc Natl Acad Sci USA 1993;90:8454–8458.

40 Clarke SD, Turini M, Jump DB: Polyunsaturated fatty acids regulate lipogenic and peroxisomal gene expression by independent mechanisms. Prostaglandins Leukotrienes Essen Fatty Acids 1997;57: 65–69.

41 Foufelle F, Lepetit N, Bosc D, Delzenne N, Morin J, Raymondjean M, Ferre P: Dnase 1 hypersensitivity sites and nuclear protein binding on the fatty acid synthase gene: Identification of an element with properties similar to known glucose-responsive elements. Biochem J 1995;308:521–527.

42 Soncini M, Yet SF, Moon Y, Chun JY, Sul HS: Hormonal and nutritional control of the fatty acid synthase promoter in transgenic mice. J Biol Chem 1995;270:3039–3043.

Mineral and Electrolyte Metabolism

Miner Electrolyte Metab 1997;23:135–139

Maret G. Traber

Department of Molecular and Cell Biology
University of California, Berkeley, Calif., USA

Cellular and Molecular Mechanisms of Oxidants and Antioxidants

Key Words
Kidney
NF-κB
Peroxide
Superoxide
Vitamin E

Abstract
This report summarizes the major categories of oxidants and antioxidants. The mechanisms for modulating the most potent antioxidant in plasma, vitamin E, are described. The roles of oxidants and antioxidants in the redox-sensitive signal transduction pathway of NF-κB are discussed.

Introduction

Free radicals and reactive oxygen species (ROS) are now implicated in virtually every chronic disease. Enthusiasm for mechanisms involving oxidants in the etiology of chronic diseases may be overstated, but evidence is accumulating that oxidants play a significant role in chronic inflammation, atherosclerosis, and renal failure. Furthermore, the evidence for the protective effects of antioxidants in chronic disease, and even aging, is growing.

This report will briefly summarize the major categories of oxidants and antioxidants, examine the molecular mechanisms for modulating antioxidant levels, and illustrate the role of oxidants and antioxidants in signal transduction pathways.

Free Radicals and ROS

A free radical has one or more unpaired electrons and can be formed either by loss (oxidation) or gain (reduction) of a single electron by a nonradical [1]. In general, a molecule with an unpaired electron has the tendency to react very rapidly with other molecules. The term ROS refers to all free radical and nonradical compounds that may cause oxidative injury [2]. These include oxygen (O_2), singlet oxygen, superoxide radical ($O_2^{\cdot-}$), hydroxyl radicals ($OH^{\cdot}$) and hydrogen peroxide (H_2O_2).

Transition metals can catalyze the production of hydroxyl radicals, the most reactive ROS. Iron and copper are good examples of metals that act as catalysts in redox reaction. For example, the Fenton reaction: iron (Fe^{2+}) plus hydrogen peroxide (H_2O_2) can create hydroxyl radicals ($OH^{\cdot}$).

$$Fe^{2+} + H_2O_2 \rightarrow Fe^{3+} + OH^{\cdot} + OH^-$$

Protection against Free Radical Damage

There are various steps that can be taken to decrease free radical damage. Free radical reactions do not occur in the absence of O_2, but sealing living systems under nitrogen is not feasible. Instead, organisms have evolved various mechanisms for protection, including both antioxidant enzymes and the utilization of low-molecular-weight antioxidants.

Antioxidant enzymes include:

(1) superoxide dismutase (CuZn SOD, Mn SOD): $O_2^{\cdot-} + O_2^{\cdot-} + 2H^+ \rightarrow H_2O_2 + O_2$;

(2) catalase: $H_2O_2 \rightarrow O_2 + 2H_2O$;

(3) glutathione (GSH) peroxidase (selenium present in the active site): $GSH + H_2O_2 \rightarrow GSSG + 2H_2O$;

(4) phospholipid hydroperoxide GSH peroxidase [3]: $GSH + LOOH \rightarrow GSSG + LOH$.

Antioxidant enzymes protect cellular components from endogenously derived oxidants. For example, leakage of electrons from the electron transport chain gives rise to superoxide, but the presence of SOD in the mitochondria prevents its escape into the cytosol. Instead, peroxide is produced, which can diffuse into the cytosol, where it may influence signal transduction pathways (see below). Additionally, GSH peroxidase uses GSH to decompose peroxides by converting them to nonradical products and thereby regulates the level of cellular peroxides. Importantly, GSH peroxidase and phospholipid hydroperoxide GSH peroxidase regulate cellular lipid hydroperoxides.

There are several small molecules that provide antioxidant protection in vivo. These include: ascorbic acid (vitamin C), GSH (substrate for GSH peroxidases), urate, β-carotene (singlet oxygen quencher), and vitamin E.

GSH functions as a redox mediator in antioxidant and conjugation reactions, as a key substance in the uptake and membrane

KARGER

E-Mail karger@karger.ch
Fax +41 61 306 12 34
http://www.karger.ch

0378–0392/97/0236–0135 $12.00/0

Maret G. Traber, PhD
Department of Molecular and Cell Biology, University of California
Berkeley, CA 94720-3200 (USA)
Tel. +1 510 642 7023, fax +1 510 642 8313
E-Mail: mgtraber@uclink4.berkeley.edu

transfer of γ-glutamylamino acids, and as a storage form of cysteine [4]. Cellular GSH concentrations are regulated both by the synthesis and efflux of GSH, as well as by the synthesis of cystine transporters [5, 6] and γ-glutamylcysteine synthetase [7].

Other antioxidants, such as flavanoids, have antioxidant activity in vitro, but their utility in vivo remains controversial. Perhaps the most important in vivo antioxidant is vitamin E, not only because of its superior ability as a chain-breaking antioxidant, but also because vitamin E interacts with other antioxidants [8].

Antioxidant Function of Vitamin E

There is overwhelming evidence that vitamin E, in vivo, functions as a chain-breaking antioxidant that prevents the propagation of free radical damage in biological membranes [9–12]. Vitamin E is a potent peroxyl radical scavenger and especially protects polyunsaturated fatty acids (PUFAs) within phospholipids of biological membranes and in plasma lipoproteins [11]. When lipid hydroperoxides are oxidized to peroxyl radicals (ROO˙), these react 1,000-fold faster with vitamin E (Vit E-OH) than with PUFAs (RH) [8]. The phenolic hydroxyl group of tocopherol reacts with an organic peroxyl radical to form the corresponding organic hydroperoxide and the tocopheroxyl radical (Vit E-O˙) [13]:

in the presence of vitamin E:

$ROO^{\cdot} + \text{Vit E-OH} \rightarrow ROOH + \text{Vit E-O}^{\cdot}$;

in the absence of vitamin E:

$ROO^{\cdot} + RH \rightarrow ROOH + R^{\cdot}$, $R^{\cdot} + O_2 \rightarrow ROO^{\cdot}$.

Vitamin E, functioning as a chain breaking antioxidant, prevents the further autooxidation of lipids.

The tocopheroxyl radical (Vit E-O˙) formed in membranes emerges from the lipid bilayer into the aqueous domain. It is here that it reacts with vitamin C (or other reductants serving as hydrogen donors (AH)), oxidizing it and returning vitamin E to its reduced state:

$$\text{Vit E-O}^{\cdot} + AH \rightarrow \text{Vit E-OH} + A^{\cdot}.$$

Biologically important hydrogen donors, which have been demonstrated in vitro to regenerate tocopherol from the tocopheroxyl radical, include ascorbate (vitamin C) and thiols [14], especially GSH [15–18]. Subsequently, the vitamin C radical can be reduced by way of metabolic processes. This phenomenon has led to the idea of 'vitamin E recycling', where the antioxidant function of oxidized vitamin E is continuously restored by other antioxidants [8]. This antioxidant network depends upon the supply of aqueous antioxidants and the metabolic activity of cells.

Lipoprotein Transport of Vitamin E

Over 40 years ago, vitamin E was shown to be transported in the circulation in plasma lipoproteins [19, 20]. Dietary vitamin E is absorbed and secreted in chylomicrons. These serve to deliver vitamin E to the liver which repackages it in very low density lipoproteins (VLDLs) for secretion into the plasma [21]. Of the various forms of vitamin E, only *RRR*-α-tocopherol is preferentially incorporated into nascent VLDL [22].

A kinetic model of plasma vitamin E transport in humans using deuterium-labeled stereoisomers of α-tocopherol demonstrated that nearly the entire plasma pool of *RRR*-α-tocopherol is replaced daily by the liver, despite the slower turnover of the circulating lipoproteins carrying the vitamin E [23]. Thus, resecretion of vitamin E by the liver is essential to maintain plasma vitamin E concentrations.

α-Tocopherol Transfer Protein (α-TTP)

The hepatic α-TTP has been isolated from rat [24, 25] and human liver [26, 27] and the amino acid sequences of both have been reported [27, 28]. All the following features of vitamin E are required for recognition and transfer by α-TTP [24]: (1) the fully methylated, intact chromanol ring with a free 6-OH group, (2) the presence of the phytyl side chain, and (3) the stereochemical configuration of the phytyl tail in the 2*R* position.

α-TTP has been localized to hepatocytes; it has not been found in any other liver cells, in other tissues or in plasma [25]. In vitro, purified α-TTP transfers α-tocopherol between liposomes and microsomes [24, 27]. This ability to recognize α-tocopherol and to transfer it suggests that α-TTP is probably responsible for the incorporation of *RRR*-α-tocopherol into VLDL, discussed above. This however, has not been demonstrated directly.

The human protein has 94% homology to the rat α-TTP [28], some homology to both the retina retinaldehyde binding protein (47.4%) and to sec14 (37.4%), a yeast phospholipid transfer protein in the Golgi complex [27], and to a few other proteins.

Genetic Defects in Vitamin E Transport

Genetic defects in α-TTP are associated with a characteristic syndrome, ataxia with vitamin E deficiency (AVED, previously called familial isolated vitamin E deficiency). AVED patients have neurologic abnormalities, similar to Friedreich's ataxia [29, 30]. They are characterized by a progressive peripheral neuropathy with a specific 'dying back' of the large-caliber axons of the sensory neurons [31].

The biochemical defect in AVED patients was demonstrated using deuterated tocopherols [32, 33]. AVED patients were unable to maintain plasma d_6-*RRR*-α-tocopherol concentrations because their ability to secrete it in VLDL was impaired. A genetic defect in α-TTP has now been identified in some of the patients who have been analyzed using deuterium-labeled tocopherols. Truncation of the terminal portion of α-TTP results in an inability to discriminate between *RRR*- and *SRR*-α-tocopherols [34]. A defective transfer function of α-TTP has also been demonstrated in vitro [35].

The identification of α-TTP and its importance in maintaining plasma concentrations of vitamin E suggest that other tissues may also have the means to regulate concentrations of vitamin E. Indeed various vitamin E binding/transfer proteins have been described [36–39], but their function remains to be established. Podda et al. [40] found that the brain contains only α-tocopherol, while the skin of the same hairless mice contained more than 15% of other forms of vitamin E. The range of vitamin E forms in various tissues suggests that tissue vitamin E may be regulated, but thus far this is mere speculation.

Vitamin E and Atherosclerosis

Vitamin E is the most potent lipid-soluble antioxidant in plasma [11, 41]. Vitamin E supplements may protect against oxidative damage and may have a protective role in numerous pathologies. For example, it is likely that oxidation of low-density lipoproteins (LDLs) is involved in the formation of atherosclerotic lesions [42] and that prevention of LDL oxidation may reduce the risk of heart disease. Vitamin E supplementation decreases in vitro susceptibility of LDLs to oxidation [43]. Epidemiologic evidence documents that vitamin E intakes greater than 100 IU reduce the risk of heart disease [44, 45]. Data from the Cambridge Heart Antioxidant Study (a double-blind, placebo-controlled study) revealed that vitamin E supplementation in patients with known heart disease reduces the risk of nonfatal myocardial infarction by 75% (supplemented vs. control, 14 vs. 41, respectively; $p<0.005$) [46]. Furthermore, Regnstrom et al. [47] reported that low LDL vitamin E concentrations were inversely correlated with a high stenosis score, suggesting that low vitamin E contributed to the severity of coronary artery disease.

Vitamin E may also play a role in preventing activation of redox-sensitive signal transduction pathways, which may be important for inflammatory responses and for abnormal cell proliferation.

Redox Regulation of Signal Transduction

One signal transduction pathway where oxidants and antioxidants regulate activation is in the nuclear factor (NF)-κB pathway. NF-κB is a transcription factor that binds to defined DNA motifs (consensus: 5′-GGGPuNNPyPyCC-3′) enhancing the transcription of a variety of genes [48, 49]. The inactive form of NF-κB is a trimeric protein consisting of p65, p50 and IκB subunits [48]. The NF-κB complex is activated by phosphorylation, ubiquitination, and consecutive proteasomal degradation of the IκB component [49, 50]. The removal of IκB enables the p50/p65 heterodimer to migrate to the nucleus where transactivation is induced by specific DNA binding, resulting in transcriptional activation of various genes [48, 49, 51]. In response to stimuli [such as tumor necrosis factor (TNF)-α, lymphotoxin, interleukin (IL)-1, lipopolysaccharide, protein kinase C activators (such as phorbol myristate) and inhibitors of protein phosphatases 1 and 2A], IκB dissociates and is degraded. NF-κB is also activated by oxidative stressors, such as hydrogen peroxide, lipid hydroperoxides, and ultraviolet light [48, 52]. Although NF-κB activation is facilitated by these oxidative reactions, neither the regulating proximal oxidant nor the molecular target prone to oxidation has been unequivocally established. Surprisingly, binding of activated NF-κB to DNA requires reducing conditions in the nucleus [53].

NF-κB activation is abrogated by a variety of thiols (N-acetyl-*L*-cysteine, *L*-cysteine, 2-mercaptoethanol, aurothioglucose), α-lipoic acid (which is intracellulary reduced to generate a dithiol), other sulfur-containing scavengers (pyrrolidine dithiocarbamate, diethyldithiocarbamate, disulfiram), phenolic antioxidants (butylated hydroxyanisol, nordihydroguaretic acid, α-tocopherol) and chelators (which prevent transition-metal-mediated free radical chain propagation, such as orthophenanthroline or deferrioxamine), as reviewed by Meyer et al. [54].

The key steps and regulatory sites in NF-κB activation remain under intense investigation because of the potential role of NF-κB in many inflammatory diseases.

Implications for Patients with Kidney Disease

Hyperlipidemia and lipoprotein abnormalities are often encountered in patients with the nephrotic syndrome or chronic renal disease, in those undergoing hemodialysis and in those with a renal transplant, as reviewed by Groene et al. [55]. They noted that glomerular lipid deposition is observed in genetic diseases (such as Fabry's disease, lecithin:cholesterol acyltransferase activity deficiency, or arteriohepatic dysplasia) and in diseases with acquired disturbances of lipid metabolism (such as nephrotic syndrome or cholestatic liver disease).

Lipid deposition, mononuclear cell infiltration, and accumulation of mesangial matrix components are early events in the development of glomerulosclerosis [56]. In vitro mesangial cells oxidize LDL, which then causes cytotoxic injury. This suggests that oxidized lipoproteins in the glomerulus may promote inflammation, cell injury, and sclerosis [56]. Jackson et al. [57] stated that 'cardiovascular disease is the major cause of mortality in patients receiving hemodialysis for chronic renal failure.' They found that dialysis caused a depletion of key antioxidants (vitamin E, ascorbate, and urate) and suggested that antioxidant depletion may lead to accelerated atherogenesis.

One protective mechanism that is important in the kidney is GSH. Visarius et al. [58] measured cellular uptake and metabolism of exogenous GSH in freshly isolated proximal tubular cells from rat kidney. They reported that GSH uptake, rather than reduction of GSSG (oxidized GSH)or synthesis, appears to be the primary mechanism in tubular cells to maintain intracellular thiol redox status. The ability of these cells to rapidly take up and use GSH for detoxification suggests that in vivo this may be an important protective mechanism against chemical injury [58]. A GSH transporter has been described in hepatocytes [59]. Although the transporter is involved in cannalicular secretion of GSH into bile, the antibody to the protein cross-reacted with proteins in brain, intestine, lung and kidney; thus, a GSH transporter could function in renal cells to obtain liver-synthesized GSH.

Ischemia/reperfusion injury to the kidney is probably mediated by free radicals [60]. Xanthine-oxidase-dependent production of superoxide is the mechanism frequently cited as the cause of injury to the kidney. Indeed, xanthine oxidase is present in the proximal tubules of the kidney [61]. The conversion of xanthine dehydrogenase, the non-oxygen-metabolite-producing form, to xanthine oxidase increases during renal ischemia [62]. The superoxide produced by xanthine oxidase could cause injury or it could react with nitric oxide (NO, also called endothelium-dependent relaxation factor) to form peroxynitrite, a far more dangerous free radical than either of the reactants [63].

Short-term dietary-induced hypercholesterolemia has also been found to cause either increased vascular tone in various blood vessels, such as coronary arteries and renal vasculature, or impaired relaxation to endothelial-dependent vasodilators [64]. In the kidney, renal blood flow, glomerular filtration rates, and ultrafiltration coefficients are decreased and glomerular capillary pressure is elevated [64]. Vasoconstrictor prostaglandins and accelerated inactivation of NO by reactive oxygen species appear to play important roles in the impaired relaxation response [64]. Indeed, NO is postulated to play a protective role in renal defense against ischemic insult and its sequelae [65]. And inducible NO synthase (iNOS) is localized in tubules and may preserve tubular integrity during hypoxia by increasing NO production [65].

NF-κB plays a key role in modulating cellular responses in the kidney. Kunz et al. [66] demonstrated that cyclosporin A, a potent immunosuppressive drug, inhibits IL-1β-dependent iNOS expression by decreasing iNOS protein and mRNA transcription, as a result of reduced NF-κB DNA binding. In contrast, FK506 activates transcription factor NF-κB and thus induces IL-6 production in the kidney [67]. IL-6 produces renal abnormalities, such as mesangioproliferative glomerulonephritis [67]. Expression of TNF-α, a stimulator of NF-κB, has been observed in proximal tubular epithelial cells in response to chemicals, immune complexes, bacterial lipopolysaccharides, and advanced glycosylation end products. TNF-α release is amplified by complement proteins (C5b-9), platelet-activating factor, and ROS, and is abrogated by prostaglandins or IL-6 and IL-10 [68].

Conclusion

Oxidants and antioxidants are no longer only the purview of chemists. Their reactions clearly take place in vivo, and their balance dictates cellular responses. It is becoming increasingly apparent that many chronic illnesses may be caused by an imbalance in favor of oxidative stress. Clearly further research needs to be carried out in the area of renal function, renal disease, and ROS.

References

1 Halliwell B, Gutteridge JMC: Free Radicals in Biology and Medicine. Oxford, Clarendon, 1989.
2 Fuchs J: Oxidative injury in dermatopathology. New York, Springer, 1992.
3 Thomas JP, Maiorino M, Rusini R, Girotti AW: Protective action of phospholipid hydroperoxide glutathione peroxidase against membrane-damaging lipid peroxidation: In situ reduction of phospholipid and cholesterol hydroperoxides. J Biol Chem 1990;265:454–461.
4 Meister A: Metabolism and function of glutathione; in Dolphin D, Poulson R, Avrannovic O (eds): Glutathione. New York, Wiley, 1989, pp 367–474.
5 Miura K, Ishii T, Sugita Y, Bannai S: Cystine uptake and glutathione level in endothelial cells exposed to oxidative stress. Am J Physiol 1992;262: C50–58.
6 Deneke SM: Induction of cystine transport in bovine pulmonary artery endothelial cells by sodium arsenite. Biochim Biophys Acta 1992;1109:127–131.
7 Cotgreave IA, Schuppe-Koistinen I: A role for gamma-glutamyl transpeptidase in the transport of cystine into human endothelial cells: Relationship to intracellular glutathione. Biochim Biophys Acta 1994;1222:375–382.
8 Packer L: Vitamin E is nature's master antioxidant. Sci Am Sci Med 1994;1:54–63.
9 Tappel AL: Vitamin E as the biological lipid antioxidant. Vitam Horm 1962;20:493–510.
10 Burton GW, Ingold KU: Vitamin E: Application of the principles of physical organic chemistry to the exploration of its structure and function. Acc Chem Res 1986;19:194–201.
11 Burton GW, Joyce A, Ingold KU: Is vitamin E the only lipid-soluble, chain-breaking antioxidant in human blood plasma and erythrocyte membranes? Arch Biochem Biophys 1983;221:281–290.
12 Ingold KU, Webb AC, Witter D, Burton GW, Metcalfe TA, Muller DPR: Vitamin E remains the major lipid-soluble, chain-breaking antioxidant in human plasma even in individuals suffering severe vitamin E deficiency. Arch Biochem Biophys 1987; 259:224–225.
13 Burton GW, Doba T, Gabe EJ, Hughes L, Lee FL, Prasad L, Ingold KU: Autoxidation of biological molecules. 4. Maximizing the antioxidant activity of phenols. J Am Chem Soc 1985;107:7053–7065.
14 Wefers H, Sies H: The protection by ascorbate and glutathione against microsomal lipid peroxidation is dependent on vitamin E. Eur J Biochem 1988; 174:353–357.
15 McCay PB: Vitamin E: Interactions with free radicals and ascorbate. Annu Rev Nutr 1985;5:323–340.
16 Niki E: Antioxidants in relation to lipid peroxidation. Chem Phys Lipids 1987;44:227–253.
17 Sies H, Murphy ME: Role of tocopherols in the protection of biological systems against oxidative damage. Photochem Photobiol 1991;8:211–224.
18 Sies H, Stahl W, Sundquist AR: Antioxidant functions of vitamins (vitamins E and C, beta-carotene, and other carotenoids). Ann NY Acad Sci 1992; 669:7–20.
19 Lewis LA, Quaife ML, Page IH: Lipoproteins of serum, carriers of tocopherol. Am J Physiol 1954; 178:221–222.
20 McCormick EC, Cornwell DG, Brown JB: Studies on the distribution of tocopherol in human serum lipoproteins. J Lipid Res 1960;1:221–228.
21 Traber MG, Sies H: Vitamin E in humans: Demand and delivery. Annu Rev Nutr 1996;16:321–347.
22 Traber MG, Rudel LL, Burton GW, Hughes L, Ingold KU, Kayden HJ: Nascent VLDL from liver perfusions of cynomologus monkeys are preferentially enriched in *RRR*- compared with *SRR*-α tocopherol: Studies using deuterated tocopherols. J Lipid Res 1990;31:687–694.
23 Traber MG, Ramakrishnan R, Kayden HJ: Human plasma vitamin E kinetics demonstrate rapid recycling of plasma *RRR*-α-tocopherol. Proc Natl Acad Sci USA 1994;91:10005–10008.
24 Sato Y, Hagiwara K, Arai H, Inoue K: Purification and characterization of α-tocopherol transfer protein from rat liver. FEBS Lett 1991;288:41–45.
25 Yoshida H, Yusin M, Ren I, Kuhlenkamp J, Hirano T, Stolz A, Kaplowitz N: Identification, purification and immunochemical characterization of a tocopherol-binding protein in rat liver cytosol. J Lipid Res 1992;33:343–350.
26 Kuhlenkamp J, Ronk M, Yusin M, Stolz A, Kaplowitz N: Identification and purification of a human liver cytosolic tocopherol binding protein. Prot Exp Purific 1993;4:382–389.
27 Arita M, Sato Y, Miyata A, Tanabe T, Takahashi E, Kayden H, Arai H, Inoue K: Human alpha-tocopherol transfer protein: cDNA cloning, expression and chromosomal localization. Biochem J 1995;306:437–443.

28 Sato Y, Arai H, Miyata A, Tokita S, Yamamoto K, Tanabe T, Inoue K: Primary structure of alpha-tocopherol transfer protein from rat liver: Homology with cellular retinaldehyde-binding protein. J Biol Chem 1993;268:17705–17710.
29 Ben Hamida M, Belal S, Sirugo G, Ben Hamida C, Panayides K, Ionannou P, Beckmann J, Mandel JL, Hentati F, Koenig M, Middleton L: Friedreich's ataxia phenotype not linked to chromosome 9 and associated with selective autosomal recessive vitamine E deficiency in two inbred tunisian families. Neurology 1993;43:2179–2183.
30 Ben Hamida C, Doerflinger N, Belal S, Linder C, Reutenauer L, Dib C, Gyapay G, Bignal A, Le Paslier D, Cohen D, Pandolfo M, Mokini V, Novelli G, Hentati F, Ben Hamida M, Mandel JL, Koenig M: Localization of Friedreich ataxia phenotype with selective vitamin E deficiency to chromosome 8q by homozygosity mapping. Nat Genet 1993;5: 195–200.
31 Sokol RJ, Kayden HJ, Bettis DB, Traber MG, Neville H, Ringel S, Wilson WB, Stumpf DA: Isolated vitamin E deficiency in the absence of fat malabsorption – Familial and sporadic cases. Characterization and investigation of causes. J Lab Clin Med 1988;111:548–559.
32 Traber MG, Sokol RJ, Burton GW, Ingold KU, Papas AM, Huffaker JE, Kayden HJ: Impaired ability of patients with familial isolated vitamin E deficiency to incorporate α-tocopherol into lipoproteins secreted by the liver. J Clin Invest 1990; 85:397–407.
33 Traber MG, Sokol RJ, Kohlschütter A, Yokota T, Muller DPR, Dufour R, Kayden HJ: Impaired discrimination between stereoisomers of α-tocopherol in patients with familial isolated vitamin E deficiency. J Lipid Res 1993;34:201–210.
34 Traber MG: Regulation of human plasma vitamin E; in Sies H (ed): Antioxidants in Disease Mechanisms and Therapeutic Strategies. San Diego, Academic Press, 1996, vol. 38, pp 49–63.
35 Gotoda T, Arita M, Arai H, Inoue K, Yokota T, Fukuo Y, Yazaki Y, Yamada N: Adult-onset spinocerebellar dysfunction caused by a mutation in the gene for the alpha-tocopherol-transfer protein. N Engl J Med 1995;333:1313–1318.
36 Dutta-Roy A, Gordon M, Leishman D, Paterson BJ, Duthie GG, James WPT: Purification and partial characterization of an alpha-tocopherol-binding protein from rabbit heart cytosol. Mol Cell Biochem 1993;123:139–144.
37 Dutta-Roy AK, Leishman DJ, Gordon MJ, Campbell FM, Duthie GG: Identification of a low molecular mass (14.2 kDa) alpha-tocopherol-binding protein in the cytosol of rat liver and heart. Biochem Biophys Res Commun 1993;196:1108–1112.
38 Gordon MJ, Campbell FM, Duthie GG, Dutta-Roy AK: Characterization of a novel alpha-tocopherol-binding protein from bovine heart cytosol. Arch Biochem Biophys 1995;318:140–146.
39 Nalecz K, Nalecz M, Azzi A: Isolation of tocopherol-binding proteins from the cytosol of smooth muscle A7R5 cells. Eur J Biochem 1992;209:37–42.
40 Podda M, Weber C, Traber MG, Packer L: Simultaneous determination of tissue tocopherols, tocotrienols, ubiquinols and ubiquinones. J Lipid Res 1996;37:893–901.
41 Burton GW, Joyce A, Ingold KU: First proof that vitamin E is major lipid-soluble, chain-breaking antioxidant in human blood plasma. Lancet 1982; ii:327.
42 Steinberg D, Parthasarathy S, Carew TE, Khoo JC, Witztum JL: Beyond cholesterol: Modifications of low-density lipoprotein that increase its atherogenicity. N Engl J Med 1989;320:915–924.
43 Jialal I, Fuller CJ, Huet BA: The effect of α-tocopherol supplementation on LDL oxidation: A dose-response study. Arterioscler Thromb Vasc Biol 1995;15:190–198.
44 Rimm EB, Stampfer MJ, Ascherio A, Giovannucci E, Colditz GA, Willet WC: Vitamin E consumption and the risk of coronary heart disease in men. N Engl J Med 1993;328:1450–1456.
45 Stampfer MJ, Hennekens C, Manson JE, Colditz GA, Rosner B, Willet WC: Vitamin E consumption and risk of coronary disease in women. N Engl J Med 1993;328:1444–1449.
46 Stephens NG, Parsons A, Schofield PM, Kelly F, Cheeseman K, Mitchinson MJ: Randomised controlled trial of vitamin E in patients with coronary disease – Cambridge Heart Antioxidant Study (CHAOS). Lancet 1996;347:781–786.
47 Regnstrom J, Nilsson J, Moldeus P, Strom K, Bavenholm P, Tornvall P, Hamsten A: Inverse relation between the concentration of low-density-lipoprotein vitamin E and severity of coronary artery disease. Am J Clin Nutr 1996;63:377–385.
48 Baeuerle PA, Henkel T: Function and activation of NF-κB in the immune system. Annu Rev Immunol 1994;12:141–179.
49 Baldwin AS: The NF-κB and IκB proteins: New discoveries and insights. Annu Rev Immunol 1996; 14:649–681.
50 Chen Z, Hagler J, Palombella VJ, Melandri F, Scherer D, Ballard D, Maniatis T: Signal induced site-specific phosphorylation targets I kappa B alpha to the ubiquitin-proteasome pathway. Genes Dev 1995;9:1586–1597.
51 Verma IM, Stevenson JK, Schwarz EM, Van Antwerp D, Miyamoto S: Rel/NF-kappa B/I kappa B family: Intimate tales of association and dissociation. Genes Dev 1995;9:2723–2735.
52 Schulze-Osthoff K, Beyaert R, Vandevoorde V, Haegeman G, Fiers W: Depletion of the mitochondrial electron transport abrogates the cytotoxic and gene-inductive effects of TNF. EMBO J 1993;12: 3095–3104.
53 Hayashi T, Ueno Y, Okamoto T: Oxidoreductive regulation of nuclear factor kappa B: Involvement of a cellular reducing catalyst thioredoxin. J Biol Chem 1993;268:11380–11388.
54 Meyer M, Pahl HL, Baeuerle PA: Regulation of the transcription factors NF-kappa B and AP-1 by redox changes. Chem Biol Interact 1994;91:91–100.
55 Groene EF, Walli AK, Groene HJ, Miller B, Seidel D: The role of lipids in nephrosclerosis and glomerulosclerosis. Atherosclerosis 1994;107:1–13.
56 Wheeler DC, Chana RS: Interaction between lipoproteins, glomerular cells and matrix. Miner Electrolyte Metab 1993;19:149–164.
57 Jackson P, Loughrey CM, Lightbody JH, McNamee PT, Young IS: Effect of hemodialysis on total antioxidant capacity and serum antioxidants in patients with chronic renal failure. Clin Chem 1995;41:1135–1138.
58 Visarius TM, Putt DA, Schare JM, Pegouske DM, Lash LH: Pathways of glutathione metabolism and transport in isolated proximal tubular cells from rat kidney. Biochem Pharmacol 1996;52:259–272.
59 Yi JR, Lu S, Fernandez-Checa J, Kaplowitz N: Expression cloning of a rat hepatic reduced glutathione transporter with canalicular characteristics. J Clin Invest 1994;93:1841–1845.
60 Baud L, Ardaillou R: Involvement of reactive oxygen species in kidney damage. Br Med Bull 1993; 49:621–629.
61 Kurosaki M, Li Calzi M, Scanziani E, Garattini E, Terao M: Tissue- and cell-specific expression of mouse xanthine oxidoreductase gene in vivo: Regulation by bacterial lipopolysaccharide. Biochem J 1995;306:225–234.
62 Linas SL, Whittenburg D, Repine JE: Role of xanthine oxidase in ischemia/reperfusion injury. Am J Physiol 1990;258:F711–F716.
63 Pryor WA, Squadrito GL: The chemistry of peroxynitrite: A product from the reaction of nitric oxide with superoxide. Am J Physiol 1995;268:L699–L722.
64 Bank N: Renal hemodynamic consequence of hyperlipidemia. Miner Electrolyte Metab 1993;19: 165–172.
65 Peer G, Blum M, Iaina A: Nitric oxide and acute renal failure. Nephron 1996;73:375–381.
66 Kunz D, Walker G, Eberhardt W, Nitsch D, Pfeilschifter J: Interleukin 1-beta-induced expression of nitric oxide synthase in rat renal mesangial cells is suppressed by cyclosporin A. Biochem Biophys Res Commun 1995;216:438–446.
67 Muraoka KI, Fujimoto K, Sun X, Yoshioka K, Shimizu KI, Yagi M, Bose HJ, Miyazaki I, Yamamoto KI: Immunosuppressant FK506 induces interleukin-6-production through the activation of transcription factor nuclear factor (NF)-κB: Implications for Fk506 nephropathy. J Clin Invest 1996;97:2433–2439.
68 Baud L, Ardaillou R: Tumor necrosis factor in renal injury. Miner Electrolyte Metab 1995;21:336–341.

Mineral and Electrolyte Metabolism

Miner Electrolyte Metab 1997;23:140–143

Saulo Klahr

Department of Internal Medicine,
Washington University School of Medicine
at Barnes-Jewish Hospital, St. Louis, Mo., USA

Oxygen Radicals and Renal Diseases

Key Words
Acute renal failure
Glomerular disease
Hydrogen peroxide
Hydroxyl
Proteinuria
Superoxide

Abstract
Reactive oxygen metabolites (superoxide, hydrogen peroxide, hydroxyl radical, and hypochlorous acid) are important mediators of renal damage in acute renal failure and glomerular and tubulointerstitial diseases. The role of these oxygen metabolites in the above entities is discussed, and the effects of antioxidants and scavengers of O_2 radicals are considered. The role of oxygen radicals in the regulation of gene transcription is also considered.

Introduction

Accumulating evidence indicates that reactive oxygen metabolites (i.e., superoxide, hydrogen peroxide, hydroxyl radical, and hypochlorous acid) are important mediators of injury in a diverse number of tissues, including the kidney [1–3]. Reactive oxygen species (ROS) have been implicated in immune and nonimmune glomerular injury, in acute renal failure (ARF) and in tubulointerstitial disease.

Oxygen (O_2) has no unpaired electrons and is not a radical. When an electron is accepted by oxygen, the product is superoxide (O_2^-). Cells known to produce superoxide include monocytes/macrophages, neutrophils, and eosinophils. In neutrophils, superoxide is formed during therespiratory burst from the reduction of oxygen by a membrane-bound oxidase that uses reduced NADPH as a substrate. If a second electron is added to superoxide, peroxide is formed. Peroxide is not a radical. Any peroxide formed at physiological pH will immediately protonate to yield hydrogen peroxide (H_2O_2). Although this compound is usually a weak oxidant, it may interact with halide anions and the neutrophil enzyme myeloperoxidase to yield noxious hypochlorous acid and the hydroxyl radical, respectively.

There are a number of compounds that provide protection against the harmful effects of ROS. For example, superoxide can be scavenged by superoxide dismutase (SOD). Hydrogen peroxide can be degraded by catalase and glutathione peroxidase, resulting in the formation of water. Some direct scavengers of the hydroxyl radical include dimethylsulfoxide, dimethylthiourea, mannitol, and ethanol. In addition, the synthesis of hydroxyl radicals can be affected by iron chelators such as deferoxamine.

Production, Role, and Effects of ROS

ROS have a role in neutrophil-dependent or macrophage-dependent or -independent glomerular injury [2, 4] and also in the development of ARF [3] caused experimentally by the administration of endotoxin, glycerol, or gentamicin, or the institution of ischemia. The production of ROS appears to affect several metabolic pathways, including prostaglandin synthesis, cyclic nucleotide metabolism, the degradation of basement membrane, and alterations in the permeability of the vascular bed [5–7]. In addition to invading cells (macrophages, neutrophils), ROS may be produced by isolated glomeruli and by cultured mesangial cells in response to diverse stimuli. In vitro studies have revealed that a variety of particulate and soluble stimuli enhance the generation of ROS by macrophages and neutrophils. The oxidative burst can be triggered by several immune reactants such as the C3b receptor, heat-aggregated IgG, Fc receptor stimulus, immune complexes, serum-treated zymosan, and complement components. Antineutrophil cytoplasmic autoantibodies, present in the blood of patients with pauci-immune vasculitis and crescentic glomerulonephritis, can significantly increase the generation of superoxide in neutrophils. It appears that stimulated neutrophils or macrophages may serve as sources of ROS in glomerulonephritis characterized by proliferation and exudation. Resident glomerular cells may also be sources of ROS in noninflammatory forms of glomerular disease. Crescentic and glomerular cells, particularly mesangial cells, can generate ROS in response to perturbation of their cell membrane, such as that produced by phorbol myristate acetate (PMA). Normal glomeruli can generate H_2O_2 in vivo. Agents such as adriamycin that induce the nephrotic syndrome in experimental animals enhance the generation of ROS in freshly isolated glomeruli in vitro. The aminonucleoside of puromycin, which induces an entity that resembles minimal-change disease with the nephrotic syndrome, also enhances the generation of superoxide

KARGER
E-Mail karger@karger.ch
Fax +41 61 306 12 34
http://www.karger.ch

0378–0392/97/0236–0140 $12.00/0

Saulo Klahr, MD
Department of Internal Medicine, Barnes-Jewish Hospital, North Campus,
216 South Kingshighway, St. Louis, MO 63110 (USA)
Tel. +1 314 454 7107, fax +1 314 454 5110
E-Mail: sklahr@imgate.wustl.edu

anion, hydroxyl radical and H_2O_2 when added to freshly isolated glomeruli. Mesangial cells can generate increased amounts of superoxide and H_2O_2 in response to several stimuli, including opsonized zymosan, membrane attack complex, platelet-activating factor (PAF) and immune complexes. Thus, either leukocytes or macrophages in inflammatory glomerular diseases or intrinsic cells (mesangial cells) in noninflammatory diseases may be sources of ROS.

ROS and Regulation of Gene Transcription

ROS may act as signal transduction messengers. This is particularly the case for H_2O_2. To explore the mechanisms that underlie ROS-dependent disorders in biological systems, recent studies have investigated the regulation of gene expression by oxidants, antioxidants, and other determinants of the intracellular reduction-oxidation (redox) state [8]. At least two well-defined transcription factors, nuclear factor (NF)-κB, and activator protein (AP)-1, are regulated by the intracellular redox state [8–11]. The regulation of gene expression by oxidants, antioxidants, and the redox state has opened new areas of investigation with therapeutic implications. Binding sites of the redox-regulated transcription factors NF-κB and AP-1 are located in the promoter region of a large variety of genes that are directly involved in the pathogenesis of disease. Biochemical and clinical studies indicate that antioxidant therapy may be therapeutically useful. It appears that critical steps in the signal transduction cascade are sensitive to oxidants and antioxidants. Many basic events of cell regulation such as protein phosphorylation and binding of transcription factors to consensus sites on DNA are driven by physiological oxidant-antioxidant homeostasis, especially by the thioldisulfide balance. Endogenous glutathione and thioredoxin systems and the exogenous lipoate-dehydrolipoate couple are effective regulators of redox-sensitive gene expression. A variety of structurally diverse antioxidants can inhibit NF-κB activation induced by a wide array of stimuli, ranging from phorbol esters to tumor necrosis factor (TNF)-α to H_2O_2. This evidence lends support to the hypothesis that ROS are involved in a common pathway of NF-κB activation. Vitamin E derivatives inhibit TNF-α-induced NF-κB activation. α-Tocopherol succinate not only inhibited activation and translocation of NF-κB to the nucleus, but also inhibited the in vitro binding of the activated proteins to the kB DNA site [8]. Catechol derivatives with characterized antioxidant properties also inhibited TNF-α-induced NF-κB activation [8].

Role of ROS in Proteinuria

Studies in vitro have shown that ROS can trigger degradation of glomerular basement membrane (GBM) by activating metalloproteinases of infiltrating polymorphonuclear leukocytes (PMNs) and/or glomerular mesangial cells [4, 7]. In addition, administration of H_2O_2 into the renal artery induces marked glomerular morphological abnormalities without PMN infiltration [5]. Furthermore, several antioxidants substantially attenuate glomerular injury and proteinuria in animal models of puromycin aminonucleoside nephrosis, passive Heymann nephritis, and anti-GBM-antibody-induced glomerulonephritis [12, 13]. These findings suggest that ROS are involved in the development of proteinuria and glomerular injury. Studies in vivo showed that the infusion of H_2O_2 into the renal artery induces a dose-dependent increase in urine protein excretion [5]. Hydrogen-peroxide-induced proteinuria peaked over 1 h and completely normalized by 24 h after infusion. Electrophoresis revealed that the urine protein was primarily of glomerular origin. The H_2O_2-induced proteinuria was almost completely prevented by treatment with catalase or deferoxamine [5]. Thus, iron-dependent metabolites of H_2O_2 appears to be involved in the proteinuria. This study suggests that ROS cause massive reversible proteinuria by inducing a molecular site selectivity defect of the glomerular capillary wall without apparent ultrastructural abnormalities. Thus, proteinuria in a variety of renal diseases may reflect the persistence of pathogenetic ROS acting on glomeruli – as opposed to nonpersistent ROS, which, as pointed out, produce transient proteinuria.

The administration of puromycin aminonucleoside (PAN) also causes proteinuria in rats [1]. The coadministration of α-tocopherol/ascorbic acid, dimethylthiolurea or SOD to PAN-treated rats reduced proteinuria by approximately 90%, 40%, and 60%, respectively, over an observation period of 18 days [12]. Transmission electronmicroscopy in PAN-treated rats given antioxidants revealed significantly reduced foot process effacement on days 4, 5, and 10 compared with rats that received PAN alone. These results provide further evidence that ROS play a role in PAN nephrosis and indicate that the antioxidants provide protection against the changes occurring in glomerular epithelial cell foot processes.

Proteinuria resulting from neutrophil-mediated glomerular injury has been documented in experimental models of inflammatory glomerular disease, including the heterologous phase of nephrotoxic serum nephritis. Neutrophil-dependent injury also occurs after intrarenal infusion of PMA, which causes local activations of neutrophils or of cobra venom factor, producing local complement activation and consequent neutrophil chemotaxis. In each of these models, neutrophil depletion markedly diminishes or prevents proteinuria.

Studies in vitro have examined the specific role of ROS in PMN-mediated glomerular injury [4]. In these experiments, glomeruli were incubated with H_2O_2, with H_2O_2 and myeloperoxidase (MPO) or with activated PMNs. Scavengers of ROS were included in some experiments. Glomerular albumin permeability was measured and found to be increased in glomeruli incubated with H_2O_2 and MPO. Glomeruli exposed to H_2O_2 alone did not demonstrate an increase in albumin permeability. Permeability to albumin after incubation with PMA or with nonactivated PMNs was no different from that of control glomeruli. In contrast, activated neutrophils increased albumin permeability. The increase was inhibited by SOD, catalase, or taurine [4]. These results indicate that hypohalous acid, usually hypochlorous acid, derived from the H_2O_2-MPO halide system, can increase albumin permeability. In fact, hypohalous acid may be the primary mediator of the immediate increase in glomerular protein permeability induced by PMNs.

Oxidants can contribute to glomerular basement membrane damage by increasing its susceptibility to proteolytic damage [14], and by inactivating the α-1 proteinase inhibitor (the primary regulator of neutrophil elastase [15–18]), thus allowing the released elastase to more readily inflict injury to the extracellular matrix [19]. It has also been reported that lipid peroxide (linoleic acid hydroperoxide) enhances the synthesis of gelatinase by mesangial cells; thus, increased generation of matrix-degrading proteinase may contribute to proteinuria [20]. Prevention of proteinuria by catalase in neutrophil-dependent glomerular nephritides suggests that the oxidative mechanism for GBM degradation described above may be relevant to leukocyte-dependent glomerular injury. On the other hand, Kashihara et al. [21] have postulated that synthesis of glomerular heparan sulfate proteoglycans is susceptible to oxidant injury. Using an isolated perfused kidney model, these investigators demonstrated a drastic dose-dependent decrease in the de novo synthesis of proteoglycans in response to xanthine-xanthine oxidase (a system that generates ROS). Thus, direct damage from ROS occurs during the de novo synthesis of heparan sulfate proteoglycans, molecules necessary to maintain integrity of the GBM and a normal glomerular filtration rate.

Role of ROS in Animal Models of Glomerular Disease

Anti-GBM Antibody Disease

This is a well-characterized model of complement- and neutrophil-dependent glomerular injury. In this setting, treatment of animals with catalase markedly reduced the proteinuria, whereas SOD had no protective effect [22]. Other potent hydroxyl radical scavengers or iron chelators also attenuated proteinuria in this model.

PAN Model of Minimal-Change Disease in Humans

ROS may mediate glomerular injury in glomerular diseases that lack infiltrating inflammatory cells. A single intravenous injection of PAN in experimental animals causes marked proteinuria and glomerular morphological changes, with characteristics similar to those of minimal-change disease in humans. Diamond et al. [23] reported that allopurinol, an inhibitor of xanthine oxidase, and SOD were protective in PAN-induced nephrotic syndrome, suggesting a role for xanthine-oxidase-generated superoxide anions in this model of renal disease. Thakur et al. [24] reported the protective effects of two hydroxyl radical scavengers and an iron chelator, implicating the hydroxyl radical in PAN nephrotic syndrome. Recently, coadministration of α-tocopherol-ascorbic acid was found to markedly reduce proteinuria and to reduce effacement of epithelial cell foot processes in this model [25].

Passive Heymann Nephritis Model of Membranous Nephropathy

Passive Heymann nephritis induced in experimental animals by a single intravenous injection of anti-FxlA is a complement-dependent and neutrophil-independent model of glomerular disease that resembles membranous nephropathy in humans. Shah [26] found that SOD or catalase did not affect the anti-FxlA-induced proteinuria. In contrast, scavengers of hydroxyl radicals and an iron chelator, desferoxamine, markedly reduced the proteinuria. The protective effects of both the hydroxyl radical scavengers and an iron chelator suggest a role for the hydroxyl radical in passive Heymann nephritis. Thus, the hydroxyl radical very likely plays an important role in the development of membranous nephropathy in animal models. A role for lipid peroxidation in the proteinuria that occurs in passive Heymann nephritis has also been reported [27]. Clearly, then, evidence exists to indicate that ROS are important mediators in glomerular pathophysiology. However, it should be pointed out that multiple mechanisms of injury may be operative and that ROS may be one of many interactive factors in the injury produced during the natural history of glomerular disease.

Role of ROS in Tubulointerstitial Disease

Several studies implicate ROS in the pathogenesis of tubulointerstitial damage. While oxygen consumption per nephron was increased in animals with subtotal nephrectomy fed a standard protein intake, no evidence of oxidative stress was present in the kidney, suggesting commensurate increases in the antioxidant defense mechanisms [28], such as enhanced urinary clearance of oxidative products. However, rats fed a high-protein diet compared to rats fed a low-protein diet had increased rates of oxygen consumption, increased lipid peroxidation, and alterations in the glutathione redox ratio in surviving nephrons. Thus, in the remnant kidney, increased dietary protein intake may be a cause of oxidative damage.

To further evaluate the role of ROS in tubulointerstitial disease, weanling rats were fed diets deficient in selenium and vitamin E for an additional 14 weeks [29]. Vitamin E is the major lipid-soluble scavenger for ROS and lipid bilayers. Selenium is required for the activity of glutathione peroxidase, the enzyme that degrades H_2O_2. Thus, a combined deficiency of these two substances promotes the accumulation of H_2O_2 and attendant oxidizing effects. The dietary manipulation resulted in increased rates of excretion of albumin, IgG, and low-molecular-weight proteins in the urine. This occurred approximately 8–9 weeks after the initiation of the diet. Increased protein excretion was accompanied by decreases in glomerular filtration rate and cellular infiltration of the tubulointerstitial space. Interestingly, renal hypertrophy and hyperplasia occurred in rats on the deficient diet, despite blunted somatic growth. Progressive renal insufficiency was noted as the administration of the diets was prolonged. Thus, dietary deficiency of antioxidants induces tubulointerstitial infiltration in the intact kidney and a decline in renal function.

ROS exert numerous toxic effects on the kidney. An effect that has been described is the induction of collagen gene expression by increased lipid peroxidation in vitro and the inhibition of such induction by vitamin E [30]. Long-term administration of carbon tetrachloride, an agent which promotes lipid peroxidation, causes tubulointerstitial fibrosis in rats. Kidney extracts from animals given this agent induced expression of type I and IV procollagen in rat mesangial cells [31]. Thus, it is likely that H_2O_2 and other ROS generated during renal oxidative metabolism may contribute to renal fibrosis.

Role of ROS in ARF

Several studies incriminate ROS in the pathogenesis of ischemia-reperfusion injury. Increased synthesis of ROS has been described in vitro in animal models of hypoxia and hypoxia reoxygenation and in vivo in a model of ischemia reperfusion. Paller and Newmann reported that under normoxic conditions, renal proximal tubular epithelial cells produced superoxide, hydroxyl radical, and H_2O_2 and that the production of these ROS was markedly increased during reoxygenation following hypoxia [32]. In addition, increased generation of ROS resulted in lipid peroxidation and cell injury, and addition of SOD, catalase, dimethylthiole urea or desferoxamine to the cells prevented oxidant-induced lipid peroxidation, and cell injury and hypoxia reoxygenation in vitro [32]. Xanthine oxidase has also been proposed as a source of ROS during ischemic reperfusion as described above. An inhibitor of xanthine oxidase (tungsten) and other scavengers, such as flavonoids, have been reported to reduce this conversion [33], resulting in improvement of renal function [33] or cell injury. Other inhibitors of xanthine oxidase, such as allopurinol or oxypurinol, have been reported to protect against cell death in an in vitro model of hypoxia reoxygenation and also in an in vivo model of ischemia reperfusion injury. Studies in specific models of ARF, induced by drugs or other substances, have also implicated ROS in the development of injury. These models include the gentamicin-induced nephrotoxicity and the ARF induced in animals by the administration of glycerol. In this setting there is enhanced in vivo generation of H_2O_2 in the renal cortex. Pyruvate, a scavenger of H_2O_2, hydroxyl radical scavengers and iron chelators are protective in glycerol-induced ARF in rats.

Although multiple factors are responsible for injury in ARF, it is clear that ROS play an important role in the pathophysiology of this entity. The exact mechanisms at the cellular and molecular level by which ROS produce injury need to be defined. The development of targeted therapy to prevent the action or the formation of ROS may influence the progression of renal disease as well as the severity of ARF.

References

1 Diamond JR: The role of reactive oxygen species in animal models of glomerular disease. Am J Kidney Dis 1992;19:292–300.
2 Shah S: Oxidant mechanisms in glomerulonephritis. Semin Nephrol 1991;11:320–326.
3 Shah SV, Walker PD, Ueda N, Nath KA: Reactive oxygen metabolites in toxic acute renal failure; in Gonick HC (ed): Current Nephrology. St Louis, Mosby, 1995, vol 18, pp 75–95.
4 Li JZ, Sharma R, Dileepan KN, Savin VJ: Polymorphonuclear leukocytes increase glomerular albumin permeability via hypohalous acid. Kidney Int 1994; 46:1025–1030.
5 Yoshioka T, Ichikawa I, Fogo A: Reactive oxygen meta- bolites cause massive, reversible proteinuria and glomerular sieving defect without apparent ultrastructural abnormality. J Am Soc Nephrol 1991;2:902–912.
6 Oberle GP, Niemeyer J, Thaiss F, Schoeppe W, Stahl RAK: Increased oxygen radical and eicosanoid formation in immune-mediated mesangial cell injury. Kidney Int 1992;42:69–74.
7 Donovan KL, Davies M, Coles GA, Williams JD: Relative roles of elastase and reactive oxygen species in the degradation of human glomerular basement membrane by intact human neutrophils. Kidney Int 1994;45:1555–1561.
8 Sen CK, Packer L: Antioxidant and redox regulation of gene transcription. FASEB J 1996;10:709–720.
9 Schreck R, Rieber P, Baeuerle PA: Reactive oxygen intermediates as apparently widely used messengers in the activation of the NF-κB transcription factor and HIV-1. EMBO J 1991;10:2247–2258.
10 Satriano J, Schlondorff D: Activation and attenuation of transcription factor NF-kB in mouse glomerular mesangial cells in response to tumor necrosis factor-α, immunoglobulin G, and adenosine 3′:5′-cyclic monophosphate: Evidence for involvement of reactive oxygen species. J Clin Invest 1994;94:1629–1636.
11 Toledano MB, Leonard WJ: Modulation of transcription factor NF-κB binding activity by oxidation-reduction in vitro. Proc Natl Acad Sci USA 1991;88:4328–4332.
12 Ricardo SD, Bertram JF, Ryan GB: Antioxidants protect podocyte foot processes in puromycin aminonucleoside-treated rats. J Am Soc Nephrol 1994;4:1974–1986.
13 Kawamura T, Yoshioka T, Bills T, Fogo A, Ichikawa I: Glucocorticoid activates glomerular antioxidant enzymes and protects glomeruli from oxidant injuries. Kidney Int 1991;40:291–301.
14 Vissers MCM, Winterbourn CC: The effect of oxidants on neutrophil-mediated degradation of glomerular basement membrane collagen, Biochim Biophys Acta 1986;889:277–286.
15 Carp H, Janoff A: In vitro suppression of serum elastase-inhibitory capacity by reactive oxygen species generated by phagocytosing polymorphonuclear leukocytes. J Clin Invest 1979;63:793–797.
16 Weiss SJ, Regiani S: Neutrophils degrade subendothelial matrices in the presence of alpha-1-proteinase inhibitor. Cooperative use of lysosomal proteinases and oxygen metabolites. J Clin Invest 1984;73:1297–1303.
17 Carp H, Janoff A: Phagocyte-derived oxidants suppress the elastase-inhibitory capacity of alpha 1-proteinase inhibitor in vitro. J Clin Invest 1980;66:987–995.
18 Clark RA, Stone PJ, El Hag A, Calore JD, Franzblau C: Myeloperoxidase-catalyzed inactivation of alpha 1-protease inhibitor by human neutrophils. J Biol Chem 1981;256:3348–3353.
19 Donovan KL, Davies M, Coles GA, Williams JD: Relative roles of elastase and reactive oxygen species in the degradation of human glomerular basement membrane by intact neutrophils. Kidney Int 1994; 45:1555–1561.
20 Kakita N, Sasaguri Y, Kato S, Morimatsu M: Induction of gelatinolytic neutral proteinase secretion by lipid peroxide in cultured mesangial cells. Nephron 1993;63:94–99.
21 Kashihara N, Watanabe Y, Makino H, Wallner EI, Kanwar YS: Selective decreased de novo synthesis of glomerular proteoglycans under the influence of reactive oxygen species. Proc Natl Acad Sci USA 1992;89:6309–6313.
22 Rehan A, Johnson KJ, Wiggins RC, Kunkel RG, Ward PA: Evidence for the role of oxygen radicals in acute nephrotoxic nephritis. Lab Invest 1984;51: 396–403.
23 Diamond JR, Bonventre JV, Karnovsky MJ: A role for oxygen free radicals in aminonucleoside nephrosis. Kidney Int 1986;29:478–483.
24 Thakur V, Walker PD, Shah SV: Evidence suggesting a role for hydroxyl radical in puromycin aminonucleoside-induced proteinuria. Kidney Int 1988;34:494–499.
25 Ricardo SD, Bertram JF, Ryan GB: Reactive oxygen species in puromycin aminonucleoside nephrosis: In vitro studies. Kidney Int 1994;45:1057–1069.
26 Shah SV: Evidence suggesting a role for hydroxyl radical in passive Heymann nephritis in rats. Am J Physiol 1988;254:F337–F344.
27 Neale TJ, Ojha PP, Exner M, Poczewski H, Rüger B, Witztum JL, Davis P, Kerjaschki D: Proteinuria in passive Heymann nephritis is associated with lipid peroxidation and formation of adducts on type IV collagen. J Clin Invest 1994;94:1577–1584.
28 Nath KA, Croatt AJ, Hostetter TH: Oxygen consumption and oxidant stress in surviving nephrons. Am J Physiol 1990;258:F1354–F1362.
29 Nath KA, Salahudeen AK: Induction of renal growth and injury in the intact rat kidney by dietary deficiency of antioxidants. J Clin Invest 1990;86:1179–1192.
30 Houglum K, Brenner DA, Chojkier M: D-α to-copherol inhibits collagen α_1 (I) gene expression in cultured human fibroblasts: Modulation of constitutive collagen gene expression by lipid peroxidation. J Clin Invest 1991;87:2230–2235.
31 Ohyama K, Seyer JM, Raghow RA, Kang AH: A factor from damaged rat kidney stimulates collagen biosynthesis by mesangial cells. Biochim Biophys Acta 1990;1053:173–178.
32 Paller MS, Neumann TV: Reactive oxygen species and rat renal epithelial cells during hypoxia and reoxygenation. Kidney Int 1991;40:1041–1049.
33 Linas SL, Whittenburg D, Repine JE: Role of xanthine oxidase in ischemia/reperfusion injury. Am J Physiol 1990;258:F711–F716.

Mineral and Electrolyte Metabolism

Miner Electrolyte Metab 1997;23:144–150

Luca De Nicola
Roberto Minutolo
Vincenzo Bellizzi
Michele Andreucci
Antonietta La Verde
Bruno Cianciaruso

Nephrology Division,
School of Medicine-University Federico II,
Naples, Italy

Enhancement of Nitric Oxide Synthesis by *L*-Arginine Supplementation in Renal Disease: Is It Good or Bad?

Key Words
Angiotensin II
Arginine
Chronic Renal Failure
Cyclosporine
Glomerulonephritis
Hypertension
Nitrates
Nitric Oxide
Nitrites
Renal function

Abstract
L-Arginine is a important component of our diet characterized by multiple physiological and pharmacological actions. In the last decade, this amino acid has attracted major interest since it has been identified as the natural substrate of nitric oxide, and is now recognized to play a major role in the regulation of vascular tone. This review, while summarizing the knowledge of the renal actions of the *L*-arginine/nitric oxide pathway in health and renal disease, focuses on the potential therapeutic implications of the increase in nitric oxide synthesis attained by *L*-arginine supplementation.

Metabolism of *L*-Arginine (ARG)

Since the 1930s, ARG has been considered to occupy a key role in protein synthesis, biosynthesis of amino acids and their derivatives, and the urea cycle [1]. The sources of ARG are exogenous and endogenous. In humans, the daily intake of ARG with food generally varies from 1 to 5 g, 60–70% of which undergoes intestinal absorption. The endogenous sources are represented by release from muscle and synthesis in liver and kidney [2], with *L*-citrulline, derived from gut, liver, or renal tubule reabsorption, being the specific ARG substrate [2, 3]. Plasma ARG levels are maintained within the 60–120 μmol/l range, mainly by dietary intake and renal synthesis [1]. The activity of arginase, the enzyme that hydrolyzes ARG into urea and *L*-ornithine, is in fact several times higher in the liver than in the kidney, and the ARG concentration is therefore 5- to 10-fold greater in renal tissue than in the liver [4, 5].

ARG can be classified as a semiessential amino acid; indeed, while in ordinary conditions the amount of this amino acid introduced with the diet appears adequate, in the presence of increased degradation or utilization of ARG, as during growth, wound healing, sepsis, or injury, it is necessary to increase its dietary intake [6], since recent observations in healthy humans have shown that the biosynthesis of this amino acid does not increase in response to ARG-free diets [7]. Interestingly, although the kidney plays a central role in the generation of ARG, renal insufficiency is not associated with a reduction in its plasma levels [8]. On the other hand, *L*-citrulline levels significantly increase as renal function declines [8]. A likely explanation for this is that, as suggested by the studies of Tizianello et al. [9] in patients with chronic renal failure (CRF), the reduced uptake of *L*-citrulline is counterbalanced by increased synthesis of ARG in surviving nephrons and by the enhanced release of ARG by muscle; under these circumstances, therefore, increased dietary intake of this amino acid can become essential to maintain its plasma levels within the normal range.

Physiological Effects of ARG

As extensively reviewed by Barbul [1], ARG is characterized by multiple physiological effects (table 1). ARG occupies a central role in organism growth due to its multiple biosynthetic activities and hormonal secretagogue effects. The role of this amino acid in protein synthesis has led investigators to study its effects in injured individuals: ARG administration has been shown to accelerate wound healing and to decrease nitrogen losses after trauma, these effects being dependent on the enhanced synthesis of connective tissue proteins. A further important characteristic of ARG is represented by its ability, detected also in humans [10], to enhance lymphocyte blastogenesis in response to concanavalin A and phytohemagglutinin with the consequent improvement in host immunity and antitumor activity.

KARGER
E-Mail karger@karger.ch
Fax +41 61 306 12 34
http://www.karger.ch

0378–0392/97/0236–0144 $12.00/0

Bruno Cianciaruso, MD
Cattedra di Nefrologia, Fac. di Medicina
Università Federico II in Napoli
via Pansini 5, I–80131 Napoli (Italy)
Tel. +39 81 7462023, fax +39 81 7462149

Table 1. Physiological effects of ARG

Biosynthetic activity	
Substrate for	amino acid and protein synthesis
	creatine and other guanidine derivatives
	urea
	nitric oxide
↑ Synthesis of connective tissue proteins	
Male reproductive activity	
↑ Sperm counts	
↑ Sperm mobility	
Endocrine activity	
↑ Secretion of	growth hormone
	prolactin
	insulin
	glucagon
	somatostatin
	catecholamines
Immunological activity	
↑ Lymphocyte content of thymic gland	
↑ Peripheral blood lymphocyte blastogenesis to mitogens	
↑ Tumor recognition and antitumor response	

The above-mentioned effects have been well characterized in the past decades; on the other hand, recently renewed interest in ARG is related to the evidence that this amino acid is the physiological precursor of nitric oxide (NO), the main endothelial-derived relaxing factor [11]. In the presence of the enzyme NO synthase (NOS) and several cofactors, ARG is converted to *L*-citrulline and NO [12]. Two main isoforms of NOS have been identified [13, 14]. The constitutive form (cNOS) regulated by Ca^{2+} and calmodulin, was first described in endothelial cells, and in the brain, which tonically releases small short-lived (minutes) quantities of NO mainly in response to bradykinin, acetylcholine, and shear stress; NO generated by cNOS is responsible for the regulation of vascular tone and neurotransmission. The inducible form (iNOS), present in endothelial cells, macrophages, mesangial and vascular smooth muscle cells, and hepatocytes, which is Ca^{2+} and calmodulin independent is activated by endotoxin and/or cytokines and produces NO in nanomolar concentrations, more than 100-fold greater than the amount generated by cNOS, over a more prolonged period (hours). iNOS contributes to macrophage cytotoxicity. Both isoforms are widely spread throughout the different segments of the kidney (mesangium, interstitial cells, vessels, tubules and macula densa) in animals as in humans [15–17].

Systemic and Renal Actions of NO

In the last few years, several reports have been published on the diverse effects of NO, which is now considered as an important mediator of systemic homeostatic and host defense processes [13, 14, 18, 19]. The main systemic actions of NO are listed in table 2. A feasible way to assess in vivo NO activity is by measuring urinary levels of cGMP, the second messenger of NO, and/or by evaluating the urinary excretion of NO_2/NO_3, the stable endproducts of NO metabolism [13, 14, 20].

Table 2. Systemic actions of NO

Maintenance of vasodilator tone
↓ Aggregation/adhesion of platelets
↓ Proliferation of vascular smooth muscle cells
↓ Atherogenesis
Neurotransmission
↑ Lung ventilation:perfusion ratio
↑ Cytotoxicity of macrophages and neutrophils
Modulation of lymphocyte proliferation

Meanwhile, experimental NOS blockade has shed some light on the role of NO in the control of renal microcirculation: early micropuncture studies have shown that NO inhibition determines profound changes in glomerular hemodynamics and tubular handling of sodium that are completely nullified by intravenous administration of ARG [21]. Specifically, acute intravenous infusion of NOS blockers in normal rats induces, in parallel with the increment of blood pressure, a marked reduction in single-nephron glomerular filtration rate (GFR) resulting from glomerular vasoconstriction and the decrement of the glomerular ultrafiltration coefficient [22, 23]. Of note, these changes resemble those observed during systemic angiotensin II (AII) infusion [24]; indeed, AII antagonists effectively prevent alterations in glomerular hemodynamics following NOS blockade in spite of an unaffected pressor response [22, 23]. Interestingly, NOS blockers also decrease proximal tubular reabsorption of Na^+; this change, prevented by AII antagonists, suggests that the inhibition of NO enhances or releases the inhibitory effect of AII on proximal reabsorption [23]. Overall, these data demonstrate that NO is crucial to the constancy of normal renal perfusion, filtration rate and tubular handling of Na^+, and, moreover, that NO acts as a physiological antagonist of AII at both the glomerular and tubular level [23]. Similar findings have been seen during the chronic NO inhibition that is associated with the development of proteinuria and glomerular sclerosis due to hypertensive microangiopathy [25]. Different studies have shown that NO also antagonizes the effects of other renal vasoconstrictors such as endothelin-1 and catecholamines [26, 27].

Further work has provided more insights into the essential role that NO plays in the physiologic control of renal function. Renal autoregulation allows renal perfusion and the filtration rate to remain constant during increases of systemic arterial pressure by two distinct mechanisms, myogenic and tubuloglomerular feedback (TGF). Braam and Koomans [28] have recently demonstrated that NO critically contributes to renal autoregulation by blunting, directly and indirectly, through AII antagonism, TGF activity. NO also influences the myogenic component of autoregulation; indeed, in vitro, glomerular afferent but not efferent arterioles synthesize and release NO that, moreover, preferentially dilates the afferent arteriole [29]. Furthermore, NO is also involved in the mechanisms of renal hyperperfusion/filtration characterizing the physiological response to acute amino acid load and pregnancy [23, 30, 31].

Finally, it is worth emphasizing that the renal effects of NO are not limited to the control of glomerular and tubular dynamics. In vitro studies have demonstrated that NO acts directly and indirectly,

Table 3. Effects of either NO increase by ARG administration (+NO) or NO inhibition (–NO) in experimental models of renal disease

Author	Model	NO	Renal effects
Schultz and Raji [35]	EIRT in normals	–	↑ renal thrombosis
Raij [36]	EIRT in pregnancy	–	↑ renal thrombosis
De Nicola et al. [37]	cyclosporine nephrotoxicity	+	afferent vasodilation, ↑ renal reserve
Amore et al. [38]	cyclosporine nephrotoxicity	+	↑ GFR, ↓ proteinuria
Reyes et al. [39]	diabetes	+	↓ proteinuria, ↓ hyperfiltration
Chen et al. [41]	salt-sensitive hypertension	+	↓ BP, ↓ renal damage
Tolins and Shultz [43]	salt-sensitive hypertension	–	↑ BP, ↓ GFR, ↑ proteinuria
Ono et al. [42]	SHR	–	↑ BP, ↑ renal damage
Schramm et al. [44]	toxic acute renal failure	+	↑ GFR
Reyes et al. [45, 46]	obstructive nephropathy	+	↓ BP, ↑ GFR, ↓ macrophage infiltration
Reyes et al. [46]	nephrotic syndrome	+	↑ GFR, ↓ macrophage infiltration
Fujihara et al. [48]	subtotal nephrectomy	–	↑ renal damage, ↓ GFR
Reyes et al. [49]	subtotal nephrectomy	+	↑ GFR, ↓ renal damage
Katoh et al. [47]	subtotal nephrectomy	+	↓ glomerular pressure
Ashab et al. [50]	subtotal nephrectomy	+	↓ BP, ↓ proteinuria, ↑ GFR
Waddington et al. [52]	nephrotoxic nephritis	–	↑ proteinuria, ↑ renal thrombosis
Ferrario et al. [51]	nephrotoxic nephritis	–	↑ glomerular pressure, ↑ proteinuria
Narita et al. [53]	ATS glomerulonephritis	–	↓ proteinuria
Narita et al. [54]	ATS glomerulonephritis	–	↓ mesangial cell lysis
Weinberg et al. [55]	lupus erythematosus	–	↓ proteinuria, ↓ renal damage

EIRT = Endotoxin-induced renal thrombosis; BP = blood pressure; SHR = spontaneously hypertensive rats; ATS = anti-Thy-1 antiserum.

through AII antagonism, as a potent antiproliferative agent for vascular smooth muscle cells, mesangial cells, and fibroblasts [32–34]; this feature implies a potential role for NO in the progression of chronic renal disease.

ARG/NO Pathway in Experimental Renal Disease

On the basis of the positive physiological actions of NO on renal function and structure, one can reasonably hypothesize that the increase of NO by ARG administration is beneficial in renal disease; accordingly, NO inhibition should be associated with the opposite, that is, detrimental effects. The data attained on this issue in most of the experimental models of renal disease are consistent with this hypothesis (table 3).

In both normal and pregnant rats, when systemic infusion of endotoxin is combined with NOS blockade, a marked glomerular thrombosis develops; this can be prevented by administration of excess ARG [35, 36]. These findings not only confirm the hemodynamic, antiaggregatory, and antiplatelet adhesion properties of NO, but also indicate that NO deficiency plays a critical role in the pathogenesis of preeclampsia.

Nephroprotective effects of NO have been detected in various models of renal disease. De Nicola et al. [37] have suggested that short-term treatment with cyclosporin A (CsA), by diminishing NO activity within the kidney because of drug-induced endothelial damage, determines an increment in the tone of preglomerular resistance and, in parallel, the loss of the normal glomerular vasodilatory response to an acute amino acid load. These hemodynamic changes are reversed by oral ARG supplementation [37]. Similarly, Amore et al. [38] have recently confirmed the beneficial renal effect of ARG administration in a chronic model of CsA nephrotoxicity. Furthermore, renal function and structure are protected by ARG administration in streptozotocin-induced diabetes mellitus [39], possibly through NO/AII antagonism [40]. Similar findings have been reported in experimental models of systemic hypertension [41–43], in uranyl-nitrate-induced acute renal failure [44], in the obstructive nephropathy resulting from bilateral ureteral ligation [45, 46], and in the puromycin-induced model of nephrotic syndrome [46]. In the latter two models, ARG decreases macrophage infiltration which is responsible, at least in part, for the GFR decrement.

In remnant rats, the main animal model of renal insufficiency, ARG administration is associated with the improvement of GFR, proteinuria, and renal morphology despite no amelioration of systemic blood pressure elevation [47–50]. The authors have ascribed the improvement of renal outcome to the capacity of NO to diminish glomerular hypertension, the major pathophysiological feature in remnant rats, and/or to the antiproliferative action of NO.

At variance with the above-mentioned models of renal disease, conflicting findings have been reported in experimental glomerulonephritis (GN). Two different groups have shown that inhibition of NO synthesis, by either NOS blockade or ARG depletion, exacerbates proteinuria in accelerated nephrotoxic nephritis [51, 52]. These results suggest that NO has an important role in limiting acute glomerular injury in this experimental model of immunological GN, possibly through attenuation of glomerular vasoconstriction/thrombosis and/or the ability of scavenging oxygen radicals [51, 52]. In contrast, in experimental models of mesangial proliferative GN and systemic autoimmune disease, the early decrement of NO synthesis, resulting from iNOS blockade in infiltrating macrophages, leads to amelioration of the renal outcome secondary to the decreasing expression of inflammatory factors (such as transforming growth factor-β1), to a reduction in both the production and deposition of matrix components, as well as to a decrease in the formation of cytotoxic compounds derived from the reaction of NO with O_2^- radicals [53–56]. These observations have led investigators to speculate that the therapeutic effects of low-protein diets in renal disease are mainly mediated by the restriction of ARG intake [53].

The discrepancies detected on the renal effects of ARG/NO in the different models of experimental GN cannot be readily explained. Nevertheless, it is possible to hypothesize that the model studied, and the stage at which observations are made, account for the differences. Indeed, immune complex GN is characterized, at the very early stages, by marked cytokine-indiced activation of the iNOS present in infiltrating leukocytes; large quantities of NO are consequently released for many hours with the enhancement of the cytotoxic activity of macrophages [57, 58]. Similarly, the activity of arginase, the enzyme that converts ARG to urea and *L*-ornithine, is also elevated at the early stages of these experimental GNs [58]. Indeed, *L*-ornithine is an important substrate for the synthesis of polyamine and proline, substances that lead to glomerulosclerosis by increasing proliferation of mesangial cells and deposition of collagen [14, 59, 60].

Therefore, detrimental effects on renal function may result from ARG administration in the acute phase of GN characterized by major leukocyte infiltration and enhanced activity of iNOS and/or arginase. On the other hand, ARG supplementation may be beneficial when the acute phase of the immunological injury is finished. Indeed, as shown in renal diseases other than GN, in the presence of endothelial damage, glomerular hypertension and/or systemic hypertension, a small increment in NO production by cNOS may protect renal function by diminishing glomerular vasoconstriction and thrombosis, as well as by counteracting mesangial cell contraction/proliferation [33, 61]. Intriguing data have been recently reported on the major role of NO in the nephroprotective effects of treatment with converting enzyme inhibitors (CEIs) in chronic renal disease; specifically, the CEI-induced increase in NO production proved to be essential to the improvement of renal function and histology detected in experimental models of CRF and hypertensive nephrosclerosis after chronic CEI administration [42, 50].

ARG/NO Pathway in Healthy Subjects and Renal Patients

Since the discovery of NO, evidence has progressively accumulated in humans on the activity of the ARG/NO pathway and the two isoforms of NOS [15–17, 62–64]. In humans, various studies, in both healthy subjects and patients, have proved the biological significance of the ARG/NO pathway in the control of vascular tone. In healthy volunteers, systemic inhibition of NO synthesis evokes a parallel increase in blood pressure and vascular resistance [65], while opposite effects ensue following ARG infusion [66, 67].

Nevertheless, while it has been proved that NO plays a major role in the physiological regulation of vascular tone, to date scanty and often contradictory information has been provided on the efficacy of ARG supplementation/NO increase in human renal disease.

In essential hypertension, the basal activity of the ARG/NO pathway, tested by acute NOS blockade, is impaired, with the degree of impairment being inversely correlated with blood pressure [68, 69]; these data suggest that NO deficiency may play a pathogen role in hypertensive patients. However, the normalization of the response to NOS blockers after antihypertensive treatment [70] and the absence of improvement after ARG treatment [71] alternatively suggest that the reduced NO production in these patients merely represents an adaptation to the increased blood pressure. Similarly, increased NO release following ARG infusion has been found to enhance diuresis and natriuresis in cirrhotic patients with ascites, but the significance of this finding remains undefined [72].

In addition, on the basis of the impressive body of evidence provided on the striking nephroprotective effects of ARG supplementation in remnant rats, investigators have focused their interest on the activity of the ARG/NO pathway in CRF patients; but, again, the results have been contradictory. In patients with end-stage renal failure, Vallance and colleagues [73–75] have demonstrated the presence of an ARG analogue, asymmetrical dimethylarginine (ADMA) released by endothelial cells, that, by competing with ARG at the level of both isoforms of NOS, acts as a potent endogenous inhibitor of NO synthesis. This group of investigators found that in end-stage CRF, plasma ADMA concentrations increase proportionally to the increment in serum creatinine, and are reduced by dialysis. These data therefore suggest a role for ADMA as a uremic toxin; it possibly contributes to the development of some of the vascular, immunological, and neurological dysfunctions characteristically present in uremia. In contrast, Noris et al. [76] and Yokokawa et al. [77] have shown that plasma from uremic patients actually induces NO generation by iNOS from cultured endothelial cells, and that such increased production of NO may be responsible for platelet dysfunction and dialysis hypotension in these patients. The discrepancies with the previous findings may be determined by the much higher levels of circulating ARG found in the patients studied by Noris et al. [76] compared with those in the study by Vallance et al. [73]; indeed, the activity of iNOS strictly depends on the availability of extracellular ARG [78]. More recent studies have further complicated this issue by showing that about 80% of the plasma samples from uremic patients and patients with moderate renal failure inhibit NOS activity, while up to 20% of the samples stimulate it [79]. Therefore, the modulation of NO synthesis by plasma of CRF patients remains controversial.

Conflicting data have also been attained on the renal effects of ARG administration in proteinuric patients with decreased renal function. In patients with chronic glomerulonephritis and a reduced GFR, acute ARG infusion significantly decreases proteinuria in the absence of any effect on systemic and renal hemodynamics [80]. Similarly, Claris-Appiani et al. [81] have shown an antiproteinuric effect of a 2-month treatment with oral ARG in pediatric patients with interstitial nephritis and CRF. However, the interpretation of the data of both studies is limited by the short follow-up and the absence of insights into the pathophysiological mechanism/s underlying the described phenomenon.

We have recently performed a randomized, double-blind study aimed at evaluating the effects on NOS activity and renal function of oral supplementation with ARG (0.2 g/kg per day) over a 6-month period in patients with moderate renal failure (inulin clearance = 20–60 ml/min) resulting from biopsy-proven primary GN; the control group (CON) comprised CRF patients with GN receiving placebo [82]. All patients were moderately proteinuric (U_{prot}/U_{creat} = 0.9 ± 0.2 in ARG and 1.2 ± 0.3 in CON). Basal mean arterial pressure was similar in both groups (95 ± 3 mm Hg in ARG and 100 ± 5 mm Hg in CON) and did not vary through the study. Plasma ARG, in the normal range (about 80 μmol/l) at baseline in both groups, significantly increased in the ARG group while it remained unchanged in CON, suggesting adherence to the prescribed ARG intake in ARG patients. At baseline, the urinary excretion of nitrates, the end products of NO metabolism, was equivalent in the two groups of patients (447 ± 40 nmol/min in ARG and 430 ± 30 nmol/min in CON); these values were strikingly similar to those detected in healthy subjects in a previous study by our group [20]. After 6 months of ARG administration, the increment in plasma ARG was paralleled by a marked increase in urinary nitrates (770 ± 30 nmol/min) while no change was observed in the controls (360 ± 60 nmol/min). In ARG patients, the increase in NO production was associated with no variation in either GFR or proteinuria. Therefore, this study provides the first evidence that in patients with moderate CRF secondary to GN, NO production is normal, and that chronic ARG supplementation, while being effective in enhancing NO generation, does not affect the renal outcome.

On the basis of these data, we may assume that in moderate CRF, the ARG/NO pathway is intact. Nevertheless, it is important to note that in our ARG patients, azotemia was increased by 30% at the end of the study, indicating additional activation of the arginase pathway of ARG metabolism. Consequently, the simultaneous activation of the two pathways may have led to the production of both NO, an antisclerotic agent, and *L*-ornithine, a prosclerotic agent; this hypothesis certainly requires further studies. Similarly, another aspect warrants further investigation; it is now well established that in GN patients with moderate CRF, 2 months treatment with CEIs effectively decreases proteinuria [83]; therefore, at variance with the experimental data [42, 50], our study suggests that in CRF patients, the increment of NO production is probably not sufficient to limit protein excretion if it is not associated with AII antagonism, at least within the initial few months of therapy.

Conclusions

An impressive body of experimental work has been conducted on the renal effects of the ARG/NO pathway. Most of the animal studies have demonstrated that the integrity of this pathway is crucial to the physiological functioning and the normal structure of both glomeruli and tubules. These findings have been further confirmed by the observation that the increased generation of NO by ARG is nephroprotective in a variety of experimental renal diseases, particularly those characterized by derangements of glomerular hemodynamics. In contrast, in humans, while it has been demonstrated that NO plays a major role in the physiological regulation of vascular tone, very few studies have addressed the critical issue as to whether the outcome of renal diseases can be favorably influenced by the ARG-induced generation of NO. The few clinical studies, moreover, have produced results that do not confirm the striking benefits indicated by the experimental work. Investigators should therefore face this reality, and try to gain in humans more insights into the activity and functions of the ARG/NO pathway.

Acknowledgment

This study was partially supported by funding to B.C. from M.U.R.S.T. 60%, 1994–1995.

References

1 Barbul A: Arginine: Biochemistry, physiology and therapeutic implications. J Parent Enter Nutr 1986; 10:227–238.
2 Levillain O, Hus-Citharel A, Morel F, Bankir L: Localization of arginine synthesis along rat nephron. Am J Physiol 1990;259:F916–F923.
3 Harper A, Yoshimura NN: Protein quality, amino acid balance, utilization and evaluation of diets containing amino acids as therapeutic agents. Nutrition 1993;9:460–469.
4 Featherstone WR, Rogers QR, Freedland RA: Relative importance of kidney and liver in synthesis of arginine by the rat. Am J Physiol 1973;224:127–129.
5 Gopalakrishna R, Nagarajan B: Effect of growth and differentiation on distribution of arginase and arginine in rat tissue. Indian J Biochem Biophys 1979;16:66–68.
6 Rose WC: The nutritive significance of the amino acids and certain related compounds. Science 1937; 86:298–300.
7 Castillo L, Chapman TE, Sanchez M, Yu YM, Burke JF, Ajami AM, Vogt J, Young VR: Plasma arginine and citrulline kinetics in adult given adequate and arginine-free diets. Proc Natl Acad Sci USA 1993;90:7749–7753.
8 Kuhlmann MK, Kopple JD: Amino acid metabolism in the kidney. Semin Nephrol 1990;10:445–457.
9 Tizianello A, De Ferrari G, Garibotto G, Guerreri G, Robaudo C: Renal metabolism of amino acids and ammonia in subjects with normal renal function and in patients with chronic renal insufficiency. J Clin Invest 1980;65:1162–1173.
10 Barbul A, Sisto DA, Wasserkrug HL: Arginine stimulates lymphocyte immune response in healthy humans. Surgery 1981;90:244–251.
11 Palmer RMJ, Ferrige AG, Moncada S: Nitric oxide release accounts for the biological activity of endothelium-derived relaxing factor. Nature 1988;327: 524–526.
12 Marletta MA: Nitric oxide synthase structure and mechanism. J Biol Chem 1993;268:12231–12234.
13 Moncada S, Higgs A: The *L*-arginine-nitric oxide pathway. N Engl J Med 1993;27:2002–2012.
14 Ketteler M, Border WA, Noble N: Cytokines and *L*-arginine in renal injury and repair. Am J Physiol 1994;267:F197–F207.
15 Nicholson AG, Haites NE, McKay NG, Wilson HM, MacLeod AM, Benjamin N: Induction of nitric oxide synthase in human mesangial cells. Biochem Biophys Res Commun 1993;193:1269–1274.
16 Bosse HM, Moser D, Bachmann S: Constitutive isoforms of nitric oxide synthase (NOS) in the kidney: Localisation in various mammalian species including man. J Am Soc Nephrol 1994;5:574.
17 McLay JS, Chatterjee P, Nicolson G, Jardine AJ, McLay NG, Ralston SH, Grabowski P, Haites NE, MacLeod AM, Hawksworth GM: Nitric oxide production by human proximal tubular cells: A novel immunomodulatory mechanism? Kidney Int 1994; 46:1043–1049.

18 Reyes AA, Karl IE, Klahr S: Role of arginine in health and in renal disease. Am J Physiol 1994;267: F331–F346.

19 Cooke JP, Singer AH, Tsao P, Zera P, Rowan RA, Billingham ME: Antiatherogenic effects of *L*-arginine in the hypercholesterolemic rabbit. J Clin Invest 1992;90:1168–1172.

20 Corso G, De Nicola L, Cianciaruso B, Dello Russo A, Napoli C, Bellizzi V: Automated enzymatic determination of urinary nitrates in humans. Curr Ther Res 1996;57:878–884.

21 Raji L, Baylis C: Glomerular actions of nitric oxide. Kidney Int 1995;48:20–32.

22 Zatz R, De Nucci G: Effect of acute nitric oxide inhibition on rat glomerular microcirculation. Am J Physiol 1991;261:F360–F363.

23 De Nicola L, Blantz RC, Gabbai FB: Nitric oxide and angiotensin II: Glomerular and tubular interaction in the rat. J Clin Invest 1992;89:1248–1256.

24 Blantz RC, Konnen KS, Tucker BJ: Angiotensin II effects upon glomerular microcirculation and ultrafiltration coefficient in the rat. J Clin Invest 1976; 57:419–434.

25 Baylis C, Mitruka B, Deng A: Chronic blockade of nitric oxide synthesis in the rat produces systemic hypertension and glomerular damage. J Clin Invest 1992;90:278–281.

26 Qiu C, Engels K, Baylis C: Endothelin modulates the pressor actions of acute systemic nitric oxide blockade. J Am Soc Nephrol 1995;6:1476–1481.

27 Thomson SC, Vallon V: α_2-Adrenoceptors determine the response to nitric oxide inhibition in the rat glomerulus and proximal tubule. J Am Soc Nephrol 1995;6:1482–1490.

28 Braam B, Koomans HA: Nitric oxide antagonizes the actions of angiotensin II to enhance tubuloglomerular feedback responsiveness. Kidney Int 1995; 48:1406–1411.

29 Edwards RM, Trizna W: Modulation of glomerular arteriolar tone by nitric oxide synthase inhibitors. J Am Soc Nephrol 1993;4:1127–1132.

30 Salazar JF, Alberola A, Nakamura T, Granger JP: Role of nitric oxide in the renal hemodynamic response to a meat meal. Am J Physiol 1994;267: R1050–R1055.

31 Danielson LA, Conrad KP: Acute blockade of nitric oxide synthase inhibits renal vasodilation and hyperfiltration during pregnancy in chronically instrumented conscious rats. J Clin Invest 1995;96: 482–490.

32 Garg UC, Hassid A: Nitric oxide-generating vasodilators and 8 bromo-cyclic guanosine monophosphate inhibit mitogenesis and proliferation of cultured vascular smooth muscle cells. J Clin Invest 1989;83:1774–1777.

33 Garg UC, Hassid A: Inhibition of rat mesangial cell mitogenesis by nitric oxide-generating vasodilators. Am J Physiol 1989;257:F60–F66.

34 Garg UC, Hassid A: Nitric oxide-generating vasodilators inhibit mitogenesis and proliferation of BALB/c3T3 fibroblasts by a cGMP independent mechanism. Biochem Biophys Res Commun 1990; 171:474–479.

35 Schultz PJ, Raji L: Endogenously synthesized nitric oxide prevents endotoxin-induced glomerular thrombosis. J Clin Invest 1992;90:1718–1725.

36 Raij L: Glomerular thrombosis in pregnancy: Role of *L*-arginine and nitric oxide pathway. Kidney Int 1994;45:775–781.

37 De Nicola L, Thomson SC, Wead LM, Brown MR, Gabbai FR: Arginine feeding modifies cyclosporine nephrotoxicity in rats. J Clin Invest 1993;92:1859–1865.

38 Amore A, Gianoglio B, Ghigo D, Peruzzi L, Porcellini MG, Bussolino F, Costamagna C, Cacace G, Picciotto G, Mazzucco G, Sena LM, Coppo R: A possible role for nitric oxide in modulating the functional cyclosporine toxicity by arginine. Kidney Int 1995;47:1507–1514.

39 Reyes AA, Karl IE, Kissane J, Klahr S: *L*-Arginine administration prevents glomerular hyperfiltration and decreases proteinuria in diabetic rats. J Am Soc Nephrol 1993;4:1039–1045.

40 Mathis KM, Banks RO: Role of nitric oxide and angiotensin II in diabetes mellitus-induced glomerular hyperfiltration. J Am Soc Nephrol 1996;7: 105–112.

41 Chen PY, St John PL, Abrahamson DR, Sanders PW: Chronic arginine supplementation prevents hypertensive nephrosclerosis and hypertension in Dahl/Rapp salt-sensitive (SS/Jr) rats. Clin Res 1992; 40:353.

42 Ono H, Ono Y, Froloich ED: ACE inhibition prevents and reverses *L*-NAME-exacerbated nephrosclerosis in spontaneously hypertensive rats. Hypertension 1996;27:176–183.

43 Tolins JP, Shultz PJ: Endogenous nitric oxide synthesis determines sensitivity to the pressor effect of salt. Kidney Int 1994;46:230–236.

44 Schramm L, Heidbreder E, Lopau K, Schaar J, De Cicco D, Götz R, Heidland A: Toxic acute renal failure in the rat: Effects of *L*-arginine and N-methyl-*L*-arginine on renal function. Nephrol Dial Transplant 1994;9(suppl4):88–93.

45 Reyes AA, Martin D, Settle S, Klahr S: EDRF role in renal function and blood pressure of normal rats and rats with obstructive nephropathy. Kidney Int 1992;41:403–413.

46 Reyes AA, Porras BH, Chasalow F, Klahr S: Administration of *L*-arginine decreases the infiltration of the kidney by macrophages in obstructive nephropathy and puromycin induced nephrosis. Kidney Int 1994;45:1346–1354.

47 Katoh T, Takahashi K, Klahr S, Reyes AA, Badr KF: Dietary supplementation with *L*-arginine ameliorates glomerular hypertension in rats with subtotal nephrectomy. J Am Soc Nephrol 1994;4: 1690–1694.

48 Fujihara CK, De Nucci G, Zatz R: Chronic nitric oxide synthase inhibition aggravates glomerular injury in rats with subtotal nephrectomy. J Am Soc Nephrol 1995;5:1498–1507.

49 Reyes AA, Purkerson ML, Karl I, Klahr S: Dietary supplementation with *L*-arginine ameliorates the progression of renal disease in rats with subtotal nephrectomy. Am J Kidney Dis 1992;20:168–176.

50 Ashab I, Peer G, Blum M, Wollman Y, Chernihovsky T, Hassner A, Schwartz D, Cabili S, Silverberg D, Iaina A: Oral administration of *L*-arginine and captopril in rats prevents chronic renal failure by nitric oxide production. Kidney Int 1995;47:1515–1521.

51 Ferrario R, Takahashi K, Fogo A, Badr K, Munger KA: Consequences of acute nitric oxide synthesis inhibition in experimental glomerulonephritis. J Am Soc Nephrol 1994;4:1847–1854.

52 Waddington S, Cook TH, Reaveley D, Jansen A, Cattell V: *L*-Arginine depletion inhibits glomerular nitric oxide synthesis and exacerbates rat nephrotoxic nephritis. Kidney Int 1996;49:1090–1096.

53 Narita I, Border WA, Ketteler M, Rouslahti E, Noble NA: *L*-Arginine may mediate the therapeutic effects of low protein diet. Proc Natl Acad Sci USA 1995;92:4552–4556.

54 Narita I, Border WA, Ketteler M, Noble NA: Nitric oxide mediates immunologic injury to kidney mesangium in experimental glomerulonephritis. Lab Invest 1995;72:17–24.

55 Weinberg JB, Granger DL, Pisetski DS, Seldin MF, Misukonis MA, Mason SN, Pippen AM, Ruiz P, Wood ER, Gilkeson GS: The role of nitric oxide in the pathogenesis of spontaneous murine autoimmune disease: Increased nitric oxide production and nitric oxide synthase expression in MRL-lpr/lpr mice, and reduction of spontaneous glomerulonephritis and arthritis by orally administered N^G-monomethyl-*L*-arginine. J Exp Med 1994;179:651–660.

56 Beckman JS, Beckman TW, Chen J, Marshall PA, Freeman BA: Apparent hydroxyl radical production by peroxynitrite: Implication for endothelial injury from nitric oxide and superoxide. Proc Natl Acad Sci USA 1990;87:1223–1227.

57 Jansen A, Cook TH, Taylor MG, Largen P, Riveros-Moreno V, Moncada S, Cattell V: Induction of nitric oxide synthase in rat immune complex glomerulonephritis. Kidney Int 1994;45:1215–1219.

58 Cook TH, Jansen A, Lewis S, Largen P, O'Donnell M, Reaveley D, Cattell V: Arginine metabolism in experimental glomerulonephritis: Interaction between nitric oxide synthase and arginase. Am J Physiol 1994;267:F646–F653.

59 Schulze-Lohoff E, Brand K, Fees H, Netzker R, Sterzel RB: Role of ornithine decarboxylase for proliferation of mesangial cells in culture. Kidney Int 1991;40:684–690.

60 Kubes P, Suzuki M, Granger DN: Nitric oxide: An endogenous modulator of leukocyte adhesion. Proc Natl Acad Sci USA 1991;88:4651–4655.

61 Schultz PJ, Schorer AE, Raij L: Effects of endothelium-derived relaxing factor and nitric oxide on rat mesangial cells. Am J Physiol 1990;258:F162–F167.

62 Vallance P, Collier J, Moncada S: Effects of endothelium-derived nitric oxide on peripheral arteriolar tone in man. Lancet 1989;ii:997–1000.

63 Vallance P, Collier J, Moncada S: Nitric oxide synthesised from *L*-arginine mediates endothelium-dependent dilatation in human veins in vivo. Cardiovasc Res 1989;23:1053–1057.

64 Vallance P, Patton S, Bhagat K, MacAllister R, Radomski M, Moncada S, Malinski T: Direct measurement of nitric oxide in human beings. Lancet 1995;345:153–154.

65 Haynes WG, Noon JP, Walker BR, Webb DJ: *L*-NMMA increases blood pressure in man. Lancet 1993;342:931–932.

66 Nakaki T, Hishikawa K, Suzuki H, Saruta T, Kato R: *L*-Arginine-induced hypotension. Lancet 1990; 336:696.

67 Kanno K, Hirata Y, Emori T, Ohta K, Eguchi S, Imai T, Marumo F: *L*-Arginine infusion induces hypotension and diuresis/natriuresis with concomitant increased urinary excretion of nitrite/nitrate and cyclic GMP in humans. Clin Exp Pharmacol Physiol 1992;19:619–625.

68 Calver A, Collier J, Moncada S, Vallance P: Effect of local intra-arterial N^G-monomethyl-*L*-arginine in patients with hypertension: The nitric oxide dilator mechanism appears abnormal. J Hypertens 1992;10:1025–1031.

69 Panza JA, Casino PR, Kilcoyne CM, Quyyumi AA: Role of endothelium-derived nitric oxide in the abnormal endothelium-dependent vascular relaxation of patients with essential hypertension. Circulation 1993;87:1468–1474.

70 Calver A, Collier J, Vallance PTJ: Effect of N^G-monomethyl-*L*-arginine (L-NMMA) on forearm blood flow (FBF) in treated hypertensive patients. Br J Clin Pharmacol 1992;35:68P.

71 Panza JA, Casino PR, Badar DM, Quyyumi AA: Effect of increased availability of endothelium-derived nitric oxide precursor on endothelium-dependent vascular relaxation in normal subjects and in patients with essential hypertension. Circulation 1993;87:1475–1481.
72 Tajiri K, Miyakawa H, Izumi N, Marumo F, Sato C: Systemic hypotension and diuresis by *L*-arginine in cirrhotic patients with ascite: Role of nitric oxide. Hepatology 1995;22:1430–1435.
73 Vallance P, Leone A, Calver A, Collier J, Moncada S: Accumulation of an endogenous inhibitor of nitric oxide synthesis in chronic renal failure. Lancet 1992;339:572–575.
74 MacAllister RJ, Whitley GS, Vallance P: Effects of guanidino and uremic compounds on nitric oxide pathways. Kidney Int 1994;45:737–742.
75 Fickling S, Leone A, Nussey S, Vallance P, Whitley GS: Synthesis of N^GN^G dimethylarginine by human endothelial cells. Endothelium 1993;1:137–140.
76 Noris M, Benigni A, Boccard P, Aiello S, Gaspari F, Todeschini M, Figliuzzi M, Remuzzi G: Enhanced nitric oxide synthesis in uremia: Implications for platelet dysfunction and dialysis hypotension. Kidney Int 1993;44:445–450.
77 Yokokawa K, Mankus R, Saklayen MG, Kohno M, Yasunari K, Minami M, Kano H, Horio T, Takeda T, Mandel AK: Increased nitric oxide production in patients with hypotension during hemodialysis. Ann Intern Med 1995;123:35–37.
78 Yui Y, Hattori R, Kosuga K, Eizawa H, Hiki K, Kawai C: Purification of nitric oxide synthase from rat macrophages. J Biol Chem 1991;266:12544–12547.
79 Arese M, Strasly M, Ruva C, Costamagna C, Ghigo D, MacAllister R, Verzetti G, Tetta C, Bosia A, Bussolino F: Regulation of nitric oxide synthesis in uremia. Nephrol Dial Transplant 1995;10:1386–1397.
80 Wolf SC, Erley CM, Kenner S, Berger ED, Risler T: Does *L*-arginine alter proteinuria and renal hemodynamics in patients with chronic glomerulonephritis and hypertension? Clin Nephrol 1995;43:S42–S46.
81 Claris-Appiani A, Ardissimo G, Coppo R, Bonaudo R, Daccò V, Bettinelli A, Tirelli A: Effect on renal function of arginine supplementation in children with chronic renal insufficiency (abstract). J Am Soc Nephrol 1993;4:766.
82 Cianciaruso B, Bellizzi V, Minutolo R, Andreucci M, Russo R, Conte G, De Nicola L: Chronic supplementation of *L*-arginine in patients with moderate renal failure secondary to chronic glomerulonephritis (abstract). J Am Soc Nephrol 1996;7:1616.
83 Maschio G, Alberti D, Janin G, Locatelli F, Mann JFE, Motolese M, Ponticelli C, Ritz E, Zucchelli P: Effect of the angiotensin-converting-enzyme inhibitor benazepril on the progression of chronic renal insufficiency. N Engl J Med 1996;334:939–945.

Mineral and Electrolyte Metabolism

Miner Electrolyte Metab 1997;23:151–156

Sistiana Aiello[a]
Marina Noris[a]
Giuseppe Remuzzi[a, b]

[a] Mario Negri Institute for Pharmacological Research, and
[b] Ospedali Riuniti di Bergamo, Bergamo, Italy

Nitric Oxide Synthesis and *L*-Arginine in Uremia

Key Words
Nitric oxide
Nitric oxide synthases
L-Arginine
Chronic renal failure

Abstract
Nitric oxide (NO), an *L*-arginine derivative, is implicated in neuronal transmission, immune response and vasodilation, and acts as a modulator of platelet function. Recent studies in the experimental model of renal mass reduction (RMR) in rats have generated the hypothesis that abnormalities in the NO synthetic pathway could play a key role in mediating the complex hemodynamic and hemostatic disorders associated with the progression of renal disease. Thus, renal NO generation is lower than normal in rats with RMR 7 days after surgery and progressively worsens with time in close correlation with signs of renal injury. This abnormality is due to a major defect in inducible NO synthase (iNOS) content in the kidney. In the same model, administration of either the NO precursor, *L*-arginine, or a NO-releasing compound reduces proteinuria, slows renal disease progression, and prolongs survival. In contrast, in the systemic circulation of uremic rats, NO is formed in excessive amounts, possibly caused by higher release from systemic vessels due to the augmented expression of both iNOS and endothelial NOS. Up-regulation of systemic NO synthesis might be a defense mechanism against uremic hypertension. On the other hand, a greater availabilty of NO to circulating cells may sustain the bleeding tendency, a well-known complication of uremia.

Nitric Oxide (NO): Biochemistry

NO is a gaseous free radical which has been implicated in a variety of physiological processes, such as vasodilation, neuronal transmission, immune response, and platelet function modulation [1].

In 1981, Furchgott and Zawadski [2] showed that the relaxation of arteries by pharmacological agents such as acetylcholine was dependent on the presence of an intact endothelium, a phenomenon that was attributed to an endothelium-derived relaxing factor (EDRF). In 1987, Palmer et al. [3] demonstrated that NO accounted for EDRF activity and that the precursor for its formation was the semiessential amino acid *L*-arginine. The synthesis of NO by vascular endothelium is responsible for vasodilator tone and is essential for the regulation of blood pressure. In vivo administration of *L*-arginine analogues, such as N^G-monomethyl-*L*-arginine (L-NMMA), that inhibit NO synthase, induces an increase in blood pressure of about 30 mm Hg in rats and rabbits, indicating that there is a continuous release of NO [4].

In the central nervous system, NO is a neurotransmitter that underlies several functions, including the formation of memory [5]. In the peripheral nervous system, there is a widespread network of nerves, previously recognized as nonadrenergic and noncholinergic, that operate through an NO-dependent mechanism to mediate some forms of neurogenic vasodilation and regulate various gastrointestinal, respiratory, and genitourinary tract functions [5].

NO was also shown to play a central role in mediating macrophage cytotoxicity during host defense and immunologic reactions. Thus, macrophages can induce oxidative injury on tumor cells and mycobacteria which is associated with the release of NO [5]. Further work demonstrated that analogues of *L*-arginine (e.g., N-nitro-*L*-arginine-methyl ester, L-NAME), which inhibit NO synthesis, could block this activity and that authentic NO could mimic the pattern of macrophage cytotoxic effects [5].

NO also modulates blood clotting and contributes to the control of platelet aggregation and adhesion. Thus, human vein endothelial cells (HUVECs) blocked platelet serotonin release and aggregation through generation of NO, and this inhibitory effect was reversed by preincubation of HUVECs with L-NMMA [6]. In addition to its antiaggregatory function, NO also modulates platelet adhesion: indeed NO reduced the adhesion of thrombin-activated platelets to the endothelium [7]. An *L*-arginine/NO pathway is also identifiable in platelets. Platelet-produced NO, by stimulating the soluble guanylate cyclase and increasing the levels of cyclic GMP, serves to autoregulate platelet function [5].

NO is derived from the amino acid *L*-arginine in an unusual reaction catalyzed by a family of enzymes, termed NO synthases (NOSs), that convert arginine and oxygen into citrulline and NO

KARGER
E-Mail karger@karger.ch
Fax +41 61 306 12 34
http://www.karger.ch

0378–0392/97/0236–0151 $12.00/0

Sistiana Aiello, Biol. Sci. D.
Mario Negri Institute for Pharmacological Research
via Gavazzeni 11, I–24125 Bergamo (Italy)
Tel. +39 35 319 888, fax +39 35 319 331
E-Mail: aiello@irfmn.mnegri.it

Table 1. Characterization of the three human NOS genes

	Isoforms		
	nNOS (NOS1)	iNOS (NOS2)	ecNOS (NOS3)
Expression	constitutive	inducible	constitutive
Human chromosomal localization	12q24.2	17q11.2-q12	7q35-q36
Cell prototype	neurons, platelets, renal epithelial cells	hepatocytes, macrophages, mesangial cells, vascular smooth muscle cells	endothelial cells
Number of exons	29	26	26
Subcellular localization	cytosolic/membrane	cytosolic	membrane
Calcium dependency	+	–	+
Calmodulin dependency	+	+	+

[8]. In order of their molecular characterization, NOS isoforms include the neuronal (nNOS, NOS1), inducible (iNOS, NOS2) and endothelial (ecNOS, NOS3) NOSs, the genes for which are located on chromosomes 12, 17, and 7, respectively [8] (table 1). nNOS and ecNOS are constitutive enzymes and are inactive until there is an increase in intracellular calcium levels [8]. These NOSs are activated by calcium-elevating agonists (e.g. acetylcholine, thrombin, histamine) and generate small quantities of NO. In contrast, iNOS is a calcium-independent isoform which can be induced by various cytokines and produces huge amounts of NO for long periods of time [8]. Originally purified from neurons, vascular endothelium, and cytokine-induced macrophages, the three isoforms are now understood to be distributed across a wide spectrum of cell types and tissues. All three enzymes have been localized in the kidney and have been implicated in the control of its function.

NOSs in the Kidney: Localization and Function

Early attempts to localize NOS activity in the kidney were based on indirect evidence for NO biosynthesis obtained by measuring nitrites and nitrates, the stable end products of NO, and cGMP levels, the NO second messenger. For these measurements, tissue slices of dissected renal zones as well as isolated glomeruli and various cell cultures were used [9–11]. These in vitro studies suggested that renal NO was not only produced by endothelial cells, but also by smooth muscle cells, mesangial cells, and tubular epithelial cells. For general histochemical detection of NOS, the nitroblue tetrazolium reaction has been used [12]. The intensity of this reaction varies with the NADPH diaphorase activity of NOS, thereby indicating NOS activity irrespective of the particular isoform. Immunohistochemical methods, using isoform-specific antibodies and in situ hybridization techniques, have been used to localize the diaphorase labeling to a particular NOS isoform [12].

All three known NOS isoforms have been detected in the kidney (fig. 1). The constitutive nNOS has been localized strictly to the cells of the macula densa of the rat and mouse kidney by different groups [12]. Thus, a strong NADPH-diaphorase reaction was observed in the macula densa of the juxtaglomerular apparatus which colocalized with immunostaining for nNOS [12]. Using RT-PCR in individual, microdissected rat nephron segments, however, Terada et al. [13] localized nNOS mRNA to the inner medullary collecting duct and, to a lesser extent, to the glomerulus, inner medullary thin limb, and cortical and outer medullary collecting duct, as well as in parts of the renal vasculature. mRNA for ecNOS has been demonstrated by RT-PCR in the glomerulus and afferent and efferent arterioles [14]. In a recent report, the distribution of ecNOS has been studied immunohistochemically in the rat and found in glomerular endothelium; efferent arteriolar endothelium showed stronger staining than afferent [12]. These two forms of NOS are crucially involved in renal pathophysiology, and local release of NO serves to control renal blood flow (RBF) and modulates the activity of tubuloglomerular feedback [15]. Baylis et al. [16] have recently shown that chronic administration of an inhibitor of NO synthesis in rats results in sustained systemic and glomerular hypertension accompanied by glomerular injury. Using a juxtamedullary rat nephron preparation, Imig et al. [17] found that NO primarily alters afferent vascular tone, thereby modifying the ability of the preglomerular vasculature to autoregulate glomerular capillary pressure.

In vivo micropuncture studies have shown that L-NAME infused into the proximal tubule or peritubular capillary augments the tubuloglomerular-feedback-mediated decrease in single-nephron glomerular filtration rate (GFR) [15]. Using isolated microperfused afferent arterioles and attached macula densa, Ito and Ren [18] demonstrated that selective inhibition of NO at the macula densa constricts the afferent arteriole when the NaCl concentration is high, but has no effect when NaCl is low. High NaCl intake increases NOS activity in the kidney [15], which contributes to the rightward shift of the tubuloglomerular feedback response curve and renal vasodilation [15]. Chronic inhibition of NO synthesis impairs sodium excretion and results in hypertension. Such effects may be due, at least in part, to inappropriate RBF, insufficient rightward shift of the tubuloglomerular feedback or both.

The mesangial cell can express inducible NOS after cytokine stimulation; this has been shown for rat, bovine, and human mesangial cells [19]. In rat mesangial cells, iNOS is induced by interleukin (IL)-1β, tumor necrosis factor-α (TNF-α), and endotoxin [19]. In human mesangial cells, iNOS is induced by the combination of IL-1β and interferon-γ [19]; this effect is augmented by TNF-α. Using the PCR technique on mRNA extracted from control and cytokine-stimulated proximal tubular cells, McLay et al. [20] have found the iNOS product, indicating that proximal tubular cells

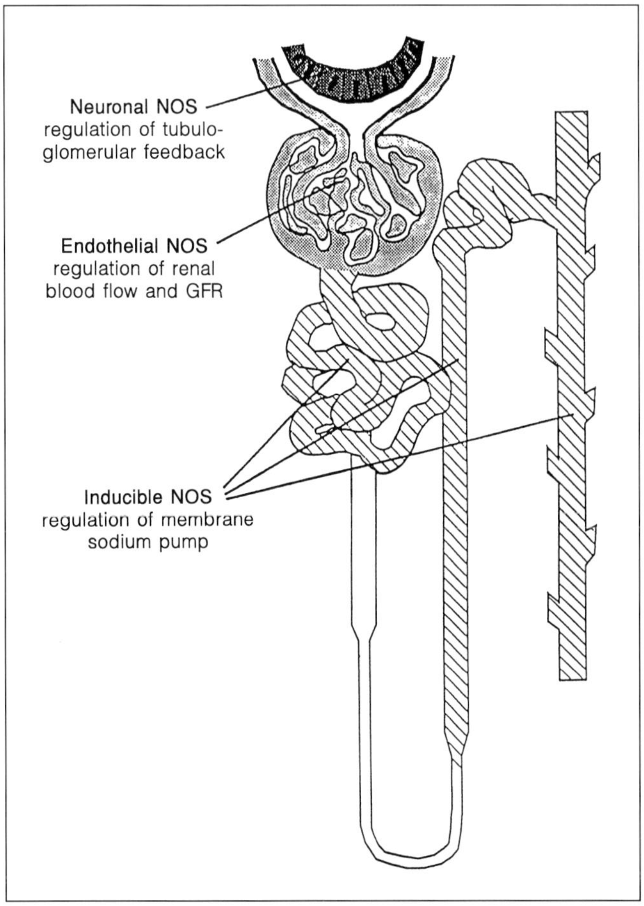

Fig. 1. Localization and biological functions of nNOSs, ecNOSs and iNOS along the nephron. GFR = Glomerular filtration rate.

express iNOS in the basal condition and respond to an immune challenge through the induction of iNOS. Other studies in the rat have shown that glomerular mesangium and afferent arterioles, as well as various segments of the nephron, express iNOS under basal conditions and that renal iNOS expression is augmented after immune activation with cytokines or endotoxin [21]. Consistently, immunoperoxidase analysis, with specific antisera to iNOS, showed the iNOS signal localized to afferent arterioles, collecting ducts, outer medullary tubuli, and medullary thick ascending limbs [12] of rat kidney. The steady-state amount of iNOS mRNA and protein in normal rat kidney is highest in tubules of the outer medulla.

In vitro studies of rat renal homogenates showed that the specific activity of NOS in the medulla was threefold that of the cortex [22]. The significance of the high expression of iNOS in normal kidney has not been completely elucidated. It has been proposed that NO production by iNOS in medullary thick ascending limbs serves to maintain and regulate medullary blood flow and oxygenation [15]. There is also evidence that NO induces natriuresis by antagonizing the effects of angiotensin II on sodium reabsorption in proximal tubules [15] and by directly inhibiting renal tubular Na^+/K^+ ATPase [23].

Data indicate that in certain models of experimental inflammatory renal disease, induction of iNOS occurs within the kidney and is mainly localized in infiltrating macrophages. The role of enhanced NO synthesis in glomerulonephritis remains to be elucidated fully, and there is evidence for both beneficial and detrimental effects depending on the model of glomerulonephritis, the timing of NOS inhibition and the agents used to inhibit NO synthesis [24].

Pathophysiology of NO in Uremia

Recent studies have led to the hypothesis that abnormalities of the NO synthetic pathway could play a key role in causing and/or mediating the complex hemodynamic and hemostatic disorders associated with the progression of renal disease. Most of these studies have focused on the experimental model of chronic renal failure (CRF) obtained in rats by extensive renal mass reduction (RMR). These animals develop systemic hypertension and exhibit a decrease in both GFR and RBF, despite an initial adaptive increase in single-nephron GFR and plasma flow per nephron [25]. Animals with RMR also develop severe proteinuria and structural changes in the kidney, including glomerulosclerosis, which eventually lead to renal insufficiency [25]. Like uremic patients, RMR rats have a bleeding tendency that manifests itself by a prolonged bleeding time [26].

The pattern of NO synthesis in experimental uremia appears to be complex. On the one hand, there is evidence that CRF is a condition of low NO formation in the kidney, and that exogenous correction of the defect markedly limits disease progression. On the other hand, other studies suggest that in the systemic circulation of uremic rats, NO is formed in excessive amounts and its overproduction leads to impaired platelet function.

The purpose of the following sections is to review more recent data from our laboratory and other laboratories concerning the pathophysiology of the *L*-arginine/NO pathway in uremia, with the aim of explaining the apparent discrepancy of less NO formed by the kidney when systemic availability is enhanced.

Renal NO

In rats with an ablated renal mass, dietary supplementation with *L*-arginine for 6 weeks reduced glomerulosclerosis; this beneficial effect appeared to result from a reduction in glomerular capillary pressure and efferent arteriolar resistance [27]. These findings were confirmed by another study, which demonstrated that oral supplementation of *L*-arginine at a much lower dose had a protective effect on urinary protein excretion and renal function in RMR rats [28]. The authors hypothesized that the RMR model is characterized by low renal NO production, and that chronic administration of *L*-arginine has beneficial effects on kidney function due to the addition of the necessary substrate for renal NOS enzymes. A recent study from our group showed that in RMR rats, renal generation of NO, measured as ex vivo conversion of ^{3}H-*L*-arginine to ^{3}H-*L*-citrulline, was markedly lower than normal [29] (fig. 2). The abnormality was already evident 7 days after surgery and progressively worsened with time, in close correlation with signs of renal

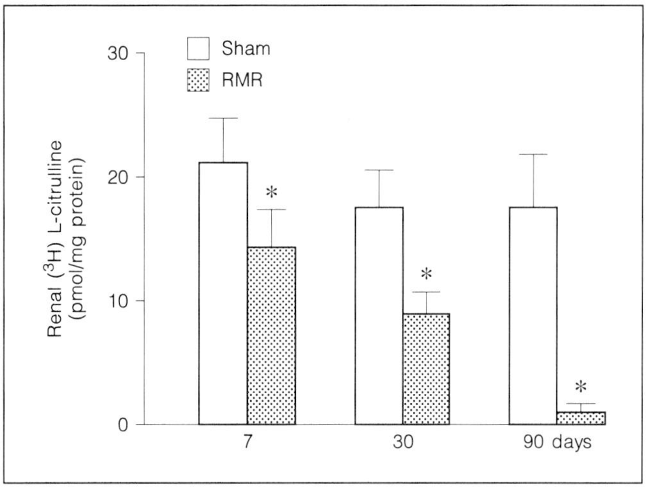

Fig. 2. Production of NO by renal tissue from rats with RMR, 7, 30, and 90 days after the surgical procedure, compared to age-matched sham-operated rats (n = 6 for each group). Data are the mean ± SE. *$p < 0.05$ vs. sham [from ref. 29].

injury, such as glomerulosclerosis and tubular damage. Consistent with these results, urinary excretion of NO_2^-/NO_3^- was significantly reduced in rats with RMR compared to control animals [30].

The kidney is a major source of arginine for the body. The largest fraction of renal arginine synthesis occurs in the proximal tubule through the consumption of filtered citrulline [31]. Chan et al. [32] studied the activity of the arginine synthetase complex in rats with advanced CRF and found that it was markedly reduced. On the other hand, in a more recent study, Boudy et al. [33] showed that total renal arginine synthesis and plasma levels are not diminished in the remnant kidney model in contrast with a significant reduction in GFR. This is possibly due to a combination of three factors: hypertrophy of the proximal tubule in remnant nephrons, hyperfiltration which increases the amount of citrulline filtrered per nephron, and increases in plasma citrulline concentrations which enhance arginine synthesis. Thus, defective substrate availability cannot be considered a main cause of impaired NO synthesis in RMR kidney.

To go deeper into the cellular basis and the biochemical nature of this abnormality, we used two histochemical approaches that could locate either NOS catalytic activity (NADPH-diaphorase) or NOS isoenzyme expression (immunoperoxidase). NADPH-diaphorase documented a progressive loss of renal NOS activity in RMR rats in all the structures examined including glomeruli (except macula densa), proximal tubules, and collecting ducts [30]. A strong, progressive decrease of iNOS immunostaining was also observed, colocalizing with the loss of the NADPH-diaphorase signal, indicating that reduced NOS activity in RMR was likely due to a decrease in iNOS enzyme content [30]. In contrast to iNOS, we found that ecNOS staining was rather similar in RMR and control kidneys. However, in RMR rats, the ecNOS signal was reduced in glomeruli affected by sclerosis, possibly because of loss of functional endothelium [30].

What are the possible causes of progressive loss of iNOS in CRF? In glomeruli soon after surgical ablation of renal mass, inflammatory mediators, such as platelet-derived growth factor (PDGF) [34] and transforming growth factor-β (TGF-β) [35], are formed in excessive amounts. PDGF and TGF-β are both potent inhibitors of NO synthesis and dose dependently block IL-1β-induced iNOS mRNA in rat mesangial cells [36]. It is therefore conceivable that elevated concentrations of PDGF and TGF-β in glomeruli of rats with RMR down-regulate glomerular iNOS. The recent observation that the potent vasoconstrictor and promitogenic peptide endothelin-1 (ET-1) inhibits IL-1β-stimulated transcription of iNOS [37] offers one possible explanation for the reduced expression of iNOS in the outer medulla of rats with RMR. Renal ET-1 gene expression and synthesis time dependently increased in the RMR model and correlated with the progressive renal injury [38]. Although the cellular origin of the increased renal synthesis of ET-1 is not yet clear, preliminary in situ hybridization studies in our laboratory in RMR rats [C. Zoja et al., personal communication] showed intense staining for ET-1 mRNA, mainly in the proximal tubular cells. One may speculate that in rats with RMR, excessive ET-1 formed by the proximal tubular epithelium down-regulates iNOS in the outer medulla.

NO in the renal vasculature and macula densa appears to control renal blood flow and tubuloglomerular feedback [15]. Besides its vasodilatory properties, NO inhibits mesangial cell proliferation [39] and extracellular matrix synthesis. Thus, dietary supplementation with *L*-arginine to rats with RMR limited the early phase of cell proliferation in the remnant glomerulus [40].

Diminished renal NO formation could therefore play a role in the hemodynamic and non-hemodynamic abnormalities reported in the model of extreme renal ablation that precede and eventually lead to the development of glomerulosclerosis and renal scarring. This possibility is further supported by a very recent study from our group showing that administration of an NO donor, molsidomine, to RMR rats reduced proteinuria, slowed the rate of renal disease progression and prolonged survival [41]. This protective effect could be documented either by giving the drug at the time of surgery or later when rats were already proteinuric and hypertensive.

Systemic NO

Contrasting data are available on the systemic *L*-arginine/NO pathway in uremia. Thus, higher than normal plasma NO_2^-/NO_3^- concentrations were found in RMR rats that paralleled the disease progression to uremia [30]. In addition, NADPH-diaphorase documented an appreciable increase of NOS activity in the aorta of uremic animals, suggesting that enhanced systemic NO in RMR rats derives, at least in part, from higher than normal synthetic activity in large blood vessels [30]. Increased diaphorase staining was accompanied by the immunohistochemistry finding of higher ecNOS and iNOS expression in the aorta of RMR rats [30]. It is also of interest that aortic ecNOS increased early during the course of the disease, while iNOS rose considerably later, when animals developed uremia.

Causes of high vascular ecNOS expression in RMR, as in other experimental models of hypertension, very likely include changes

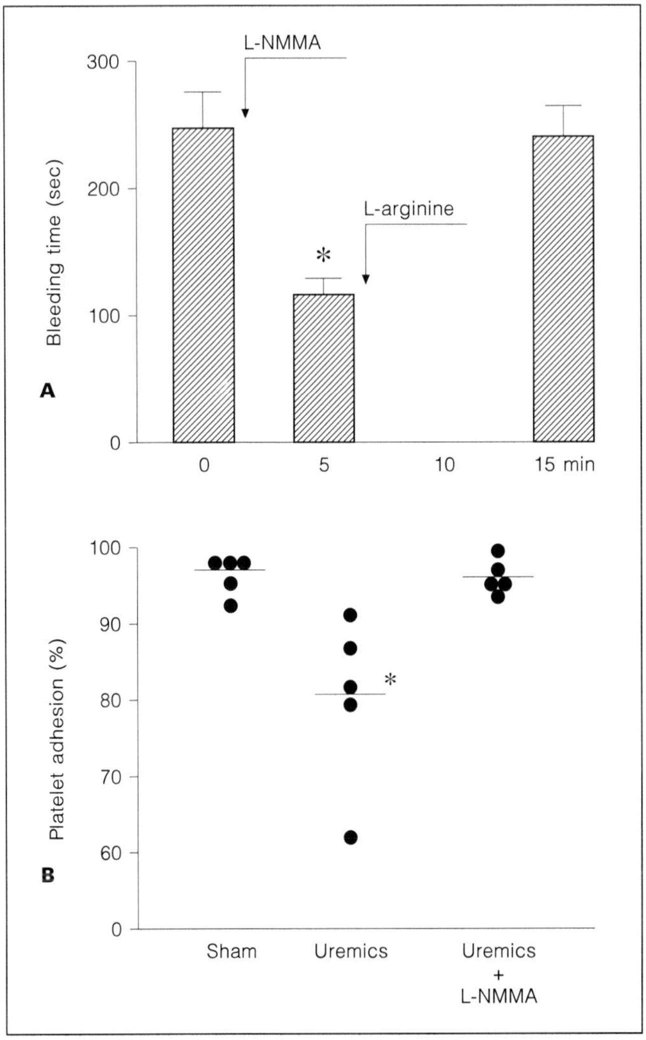

Fig. 3. A Effect of *L*-arginine (n = 6) on the shortening of bleeding time induced by L-NMMA in uremic rats. Values are the mean ± SD. *p < 0.01 vs. time 0. **B** Ex vivo platelet adhesion in sham-operated rats, uremic rats, and uremic rats given *L*-NMMA (n = 5 for each group). *p < 0.05 vs. sham-operated rats and uremic rats + L-NMMA [from ref. 26].

in shear stress, since endothelial cells appear to release NO in response to changes in blood flow, and ecNOS mRNA and the corresponding protein are induced in endothelial cells after exposure to fluid shear in a cone-and-plate apparatus [42]. A similar mechanism operates in hypertension where pulsatile stretch and distension of the arterial wall favor NO release, as indicated by recent findings that the cyclic strain induces the expression of ecNOS transcripts and protein levels in cultured endothelial cells [43].

In addition to ecNOS, increased expression of iNOS was found in systemic vessels of RMR rats which was not confined to the endothelium but was also detectable in smooth muscle. This is consistent with a previous study showing that plasma from uremic patients stimulated NO synthesis (conceivably by induction of iNOS activity) more than normal plasma [44]. The stimulatory activity of uremic plasma was attributed to TNF-α, a potent inducer of iNOS, that circulates in increased amounts in the plasma of patients with CRF who were either not dialyzed or were undergoing maintenance hemodialysis [44, 45].

The inducible NOS, due to its Km of around 30 μM, is actually dependent on the availability of *L*-arginine. Data on systemic *L*-arginine levels in uremics, as compared to healthy volunteers, are often conflicting. Some investigators have showed that *L*-arginine concentrations are increased in plasma and circulating cells from uremic patients [44, 46]. On the other hand, a recent paper reported no alterations in plasma *L*-arginine levels [47]. These conflicting results might be due to different stages of CRF, dialytic treatment or protein intake. However, it is conceivable that in uremic patients with increased levels of *L*-arginine, an increase in substrate at the local level could contribute to up-regulate NO synthesis in uremia.

Up-regulation of NO-forming enzymes might be an early defense mechanism against hypertension in uremia. However, besides its vasoactive properties, NO inhibits platelet function in vitro [6, 7]. The in vivo counterpart of this activity is the prolongation of skin bleeding time observed in healthy volunteers given NO by inhalation [48]. Bleeding, a well-known complication of uremia, is a defect which mainly involves primary hemostasis, i.e., platelet–platelet and platelet–vessel interactions [49]. We previously reported that in uremic patients affected by serious bleeding disorders, defective platelet aggregation was associated with exaggerated platelet NO synthesis [44]. Since a prolonged bleeding time in uremic animals is completely normalized by the NO antagonist L-NMMA [26] (fig. 3), it seems logical to postulate that excessive systemic NO formation may play a major role in the abnormal primary hemostasis of uremia.

Conclusion

The *L*-arginine/NO pathway plays a crucial role in the physiology and pathology of the uremic condition. In the kidney, with the progressive development of renal insufficiency, proportionately less NO is formed locally, apparently as a consequence of a time-dependent progressive reduction in the amount of the iNOS molecule expressed in the kidney. In this setting, a marked reduction in NO, in the face of continuous local generation of vasoconstrictor and promitogenic substances, may well contribute to intraglomerular capillary hypertension and cell proliferation.

In contrast, uremia appears to be associated with an abnormally large amount of NO released into the systemic circulation, as a consequence of excessive generation of the gas by systemic vessels due to up-regulation of both ecNOS and iNOS. While the increase in vascular ecNOS occurs early in response to progressive renal function deterioration, up-regulation of iNOS can only be detected in the advanced phases of the disease; moreover, it is not confined to the smooth muscle cells, but also involves the vascular endothelim. Up-regulation of NO-forming enzymes might be an early defense mechanism against the hypertension of uremia. On the other hand, greater NO availability to circulating cells may sustain the bleeding tendency, a well-known complication of uremia in experimental animals and in humans.

References

1 Moncada S, Higgs EA: Molecular mechanisms and therapeutic strategies related to nitric oxide. FASEB J 1995;9:1319–1330.

2 Furchgott RF, Zawadski JV: The obligatory role of endothelial cells in the relaxation of smooth muscle by acetylcholine. Nature 1980;288:373–376.

3 Palmer RMJ, Rees DD, Ashton DS, Moncada S: *L*-arginine is the physiological precursor for the formation of nitric oxide in endothelium-dependent relaxation. Biochem Biophys Res Commun 1988; 153:1251–1256.

4 Rees DD, Palmer RM, Moncada S: Role of endothelium-derived nitric oxide in the regulation of blood pressure. Proc Natl Acad Sci USA 1989;86: 3375–3378.

5 Moncada S, Higgs EA: The *L*-arginine-nitric oxide pathway. N Engl J Med 1993;239:2002–2012.

6 Broekman MJ, Eiroa AM, Marcus AJ: Inhibition of human platelet reactivity by endothelium-derived relaxing factor from human umbilical vein endothelial cells in suspension: Blockade of aggregation and secretion by aspirin-insensitive mechanism. Blood 1991;78:1033–1040.

7 Sneddon JM, Vane JR: Endothelium-derived relaxing factor reduces platelet adhesion to bovine endothelial cells. Proc Natl Acad Sci USA 1988;85: 2800–2804.

8 Wang Y, Marsden PA: Nitric oxide synthases: Biochemical and molecular regulation. Curr Opin Nephrol Hypertens 1995;4:12–22.

9 Brezis M, Heyman SN, Dinour D, Epstein FH, Rosen S: Role of nitric oxide in renal medullary oxygenation: Studies in isolated and intact rat kidneys. J Clin Invest 1991;88:390–395.

10 Shultz PJ, Schorer AE, Raij L: Effects of endothelium-derived relaxing factor and nitric oxide on rat mesangial cells. Am J Physiol 1990;258:F162–F167.

11 Biondi ML, Dousa T, Vanhoutte P, Romero JC: Evidence for the existence of endothelium-derived relaxing factor in the renal medulla. Am J Hypertens 1990;3:876–878.

12 Bachmann S, Mundel P: Nitric oxide in the kidney: Synthesis, localization and function. Am J Kidney Dis 1994;24:112–119.

13 Terada Y, Tomita K, Nonoguchi H, Mammo F: Polymerase chain reaction localization of constitutive nitric oxide synthase and soluble guanylyl cyclase messenger RNAs in microdissected rat nephron segments. J Clin Invest 1992;90:659–665.

14 Ujiie K, Yuen J, Hogarth L, Danziger R, Star RA: Localization and regulation of endothelial NO synthase mRNA expression in rat kidney. Am J Physiol 1994;267:F296–F302.

15 Ito S: Nitric oxide in the kidney. Curr Opin Nephrol Hypertens 1995;4:23–30.

16 Baylis C, Mitruka B, Deng A: Chronic blockade of nitric oxide synthesis in the rat produces systemic hypertension and glomerular damage. J Clin Invest 1992;90:278–281.

17 Imig JD, Roman RJ: Nitric oxide modulates vascular tone in preglomerular arterioles. Hypertension 1992;19:770–774.

18 Ito S, Ren Y: Evidence for the role of nitric oxide in macula densa control of glomerular hemodynamics. J Clin Invest 1993;92:1093–1098.

19 Pfeilschifter J, Kunz D, Muhl H: Nitric oxide: An inflammatory mediator of glomerular mesangial cells. Nephron 1993;64:518–525.

20 McLay JS, Chatterjee P, Nicolson AG, Jardine AG, MacKay NG, Ralston SH, Grabowski P: Nitric oxide production by human proximal tubular cells: A novel immunomodulatory mechanism? Kidney Int 1994;46:1043–1049.

21 Morrissey JJ, McCracken R, Kaneto H, Vehaskari M, Montani D, Klahr S: Location of an inducible nitric oxide synthase mRNA in the normal kidney. Kidney Int 1994;45:998–1005.

22 McKee M, Scavone C, Nathanson JA: Nitric oxide, cGMP, and hormone regulation of active sodium transport. Proc Natl Acad Sci USA 1994;91:12056–12060.

23 Guzman NJ, Fang MZ, Tang SS, Ingelfinger JR, Garg LC: Autocrine inhibition of Na^{+}/K^{+}-ATPase by nitric oxide in mouse proximal tubule epithelial cells. J Clin Invest 1995;95:2083–2088.

24 Cook HT, Cattell V: Nitric oxide synthetic pathway and regulation: Its potential role in immune and inflammatory reactions in the glomerulus. J Nephrol 1995;8:247–254.

25 Anderson S, Meyer TW, Brenner BM: The role of hemodynamic factors in the initiation and progression of renal disease. J Urol 1985;133:310–315.

26 Remuzzi G, Perico N, Zoja C, Corna D, Macconi D, Viganò G: Role of endothelium-derived nitric oxide in the bleeding tendency of uremia. J Clin Invest 1990;86:1768–1771.

27 Reyes AA, Purkerson ML, Karl I, Klahr S: Dietary supplementation with *L*-arginine ameliorates the progression of renal disease in rats with subtotal nephrectomy. Am J Kidney Dis 1992;20:168–176.

28 Ashab I, Peer G, Blum M, Wollman Y, Chermibovsky T, Hassner A, Schwartz D: Oral administration of *L*-arginine and captopril in rats prevents chronic renal failure by nitric oxide production. Kidney Int 1995;47:1515–1521.

29 Aiello S, Noris M, Corna D, Benigni A, Zoya C, Todeschini N, Remuzzi G: Defective renal synthesis of nitric oxide in rats with reduced renal mass (abstract). J Am Soc Nephrol 1994;5:936.

30 Zappella S, Noris M, Aiello S, Corna D, Benigni A, Zoya C, Remuzzi G: Renal and systemic nitric oxide synthase expression in rats with renal mass reduction. J Am Soc Nephrol 1996;abstract, 7: A3105.

31 Mitch W, Chesney R: Amino acid metabolism by the kidney. Miner Electrolyte Metab 1983;9:190–202.

32 Chan W, Wang M, Kopple J, Swendseid M: Citrulline levels and urea cycle enzymes in uremic rats. J Nutr 1974;104:678–683.

33 Boudy N, Hassler C, Parvy P, Bankir L: Renal synthesis of arginine in chronic renal failure: In vivo and in vitro studies in rats with 5/6 nephrectomy. Kidney Int 1993;44:676–683.

34 Floege J, Burns MW, Alpers CE, Yoshimura A, Pritzl P, Gordon K, Seifert RA, Bowen-Pope DF, Couser WG, Johnson RJ: Glomerular cell proliferation and FDGF expression precede glomerulosclerosis in the remnant kidney model. Kidney Int 1992;41:297–309.

35 Lee LK, Meyer TW, Pollock AS, Lovett DH: Endothelial cell injury initiates glomerular sclerosis in the rat remnant kidney. J Clin Invest 1995;96:953–964.

36 Ketteler M, Border WA, Noble NA: Cytokines and *L*-arginine in renal injury and repair. Am J Physiol 1994;267:F197–F207.

37 Beck KF, Mohaupt MG, Sterzel RB: Endothelin-1 inhibits cytokine-stimulated transcription of inducible nitric oxide synthase in glomerular, mesangial cells. Kidney Int 1995;48:1893–1899.

38 Orisio S, Benigni A, Bruzzi I, Corna D, Perico N, Zoja C, Benatti L, Remuzzi G: Renal endothelin gene expression is increased in remnant kidney and correlates with disease progression. Kidney Int 1993;43:354–358.

39 Garg U, Hassid A: Inhibition of rat mesangial mitogenesis by nitric oxide-generating vasodilators. Am J Physiol 1989;257:F60–F66.

40 Ingram A, Parbtani A, Thai K, Ly H, Shankland SJ, Morrissey G, Smoley JU: Dietary supplementation with *L*-arginine limits cell proliferation in the remnant glomerulus. Kidney Int 1995;48:1857–1865.

41 Corna D, Noris M, Luzzana E, Benigni A, Zoja C, Todeschini N, Remuzzi G: A nitric oxide-donor either as prophylaxis or therapy slows renal disease progression and prolongs survival in remnant kidney. J Am Soc Nephrol 1996;Abstract, 7:A3029..

42 Noris M, Morigi M, Donadelli R, Aiello S, Foppolo M, Todeschini N, Orisio S, Remuzzi G, Remuzzi A: Nitric oxide synthesis by cultured endothelial cells is modulated by flow conditions. Circ Res 1995; 76:536–543.

43 Awolesi MA, Sessa WC, Sumpio BE: Cyclic strain upregulates nitric oxide synthase in cultured bovine aortic endothelial cells. J Clin Invest 1995;96:1449–1454.

44 Noris M, Benigni A, Boccardo P, Aiello S, Gaspari F, Todeschini N, Figliuzzi M, Remuzzi G: Enhanced nitric oxide synthesis in uremia: Implications for platelet dysfunction and dialysis hypotension. Kidney Int 1993;44:445–450.

45 Descamps-Latscha B, Herbelin A, Nguyen AT, Roux-Lombard P, Zingraff J, Maynot A, Urger C, Dahmane D, de Groote D, Jungers P, Dayer JM: Balance between IL-1β, TNF-α and their specific inhibitors in chronic renal failure and maintenance dialysis. J Immunol 1995;154:882–892.

46 Metcoff J, Pederson J, Gable J, Llach F: Protein synthesis, cellular amino acids, and energy levels in CAPD patients. Kidney Int 1987;32:S136–S144.

47 Jungers P, Chauveau P, Ceballos I, Bardet J, Parvy P, Hannedouche T, Kamoun P: Plasma free amino acid alterations from early to end-stage chronic renal failure. J Nephrol 1994;7:48–54.

48 Hogman M, Frostell C, Arnberg H, Hedenstierna G: Bleeding time prolongation and NO inhalation. Lancet 1993;341:1664–1665.

49 Remuzzi G: Bleeding in renal failure. Lancet 1988; 1:1205–1208.

Mineral and Electrolyte Metabolism

Miner Electrolyte Metab 1997;23:157–160

Ralph Rabkin

Department of Medicine, Veterans Affairs Palo Alto Health Care System and Stanford University, Palo Alto, Calif., USA

Nutrient Regulation of Insulin-Like Growth Factor-I

Key Words
Hypokalemia
Insulin-like growth factor-I
Kidney
Malnutrition
Potassium
Potassium depletion
Renal hypertrophy

Abstract
Insulin-like growth factor-I (IGF-I) is a nutrient-regulated growth factor. In malnutrition, serum IGF-I levels fall despite normal or elevated growth hormone (GH) levels and this is caused by resistance to GH, defects in IGF-I gene transcription and translation, and mRNA instability. Because IGF-I production is sensitive to protein and energy deficiencies, serum IGF-I levels may be valuable in the assessment of the nutritional status of patients with wasting diseases, including advanced renal failure. It should be recognized, however, that serum or tissue IGF-I peptide levels do not always reflect IGF-I gene expression. For example, in the K-deficient rat, kidney IGF-I levels are increased even though IGF-I mRNA levels are low. This appears to be due in part to increased sequestration by IGF-binding proteins and by decreased kidney IGF-I degradation. The increase in kidney IGF-I may contribute to the exaggerated renal growth that occurs in hypokalemia.

Introduction

Insulin-like growth factor-I (IGF-I) is a potently anabolic peptide growth factor that stimulates amino acid transport, protein synthesis, and body growth [1]. Administered to normal humans it increases renal blood flow, glomerular filtration rate and tubular phosphate transport, and in animals it also promotes renal growth [2]. This growth factor is produced in tissues throughout the body and it is present in the circulation with the liver as the main source [3]. IGF-I production is regulated by growth hormone (GH) through a process that requires an adequate nutrient intake [1]. There is no significant storage of IGF-I in the IGF-I producing cells; however, in the circulation, IGF-I forms tight complexes with high-affinity IGF-binding proteins (IGFBPs) and these complexes serve as a slow-release form of the growth factor [3]. A similar process occurs in tissues. Currently, six IGFBPs have been identified. The most abundant in the circulation is IGFBP-3 which forms a ternary complex with IGF-I and a circulating acid-labile subunit. Lesser amounts of IGFBP-1, -2 and -4 (24- to 32-kD binding proteins) can also be easily detected and IGF-I complexed to these smaller proteins passes more readily into the tissues [4]. Altogether >97% of the circulating IGF-I is complexed to IGFBPs and this limits the bioavailability of the growth factor [5]. In the tissues, IGFBPs are produced in an organ-specific pattern and here they generally limit the activity of IGF-I, but in some cell types, IGFBPs may actually promote IGF-I action [3].

Effect of Malnutrition on Circulating IGF-I Levels

In malnourished humans, even though the plasma GH levels are usually normal or even elevated, there is a marked fall in serum IGF-I levels [1]. This decrease occurs earlier and to a greater extent than the decline in serum albumin or transferrin levels which are commonly used to monitor nutritional status [6]. Because of the sensitivity of IGF-I as a nutritional marker, several workers have examined the value of the serum IGF-I level as a monitor of the nutritional status of patients on maintenance hemodialysis [7, 8]. In general, low serum IGF-I levels correlate significantly with midarm circumference, triceps skinfold thickness, and prealbumin levels. Thus, anthropometric measurements together with biochemical measurements of serum IGF-I, albumin, and transferrin may be useful in identifying malnourished dialysis patients. However, it should be recognized that the use of serum IGF-I levels as a nutritional marker has several pitfalls. First, there is considerable variation in the range of normal values, and unless the baseline values of a particular individual are known, a significant fall in level may pass unrecognized. Second, the serum IGF-I levels are markedly affected by the circulating IGFBPs which are also modulated by nutrients and may also be affected by the underlying disease [9]. For example, in advanced chronic renal failure, IGFBP-1, -2, and -3 levels are usually elevated and this limits the passage of IGF-I into the extravascular compartment [5]. Nevertheless, monitoring serum IGF-I levels can be of particular value in following the efficacy of treating malnutrition, because serum IGF-I levels increase promptly when nutrients are replaced [10].

KARGER
E-Mail karger@karger.ch
Fax +41 61 306 12 34
http://www.karger.ch

0378–0392/97/0236–0157 $12.00/0

Ralph Rabkin, MD
VAMC Palo Alto (111R), 3801 Miranda Avenue
Palo Alto, CA 94304 (USA)
Tel. +1 415 858 3985, fax +1 415 858 3975
E-Mail rabkin@leland. stanford.edu

To evaluate the relative effects of energy and protein intake on plasma IGF-I levels, Isley et al. [11] studied the effect of calorie and of protein refeeding in normal volunteers after a 5-day fast. From the response, they concluded that both protein and calories are required to restore plasma IGF-I levels to normal. Recently, Fouque et al. [12] showed that if the calorie intake of patients with advanced chronic renal failure is maintained at a normal level, modest reductions in protein intake (0.6 g/kg per day) do not affect serum IGF-I levels. Thus the practice of modest protein restriction for the treatment of advanced chronic renal failure appears to be nutritionally safe.

Pathogenic Mechanisms for the Fall in Plasma IGF-I Levels in Malnutrition

There are several mechanisms accounting for the fall in serum IGF-I levels in malnutrition, which in turn may increase catabolism. In humans this does not appear to be due to a GH deficiency, because circulating GH levels are either normal or elevated and this indicates that resistance to GH is present [1]. Indeed, in starved rats, liver GH receptor binding is reduced and in patients with anorexia nervosa, levels of the circulating GH-binding protein, which corresponds to the ectodomain of the GH receptor, are reduced [9]. On the other hand, there is little or no change in GH receptor binding in protein-restricted rats and this suggests that a postreceptor defect also plays a role [13]. Undernutrition also affects the production of IGF-I at a cellular level by impairing the transcription and translation of the IGF-I gene and by reducing IGF-I mRNA stability [14]. Impaired translation is illustrated by a study of the response of hypophysectomized protein-restricted rats to GH therapy [15]. In these rats, the administration of pharmacological doses of GH overcame the resistance to GH and normalized the low hepatic total IGF-I mRNA levels, but not the serum and hepatic IGF-I peptide levels. Another mechanism contributing to the low serum IGF-I levels is related to the profound changes in the serum IGFBP profile that occur in malnutrition and that lead to an increase in the serum clearance of IGF-I. Studying protein-restricted rats, Thissen et al. [4] noted that a larger portion of the serum IGF-I circulates in a 30-kD IGFBP complex than in the 150-kD complex, predominant in protein-replete animals. Because of the difference in complex size, the 30kD moeity is cleared more rapidly, contributing to a lower serum IGF-I level.

Effect of Nutrient Deficiency on the Renal IGF-I Axis

As with other tissues, renal IGF-I production is nutrient sensitive and this has been recently reviewed [2]. High protein intake causes a transient increase, while low protein intake results in a persistent decrease in kidney IGF-I levels, as does fasting [16–18]. The implications of these changes are noteworthy, for it has been suggested that the increase in kidney IGF-I expression accompanying a high protein intake may play a role in mediating protein-induced renal hypertrophy [16]. Conversely, reduced IGF-I levels may account for the attenuated compensatory renal growth that occurs in protein-restricted uninephrectomized rats. Recently, we examined the affect of K depletion on the renal IGF-I axis [20]. This was prompted by the paradoxical effect that K depletion has on body and kidney growth and the report of transient increases in kidney IGF-I levels in kaliopenic rats [21].

In the rat, K depletion leads to severely attenuated body growth and muscle wasting which can only partly be attributed to a reduced intake of food [22]. In contrast, kidney growth accelerates and this is due to cellular hypertrophy and regional hyperplasia [23]. Prolonged hypokalemia leads to interstitial fibrosis and an increase in the prevalence of renal cysts has been documented in kaliopenic patients [24, 25]. Since the serum IGF-I levels fall in young K-depleted rats, it has been suggested that IGF-I deficiency contributes to the retarded body growth [26]. Indeed, chronic K deficiency due to gastrointestinal losses and reduced intake in protein-calorie-deficient children, as is prevalent in many underdeveloped countries, may also adversely effect growth of these children [22]. Treating kaliopenic rats with IGF-I or GH fails to promote normal growth, indicating that there is also resistance to IGF-I [27, 28]. On the other hand, it has been suggested that IGF-I may play an early role in K-induced nephromegaly as there is a short-lived increase in kidney IGF-I content [21]. To explain these paradoxes, we tested the hypothesis that K deficiency alters local IGF-I systems and target cell responsiveness in a tissue-specific manner.

In this study we measured the expression of the genes for the components of the local IGF-I system and the protein products in the kidney and gastrocnemius muscle of K-depleted and control rats pair-fed a normal diet [20]. Kaliopenia was achieved by placing male rats on a 0.01% K diet for 8 days. The characteristics of these rats are shown in table 1. As anticipated, serum K levels fell, weight gain was attenuated and gastrocnemius weight was reduced compared to control rats eating the same quantity of food but containing 1% K. Of note, the kidney of the kaliopenic rats were 25% heavier. Examination of the kidney IGF-I axis revealed several important alterations. Despite similar serum IGF-I levels, there was a twofold increase in the kidney IGF-I levels in the kaliopenic rats compared to controls (491 ± 61 vs. 227 ± 19 ng/g wet weight; $p < 0.01$). This was especially remarkable as the kidney IGF-I mRNA levels were reduced by half compared to the controls (6 ± 0.5 vs. 3 ± 0.2 arbitrary densitometry units; $p < 0.01$). GH and IGF-I receptor mRNA levels were unchanged as was IGF-I receptor binding. In muscle, the IGF-I mRNA levels were also significantly reduced compared to pair-fed controls (8 ± 0.4 vs. 4 ± 0.3 arbitrary densitometry units). But in contrast to the kidney, the muscle IGF-I peptide levels were also reduced (20 ± 4 vs. 54 ± 5 ng/g weight; $p > 0.01$). GH receptor mRNA levels were decreased significantly.

There were also changes in the serum and kidney IGFBP profiles, which provide explanations for the increase in kidney IGF-I levels in the presence of low kidney IGF-I mRNA levels. In serum of kaliopenic rats there was a 50% increase in the 30- to 34-kD IGFBPs that can be expected to increase the proportion of circulating IGF-I complexed to low-molecular-weight IGFBPs. Such a change will

Table 1. Serum potassium, body, kidney and muscle weight after 8 days of K restriction or pair-feeding

Group	Serum K mEq/l	Body weight change g	Kidney weight g	Muscle weight g
C	4.0 ±0.09	34 ±4	0.81 ±0.02	1.34 ±0.03
KD	2.6** ±0.13	10** ±2	1.02** ±0.03	1.19** ±0.01
CT	3.9 ±0.08	33 ±2	0.95* ±0.01	1.3 ±0.02
KDT	2.3 ±0.07	8 ±2	1.20*** ±0.04	1.22 ±0.01

Control (C) rats were on a 1% K diet and were pair-fed with K-depleted (KD) rats on a 0.01% K diet. Treated rats (T) received IGF-I by subcutaneous minipump for 8 days. Food consumption was similar in all groups. Average entry body weight was 179 g. Kd vs. C, **$p<0.01$; CT vs. C, *$p<0.05$; KDT vs. KD, ***$p<0.01$. Muscle = Gastrocnemius.

lead to an increase in the passage of IGF-I into tissues, especially kidney [4]. In the kidney of kaliopenic rats, there was a threefold increase in IGFBP-1 mRNA levels and this was accompanied by a 56% increase in the immunoreactive IGFBP-1 associated with kidney plasma membranes. This could conceivably produce an increase in the trapping of IGF-I in the kidney. In muscle there was no increase in IGFBP expression but IGFBP-5 levels fell. We also measured the IGF-I degrading activity of kidney plasma membranes and found that it was significantly depressed by 30% (23 ± 2 vs. 16 ± 1.4%/h) Thus, increased delivery of IGF-I to the kidney together with an increase in the trapping of IGF-I by IGFBP-1 and diminished degradation could account for the accumulation of IGF-I in the kidney of kaliopenic rats.

Since, as mentioned earlier, resistance to the body-growth-promoting action of IGF-I is a feature of K deficiency, we treated K-deficient rats with IGF-I to determine whether the kidney remains responsive to IGF-I. In K-deficient rats treated with IGF-I for 8 days, renal growth was increased even further (table 1). Kidney weight in these rats exceeded that of control rats consuming a similar quantity of food by 48% and that of untreated K-deficient rats by 18%. IGF-I-treated control rats also increased kidney size by a similar proportion, 17%. Thus in K deficiency, renal sensitivity to IGF-I is retained.

From this study we conclude that K is a profound modulator of the IGF-I system and that tissue-distinct changes in this system occur in K deficiency that may contribute to the altered tissue and body growth. We found after 8 days of severe K restriction that body and muscle growth were attenuated while the kidneys exhibited exaggerated growth. In muscle, GH receptor and IGF-I mRNA levels, IGF-I peptide content, and IGFBP levels were reduced. In kidney, IGF-I peptide levels were increased despite low IGF-I mRNA levels, IGFBP-1 mRNA and protein levels were also increased and responsiveness to IGF-I was retained. Accordingly, we propose that the local accumulation of IGF-I that occurs in the IGF-I-sensitive kidneys of K-deficient rats may, together with other mediators, play a role in promoting the exaggerated renal growth.

It is apparent from the foregoing discussion that nutrients are powerful regulators of the IGF-I system and that tissues can vary dramatically in their response to nutrient deprivation. Furthermore, the response to altered nutrient intake as seen in K deficiency, can lead to discordant changes between IGF-I gene and protein expression that again may vary in a tissue-specific manner. Finally, it is important to keep in mind that when IGF-I is employed as a therapy, e.g., for the treatment of malnourished dialysis patients [29], it is essential that an adequate intake of nutrients be maintained to avoid IGF-I resistance and to optimize the therapeutic response.

Acknowledgement

This report was supported by funds from the Research Service of the USA Department of Veterans Affairs.

References

1 Thissen JP, Ketelslegers JM, Underwood LE: Nutritional regulation of the insulin-like growth factors. Endocr Rev 1994;15:80–101.

2 Feld S, Hirschberg R: Growth hormone, the insulin-like growth factor system, and the kidney. Endocr Rev 1996;17:423–480.

3 Jones JI, Clemmons DR: Insulin-like growth factors and their binding proteins: Biological actions. Endocr Rev 1995;16:3–34.

4 Thissen JP, Davenport ML, Pucilowska JB, Miles MV, Underwood LE: Increased serum clearance and degradation of ^{125}I-labeled IGF-I in protein-restricted rats. Am J Physiol 1992;262:E406–E411.

5 Rabkin R, Fervenza FC, Maidment H, Ike J, Hintz R, Lui F, Bloedow DC, Hoffman AR, Gesundheit N: Pharmacokinetics of IGF-I in advanced chronic renal failure. Kidney Int 1996;49:1134–1140.

6 Unterman TG, Vazquez RM, Sla AJ, Martyn PA, Philips LS: Nutrition and somatomedin. XII. Usefulness of somatomedin-C in nutritional assessment. Am J Med 1985;78:228–234.

7 Jacob V, Le Carpentier JE, Salzano S, Naylor V, Wild G, Brown CB, El Nahas AM: IGF-I, a marker of undernutrition in hemodialysis patients. Am J Clin Nutr 1990;52:39–44.

8 Sanaka T, Shinobe M, Ando M, Hizuka N, Kawaguchi H, Nihei H: IGF-I as an early indicator of malnutrition in patients with end-stage renal disease. Nephron 1994;67:73–81.

9 Ketelslegers JM, Dominique M, Maes M, Underwood LE, Thissen JP: Nutritional regulation of the growth hormone and insulin-like growth factor-binding proteins. Horm Res 1996;45:252–257.

10 Clemmons DR, Underwood LE, Dickerson RN, Brown RO, Hak LJ, MacPhee RD, Heizer WD: Use of plasma somatomedin-C/insulin-like growth factor-1 measurements to monitor the response to nutritional repletion in malnourished patients. Am J Clin Nutr 1985;41:191–198.

11 Isley WL, Underwood LE, Clemmons DR: Dietary components that regulate serum somatomedin-C concentration in humans. J Clin Invest 1983;71:175–182.

12 Fouque D, Le Bouc Y, Laville M, Combarnous F, Joly MO, Raton P, Zech P: Insulin-like growth factor-1 and its binding proteins during a low-protein diet in chronic renal failure. J Am Soc Nephrol 1995;6:1427–1433.

13 Thissen JP, Triest S, Maes M, Underwood LE, Ketelshlegers JM: The decreased plasma concentrations of insulin-like growth factor-I in protein-restricted rats is not due to decreased number of growth hormone receptors on isolated hepatocytes. J Endocrinol 1990;124:159–165.

14 Straus DS: Nutritional regulation of hormones and growth factors that control mammalian growth. FASEB J 1994;8:6–12.

15 Thissen JP, Triest S, Moats-Staats BM, Underwood LE, Mauerhoff T, Maiter D, Ketelslegers JM: Evidence that pretranslational defects decrease serum insulin-like growth factor-I concentrations during dietary protein restriction. Endocrinology 1991;129:429–435.

16 Chin E, Bondy CA: Dietary protein-induced renal growth: Correlation between renal IGF-I synthesis and hyperplasia. Am J Physiol 1994;266:C1037–C1045.

17 Murray BM, Campos SP, Schoenl M, MacGillivray MH: Effect of dietary protein intake on renal growth: Possible role of insulin-like growth factor-I. J Lab Clin Med 1993;122:677–685.

18 Kobayashi S, Nogami H, Ikeda T: Growth hormone and nutrition interact to regulate expressions of kidney IGF-I and IGFBP mRNAs. Kidney Int 1995;48:65–71.

19 El Nahas M, Le Carpentier JE, Bassett AH, Hill DJ: Dietary protein and insulin-like growth factor-I content following unilateral nephrectomy. Kidney Int 1989;36:S15–S19.

20 Hsu FW, Tsao I, Rabkin R: The IGF-I axis in kidney and skeletal muscle of K deficient rats. Kidney Int, in press.

21 Flyvbjerg A, Marshall SM, Frystyk J, Rasch R, Bornfeldt KE, Arnqvist H, Jensen PK, Pallesen G, Ørskov H: Insulin-like growth factor I in initial renal hypertrophy in potassium-depleted rats. Am J Physiol 1992;262:F1023–F1031.

22 Dørup I: Magnesium and potassium deficiency. Acta Physiol Scand 1994;150:7–55.

23 Tobak GF: Phosphatidylcholine metabolism during renal growth and regeneration. Am J Physiol 1984;246:F249–F259.

24 Tolins JP, Hostetter MK, Hostetter TH: Hypokalemic nephropathy in the rat. J Clin Invest 1987;79:1447–1458.

25 Torres VE, Young WF Jr, Offord KP, Hattery RR: Association of hypokalemia, aldosteronism, and renal cysts. N Engl J Med 1990;322:345–351.

26 Flyvbjerg A, Dørup I, Everts ME, Ørskov H: Evidence that potassium deficiency induces growth retardation through reduced circulating levels of growth hormone and insulin-like growth factor I. Metab Clin Exp 1991;40:769–775.

27 Dørup I, Fkyvbjerg A, Everts ME, Ørskov H: Effects of growth hormone on growth and muscle Na^+-K^+ pump concentration in K^+-deficient rats. Am J Physiol 1992;262:E511–E517.

28 Dørup I, Fkyvbjerg A: Effects of IGF-I infusion on growth and muscle Na^+-K^+ pump concentration in K^+-deficient rats. Am J Physiol 1993;264:E810–E815.

29 Kopple JD, Ding H, Gao XL: Altered physiology and action of insulin-like growth factor-I in skeletal muscle in chronic renal failure. Am J Kidney Dis 1995;26:248–255.

Mineral and Electrolyte Metabolism

Miner Electrolyte Metab 1997;23:161–165

Stefan Greiber[a]
Christoph Wanner[b]

Division of Nephrology,
Department of Medicine, University of
[a] Freiburg and
[b] Würzburg, Germany

Lipoprotein(a) in Nephrotic Syndrome and End-Stage Renal Disease

Key Words
DNA synthesis
Fibronectin
Lipoprotein(a)
Oxygen radicals
Protein synthesis

Abstract
Lipoprotein(a) [Lp(a)] may be elevated in patients with the nephrotic syndrome and patients on hemodialysis or continuous ambulatory peritoneal dialysis. High levels of Lp(a) are due to proteinuria or an activated acute-phase response. Serum concentrations greater than 30 mg/dl are independently associated with coronary heart disease. Data from cell culture studies suggest that it is not uptake of Lp(a) by mesangial cells but trapping by matrix proteins that contributes to the generation of glomerular apo(a) deposits. Lp(a) alters mesangial cell DNA synthesis and stimulates the generation of reactive oxygen species. Prolonged exposure to Lp(a) causes mesangial cell death in vitro culture experiments. Lp(a) does not alter autocrine transforming growth factor-β transcription in human mesangial cells and has, unlike low-density lipoprotein, no effect on the production of the extracellular matrix protein fibronectin. Future cell culture studies on the role of Lp(a) in renal disease have to address whether Lp(a) induces cell death via apoptosis and to what extent the generation of oxygen radicals is involved in this process.

Introduction

Lipoprotein(a) [Lp(a)] is composed of lipids and two major protein components, apolipoprotein B-100 (apoB-100) and apolipoprotein(a) [apo(a)] [1]. Apo(a) exhibits size polymorphism, with isoprotein size ranging from 420 to 840 kD [2]. Individual isoform patterns are genetically determined and the plasma Lp(a) concentration correlates inversely with the size of the apo(a) protein isoform [1]. The concentrations of Lp(a) vary from individual to individual ranging from <0.1 to >300 mg/dl. Lp(a) is an independent risk factor for atherosclerotic vascular disease. High plasma concentrations (>30 mg/dl) are associated with the risk of premature coronary and cerebrovascular atherosclerosis, saphenous vein bypass graft stenosis, and restenosis after coronary balloon angioplasty [3–5]. Currently, only a few metabolic states, clinical disorders, or inherited diseases are known to influence the strong genetically determined plasma concentration of this unique lipoprotein. Abnormalities of Lp(a) metabolism in patients with renal disease have provoked two principal questions. Do elevated Lp(a) levels in patients with kidney disease contribute to an increased risk for atherosclerosis disease? Do elevated Lp(a) serum levels accelerate glomerular injury and if so, what are the likely mechanisms?

Elevated Plasma Concentrations in Patients with Renal Disease

Sixty percent of patients with the nephrotic syndrome [6], and 30–70% of patients on hemodialysis (HD) or continuous ambulatory peritoneal dialysis (CAPD) have plasma Lp(a) levels above 30 mg/dl (for controls this is <20%) [7–10]. On average, Lp(a) levels are elevated sixfold in patients with severe nephrotic syndrome, fourfold in CAPD, and twofold in patients receiving hemodialysis as compared to healthy controls (fig. 1). In HD patients, an elevation of Lp(a) plasma levels has preferentially been identified in those patients exhibiting high-molecular-weight isoforms [11]. In these patients, levels decrease following renal transplantation [12, 13].

Why Do Some Patients with Renal Disease Develop High Plasma Lp(a)?

In patients with the nephrotic syndrome, the loss of a regulatory substance and/or hypoalbuminemia may stimulate hepatic synthesis of albumin and other liver-derived proteins, including apolipoproteins. This mechanism may in part be applicable in CAPD patients as well. In HD patients, Kario et al. [14] demonstrated a close relationship between high Lp(a) levels and acute-phase proteins as demonstrated by correlations with C-reactive protein, sialic acid and with interleukin (IL)-6. Since seven IL-6-responsive element

KARGER
E-Mail karger@karger.ch
Fax +41 61 306 12 34
http://www.karger.ch

0378–0392/97/0236–0161 $12.00/0

C. Wanner, MD
Department of Medicine, Division of Nephrology, University of Würzburg
Josef-Schneider-Strasse 2
D-97080 Würzburg (Germany)
Tel. +49 931 201 5330, fax +49 931 201 3502

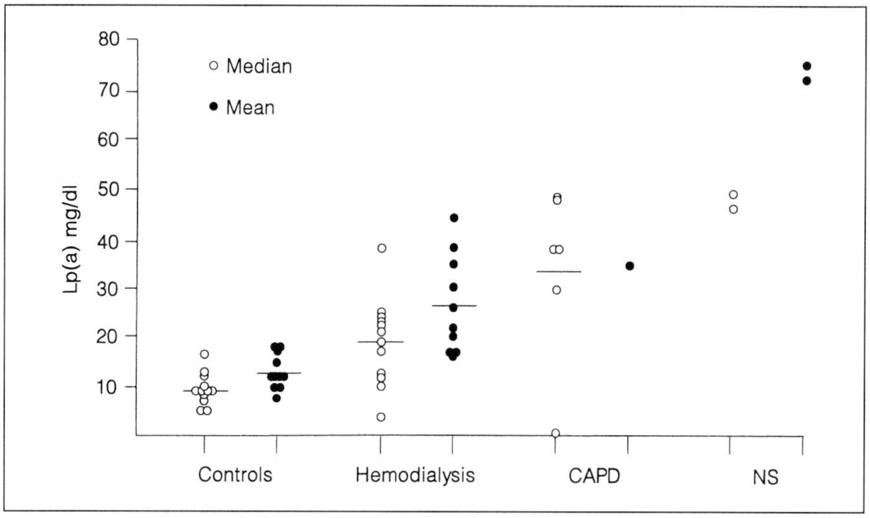

Fig. 1. Mean and median Lp(a) levels from 27 studies for nephrotic syndrome patients (NS), patients receiving hemodialysis or CAPD and the respective control group. The studies included 1,861 hemodialysis patients, 417 CAPD patients, 156 patients with nephrotic syndrome and 3,064 control subjects. Each circle represents one study. Considerable variability in results is apparent. It can best be explained by potential difficulties in accurately measuring Lp(a). Because of the high degree of polymorphism, it is impossible to match the isoform of the assay calibrator with that of the unknown samples. [Modified according to ref. 48.]

sequence motifs can be identified in the 5′-flanking regulatory regions of the apo(a) gene on chromosome 6, it is likely that apo(a) responds as an acute-phase reactant [15]. Furthermore, most or all of serum apo(a) is derived from the liver and all factors which may interfere with liver synthesis can influence hepatic production. It also appears that serum Lp(a) levels vary and are not constantly elevated over time in dialysis patients.

Is There an Increased Risk for Atherosclerosis Disease in Dialysis Patients?

Data by Cressmann et al. [16] obtained in 129 HD patients during a follow-up period of 48 months showed that the patients with the highest plasma Lp(a) levels had the highest incidence of cardiovascular disease. Serum Lp(a) concentrations and the presence of a previous clinical event were the only independent contributors to the risk of a clinical event during the period of follow-up. Recently, Kronenberg et al. [17] demonstrated that HD patients with low-molecular-weight (LMW) phenotypes had significantly more carotid arterial sites affected by atherosclerosis than those with high-molecular-weight phenotypes. These authors suggested that the apo(a) phenotype rather than the Lp(a) plasma concentration is a better predictor of carotid atherosclerosis in end-stage renal disease patients and may be used for assessment of general atherosclerosis risk in this patient group. In agreement with these findings, LMW isoforms were also more frequent in a Spanish HD patient population [8] as well as in our series of CAPD patients who were affected by cardiovascular disease (fig. 2) [18].

Why Is the Risk for Progressive Vascular Damage Enhanced?

Many individuals with genetically determined high levels of Lp(a) never develop coronary heart disease. In part of the world where there are few other predisposing factors, atherosclerosis is a rare disease. From a clinical viewpoint, one can speculate that Lp(a) represents an 'innocent bystander' which becomes atherogenic under certain clinical circumstances or under the influence of a trigger. Because risk factors for coronary artery disease are usually additive, and dialysis patients express many of them, physicians would do well to control the known risk factors. Another trigger could be a condition in which the endothelium or endothelium-dependent processes are activated. The rate of restenosis after coronary angioplasty is especially high in patients with high Lp(a) [5]. Dialysis patients do better after a coronary bypass intervention than following angioplasty. Another potential candidate is plasma homocysteine. Homocysteine levels were enhanced in dialysis patients. Harpel et al. [19] recently provided evidence that homocysteine stimulated a 20-fold increase in the affinity between Lp(a) and plasmin-modified fibrin, and a 4-fold increase with unmodified fibrin. Interestingly, Smith and Crosbie [20] have reported extensive colocalization of Lp(a) with fibrin in plaques.

Lipids as Modulators of Progressive Renal Disease: A Role for Lp(a)?

Lipid abnormalities are considered to contribute to glomerular injury during the course of renal disease [21]. Possible mechanisms

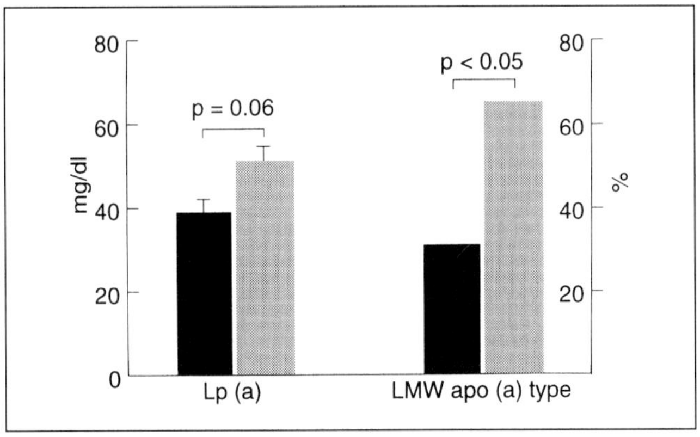

Fig. 2. Serum Lp(a) concentrations and frequency distribution of apo(a) isoforms in 64 CAPD patients with (▨, n=26) and without (■, n=38) coronary artery disease. LMW=All subjects with low-molecular-weight isoforms (F, B, S1, S2). [Adapted and modified from ref. 18.]

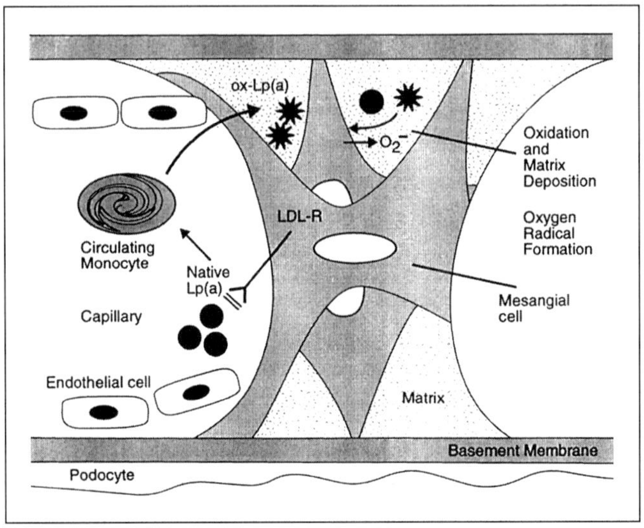

Fig. 3. Schematic drawing of Lp(a) handling by the glomerulus. After permeation of the endothelial-mesangial interface, native Lp(a) may be modified and deposited in its native or oxidized form. Matrix-associated oxidized apo(a) stimulates mesangial cells to produce oxygen-derived radicals or may cause programmed cell death.

include mesangial matrix accumulation, focal and segmental glomerulosclerosis, and monocyte infiltration into preinjured glomeruli [22]. In a recent study, high levels of Lp(a) in the serum of patients with the nephrotic syndrome were correlated with localization of apo(a) in the mesangium [23]. Patients with mesangial Lp(a) deposits also displayed a more severe course of renal disease. However, this study did not determine whether Lp(a) deposits just passively follow the clinical course of the disease, or whether they induce changes in cellular functions that could be relevant to the progression of glomerular disease.

Interaction of Native and Oxidized Lp(a) with Human Mesangial Cells

It remained unclear whether glomerular deposition of Lp(a) requires receptor-mediated pathways and how specific cellular uptake and deposition occur. Since apoB-100 is the main protein component of the Lp(a) particle [24], it has been speculated that the low-density lipoprotein (LDL) receptor may be involved in its clearance. Binding of Lp(a) to the LDL receptor has been reported in fibroblasts and hepatic tissue, although its binding was significantly less than that of LDL [25–27]. Human mesangial cells possess LDL receptors, capable of specific and high-affinity uptake of apoB- and apoE-containing lipoproteins [28]. We investigated the uptake of native and oxidized Lp(a) by cultured human mesangial cells [29] and demonstrated minimal uptake due to the low affinity for the LDL receptor. Scavenger receptor activity which might allow uptake of Lp(a) could not be demonstrated in mesangial cells under our cell culture conditions [29].

In the light of poor cellular uptake, the question of the origin of glomerular deposits remains unanswered. However, there was considerable Lp(a) binding to mesangial matrix, regardless of its oxidative status [30]. Thus, glomerular deposits of Lp(a), found in various forms of renal disease, probably represent matrix-bound rather than internalized Lp(a). The pathogenetic consequences of matrix-bound Lp(a) in kidney could be as follows: Matrix-bound Lp(a) enhances migration and adherence of monocytes which in turn contribute to glomerular injury directly or through oxidation of lipoproteins. A schematic drawing is given in figure 3. This concept is supported by a recent report, where Lp(a) stimulated leukocyte adherence to endothelial cells in vitro [30], although without interfering with the expression of adhesion molecules such as ELAM-1, ICAM-1, or VCAM-1. Alternatively, matrix-bound lipoproteins could interfere with cell-matrix interactions, thereby inducing cell apoptosis [31].

Effect of Lp(a) on Mesangial Cell Proliferation

Focal sclerosis in human minimal-change disease [32] is preceded by mesangial cell proliferation. In animal models of glomerular or tubular injury, suppression of cellular proliferation prevented the development of sclerosis [33].

Stimulation of vascular smooth muscle proliferation has recently been proposed as the mechanism by which Lp(a) accelerates the progression of atherosclerotic disease [34]. We and others have shown that high levels of Lp(a) (above 10 µg/ml) suppress DNA synthesis and are ultimately cytotoxic to human and rat mesangial cells [29, 35, 36]. There is a controversy about the effects of low Lp(a) concentrations. In studies with rat mesangial cells, we demonstrated induction of DNA synthesis and elevated mRNA levels of immediate-early response genes c-*fos* and c-*myc* [35]. In studies with human mesangial cells, we could not detect induction of DNA synthesis; in fact, Lp(a) even inhibited the stimulatory effects of platelet-derived growth factor on DNA synthesis [29]. Other researchers however detected a small stimulatory effect of Lp(a) (5 μg/ml) on DNA synthesis, an effect apparently mediated by phospholipase C activation by a pertussis-toxin-sensitive G protein [36]. Since enhanced DNA synthesis and increased expression of immediate-early genes may very well occur via a stress response pathway, as has been shown after reperfusion injury [37] or oxidative stress [38], the general effect of Lp(a) on mesangial cells seems to be a noxious one, ultimately leading to cell death.

Lp(a) and the Generation of Extracellular Matrix

The excess generation of extracellular matrix is a critical step for the development of glomerulosclerosis. LDL cholesterol can apparently directly interfere with matrix generation in cultured human mesangial cells, since it stimulates fibronectin production, presumably by increasing cellular fibronectin mRNA levels [39]. The generation of excess extracellular matrix proteins in the glomeruli of diabetic rats has been attributed to increased expression of transforming growth factor-β (TGF-β) [40], most likely due to increased transcription of TGF-β mRNA [41]. However, we assessed secretion and cellular levels of the extracellular matrix protein fibronectin and could not detect an increase in fibronectin protein or RNA after stimulation of mesangial cells with Lp(a) for up to 72 h [42]. Furthermore, Lp(a) generally inhibited protein synthesis. Consequently, Lp(a) also had no effects on cellular TGF-β mRNA levels [29].

Lp(a) and the Generation of Reactive Oxygen Species

Several lines of evidence suggest that Lp(a) induces the generation of reactive oxygen species in glomerular cells. In rat mesangial cells, the effects of Lp(a) on DNA synthesis were partially inhibited by the oxygen radical scavengers catalase and superoxide dismutase [35]. Measurement of cytochrome C reduction provided direct evidence for increased O_2^- generation in Lp(a)-stimulated mesangial cells. In intact rat glomeruli, Lp(a) increases O_2^- production via a protein-kinase-C-sensitive pathway [43]. Lp(a)-induced generation of oxygen radicals has also been demonstrated in rat juxtaglomerular cells [44] and in rabbit renal arteries [45], suggesting a general mechanism. The numerous adverse effects of reactive oxygen species on glomerular cell function have been reviewed elsewhere [46, 47]. In the light of the toxic effects of Lp(a) on cultured glomerular cells, future studies have to address whether Lp(a) induces cell death via specific pathways, i.e., apoptosis, and to what extent the generation of oxygen radicals is involved in this process.

References

1 Utermann G: The mysteries of lipoprotein (a). Science 1989;246:904–910.
2 Gaubatz JW, Ghanem KI, Guevara JJ, Nava ML, Patsch W, Morrisett JD: Polymorphic forms of human apolipoprotein[a]: Inheritance and relationship of their molecular weights to plasma levels of lipoprotein[a]. J Lipid Res 1990;31:603–613.
3 Rader DJ, Hoeg JM, Brewer HB Jr: Quantitation of plasma apolipoproteins in the primary and secondary prevention of coronary artery disease. Ann Intern Med 1994;120:1012–1025.
4 Jürgens G, Taddei-Peters WC, Költringer P, Petek W, Chen Q, Greiberger J, Macomber PF, Butman BT, Stead AG, Ransom JH: Lipoprotein(a) serum concentration and apolipoprotein(a) phenotype correlate with severity and presence of ischemic cerebrovascular disease. Stroke 1995;26:1841–1848.
5 Desmarais RL, Sarembock IJ, Ayers CR, Vernon SM, Powers ER, Gimple LW: Elevated serum lipoprotein(a) is a risk factor for clinical recurrence after coronary balloon angioplasty. Circulation 1995;91: 1403–1409.
6 Wanner C, Rader D, Bartens W, Krämer J, Brewer HB, Schollmeyer P, Wieland W: Elevated plasma lipoprotein(a) in patients with the nephrotic syndrome. Ann Intern Med 1993;119:263–269.
7 Kronenberg F, König P, Neyer U, Auinger M, Pribasnig A, Lang U, Reitinger J, Pinter G, Utermann G, Dieplinger H: Multicenter study of lipoprotein(a) and apolipoprotein(a) phenotypes in patients with end-stage renal disease treated by hemodialysis or continuous ambulatory peritoneal dialysis. J Am Soc Nephrol 1995;6:110–120.
8 Auguet T, Senti M, Rubies-Prat J, Pelegri A, Pedro-Botet J, Nogues X, Romero R: Serum lipoprotein(a) concentrations in patients with chronic renal failure receiving hemodialysis: Influence of apolipoprotein(a) genetic polymorphism. Nephrol Dial Transplant 1993;8:1099–1103.
9 Hirata K, Kikuchi S, Saku K, Jimi S, Zhang B, Naito S, Hamaguchi H, Arakawa K: Apolipoprotein(a) phenotypes and serum lipoprotein(a) levels in maintenance hemodialysis patients with/without diabetes mellitus. Kidney Int 1993;44: 1062–1070.
10 Shoij T, Nishizawa Y, Nishitani H, Yamakawa M, Morii H: High serum concentrations in uremic patients treated with continuous ambulatory peritoneal dialysis. Clin Nephrol 1992;38:271–276.
11 Dieplinger H, Lackner C, Kronenberg F, Sandholzer C, Lhotta K, Hoppichler F, Graf H, König P: Elevated plasma concentrations of lipoprotein(a) in patients with end-stage renal disease are not related to the size polymorphism of apolipoprotein(a). J Clin Invest 1993;91:397–401.
12 Black IW, Wilcken DEL: Decreases in apolipoprotein(a) after renal transplantation: Implications for lipoprotein(a) metabolism. Clin Chem 1992;38: 353–357.
13 Kronenberg F, König P, Lhotta H: Apolipoprotein(a) associated decrease in lipoprotein(a) plasma concentrations after renal transplantation. Arterioscler Thromb 1994;14:1399–1404.
14 Kario K, Matsuo T, Kobayashi H, Matsuo M, Asada R, Koide M: High lipoprotein(a) levels in chronic hemodialysis patients are closely related to the acute phase reaction. Thromb Haemost 1995; 74:1020–1025.
15 Wade DP, Clarke JG, Lindahl GE, Liu AC, Zysow BR, Meer K, Schwartz K, Lauw RM: Genetic influences on lipoprotein(a) concentration. Biochem Soc Trans 1993;21:499–501.
16 Cressman MD, Heyka RJ, Paganini EP, O'Neil J, Skibinski CI, Hoff HF: Lipoprotein(a) is an independent risk factor for cardiovascular disease in hemodialysis patients. Circulation 1992;86:475–482.
17 Kronenberg F, Kathrein H, König P, Neyer U, Sturm W, Lhotta K, Gröchenig E, Utermann G, Dieplinger H: Apolipoprotein(a) phenotypes predict the risk for carotid atherosclerosis in patients with end-stage renal disease. Arterioscler Thromb 1994;14:1405–1411.
18 Wanner C, Bartens W, Walz G, Nauck M, Schollmeyer P: Protein loss and genetic polymorphism of apolipoprotein(a) modulate serum lipoprotein(a) in CAPD patients. Nephrol Dial Transplant 1995;10: 75–81.
19 Harpel PC, Chang VT, Borth W: Homocysteine and other sulfhydryl compounds enhance the binding of lipoprotein(a) to fibrin: A potential biochemical link between thrombosis, atherogenesis and sulfhydryl compound metabolism. Proc Natl Acad Sci USA 1992;89:10193–10197.

20 Smith EB, Crosbie L: Does lipoprotein(a) [Lp(a)] compete with plasminogen in human atherosclerotic lesions and thrombi? Atherosclerosis 1991;89: 127–136.
21 Neugarten J, Schlondorff D: Lipoprotein interactions with glomerular cells and matrix; in Keane WF, Stein JH (eds): Lipids and Renal Disease. New York, Churchill Livingstone, 1991, pp 173–206.
22 Keane WF, Kasiske BL, O'Donnell MP: Hyperlipidemia and the progression of renal disease. Am J Clin Nutrition 1988;47:157–160.
23 Sato H, Suzuki S, Ueno M, Shimada H, Karasawa R, Nishi S, Arakawa M: localization of apolipoprotein(a) and B-100 in various renal diseases. Kidney Int 1993;43:430–435.
24 Snyder LM, Polacek D, Scanu AM, Fless GM: Comparative binding and degradation of lipoprotein(a) and low density lipoprotein by human monocyte derived macrophages. J Biol Chem 1992; 267:339–346.
25 Floren CH, Albers JJ, Bierman EL: Uptake of Lp(a) lipoprotein by cultured fibroblasts. Biochem Biophys Res Commun 1981;102:636–639.
26 Havekes L, Vermeer BJ, Brugman T, Emeis J: Binding of Lp(a) to the low density lipoprotein receptor of human fibroblasts. FEBS Lett 1981;132:169–173.
27 Kostner GM: Interaction of Lp(a) and of Apo(a) with liver cells. Arterioscler Thromb 1993;13:1101–1109.
28 Krämer A, Nauck M, Pavenstädt P, Schwedler S, Wieland H, Schollmeyer P, Wanner C: Receptor-mediated uptake of IDL and LDL from nephrotic patients by glomerular cells. Kidney Int 1995;44: 1341–1451.
29 Krämer-Guth A, Greiber S, Pavenstädt H, Quaschning T, Winkler K, Schollmeyer P, Wanner C: Interaction of native and oxidized lipoprotein(a) with human mesangial cells and matrix. Kidney Int 1996;49:1250–1261.
30 Ramaswamy C, Steadman R, Jordan N, Westwick J, Williams JD: lipoprotein(a) stimulates leukocyte adherence to endothelial cells in vitro (abstract). J Am Soc Nephrol 1995;6:912.
31 Sugiyama H, Kashihara T, Sekikawa Y, Maeshima Y, Okamoto K, Kanao K, Makino H, Ota Z: Cell-matrix interactions modulate cell apoptosis (abstract). J Am Soc Nephrol 1995;6:912.
32 Fogo A, Hawkins EP, Berry PL, Glick AD, Chiang ML, MacDonnel RC, Ichikawa I: Glomerular hypertrophy in minimal change disease predicts subsequent progression to focal glomerular sclerosis. Kidney Int 1990;38:115–123.
33 Eng E, Floege J, Young BA, Couser WG, Johnson RJ: Does extracellular matrix expansion in glomerular disease require mesangial proliferation? Kidney Int 1994;45(suppl 45):S45–S47.
34 Grainger DJ, Kirschenlohr HL, Metcalfe JC, Weissberg PL, Wade DP, Lawn RM: Proliferation of human smooth muscle cells promoted by lipoprotein(a). Science 1993;260:1655–1658.
35 Greiber S, Krämer-Guth A, Pavenstädt H, Gutenkunst M, Schollmeyer P, Wanner C: Effects of lipoprotein(a) on mesangial cell proliferation and viability. Nephrol Dial Transplant 1996;11:778–785.
36 Mondorf UF, Bender M, Olbrich HG, Herrero M, Weisensee D, Scheuermann E, Gross W: Phospholipase C activation by lipoprotein(a) in cultured human mesangial cell (abstract). J Am Soc Nephrol 1996;7:1682.
37 DiMari J, Megyesi J, Udarhelyi N, Price P, Safirstein R: Inhibition of immediate early gene response during reperfusion injury ameliorates renal failure (abstract). J Am Soc Nephrol 1995;6:977.
38 Rao GN, Lassegue B, Griendling KK, Alexander RW: Hydrogen peroxide stimulates transcription of c-*jun* in vascular smooth muscle cells: Role of arachidonic acid. Oncogene 1993;8:2759–2764.
39 Rovin BH, Tan LC: LDL stimulates mesangial fibronectin production and chemoattractant expression. Kidney Int 1993;43:218–225.
40 Yamamoto T, Nakamura T, Noble NA, Rouslati E, Border WA: Expression of transforming growth factor beta is elevated in human and experimental diabetic nephropathy. Proc Natl Acad Sci USA 1993;90:1814–1818.
41 Isaka Y, Fujiwara Y, Ueda N, Kaneda Y, Kamada T, Iamai E: Glomerulosclerosis induced by in vivo transfection of transforming growth factor-beta or platelet derived growth factor into the rat kidney. J Clin Invest 1993;92:2597–2601.
42 Greiber S, Daxer C, Schollmeyer P, Wanner C: Influence of lipoprotein(a) on fibronectin production in cultured human mesangial cells (abstract). J Am Soc Nephrol 1996;7:1852.
43 Kreusel M, Greiber S, Schollmeyer P, Wanner C: Lipoprotein(a) und O_2^--Produktion in isolierten Glomeruli (abstract). Nieren Hochdruckkrankheiten 1995;24:532.
44 Galle J, Stunz P, Schollmeyer P, Wanner C: Oxidized LDL and lipoprotein(a) stimulate renin secretion of juxtaglomerular cells. Kidney Int 1995;47: 45–52.
45 Galle J, Bengen J, Schollmeyer P, Wanner C: Impairment of endothelium-dependent dilatation of rabbit renal arteries by oxidized lipoprotein(a): Role of oxygen-derived radicals. Circulation 1995;92: 1582–1589.
46 Nath KA, Fischereder M, Hostetter THL: The role of oxidants in progressive renal injury. Kidney Int 1994;45:S111–S115.
47 Shah SV: The role of reactive oxygen metabolites in glomerular disease. Annu Rev Physiol 1995;57: 245–262.
48 Wanner C: Clinical significance of lipoprotein(a) determination in patients with renal disease. J Nephrol 1996;9:157–159.

Mineral and Electrolyte Metabolism

Miner Electrolyte Metab 1997;23:166–169

William F. Keane

Department of Medicine, Division of Nephrology, Hennepin County Medical Center, Minneapolis, Minn., USA

Derangement of Lipid Metabolism and Its Management in Renal Transplant Patients

Key Words
ESRD
Transplant hyperlipidemia
Cardiovascular disease
HMG CoA reductase inhibitors

Abstract
Hyperlipidemia is a common clinical finding in postrenal transplant patients. Increasing clinical data suggest that the dyslipidemia seen in these patients is associated with the development of atherosclerotic cardiovascular disease. In addition, recent data suggest that disordered lipid metabolism may be a risk factor for chronic allograft loss. The precise mechanisms whereby dyslipidemia contributes to vascular injury are unknown. No long-term interventional studies of lipid-lowering therapies have been conducted in this patient population. However, a number of short-term studies have shown that the newer antilipemic agents can be safely used in renal transplant patients. Thus, it appears that antilipemic therapy is warranted in renal transplant patients with an atherogenic lipid profile and multiple risk factors.

Hyperlipidemia is a common clinical complication of renal transplantation [1–3]. Considerable progress has been made during the past decade not only in understanding the causes but also in the management of hyperlipidemia in the posttransplantation patient. These lipid abnormalities have been recently associated with an increased risk for the development of cardiovascular disease and for chronic allograft loss, both of which are frequently seen in this population of patients. In this review, the mechanisms and management of lipoprotein abnormalities after renal transplantation will be examined.

Hyperlipidemia in Renal Transplant Patients

Increased total cholesterol and triglycerides are common after renal transplantation. Elevated serum low-density lipoprotein (LDL) cholesterol as well as apolipoprotein (apo)B levels are common [4–8]. On the other hand, serum high-density lipoprotein levels have been reported to be low, normal, or high, while apoA levels, in general, have been considered normal [5, 9–12]. Most studies have reported normal lipoprotein(a) [Lp(a)] in patients who were not receiving cyclosporin A [13–17], while renal transplant patients treated with cyclosporin A alone have demonstrated elevated Lp(a) levels [18, 19]. Elevated lipid peroxidation products, increased LDL susceptibility to oxida- tion, and decreased antioxidant levels have been observed in renal transplant patients treated with cyclosporin A [20–22]. It has also been demonstrated that LDL susceptibility to oxidation decreased upon conversion of cyclosporine to azathioprine [23]. Since oxidative modification of LDL has been identified as important in the pathogenesis of atherosclerosis in the general population, these findings may be of considerable importance in understanding the increased prevalence of vascular disease in the posttransplant patient.

Mechanisms of Hyperlipidemia after Kidney Transplantation

The mechanisms of hyperlipidemia after renal transplantation have been the subject of investigation during the past decade. Several possible causes have been identified including immunosuppressive therapy, genetic predisposition, age, gender, body habitus, renal function, proteinuria, diabetes, and antihypertensive therapy [4, 24–26]. In addition, posttransplant hyperlipidemia is related to immunosuppressive therapy in a dose-dependent fashion [1–3]. The relative contribution of these factors to the development of posttransplantation hyperlipidemia is unclear.

The altered metabolism of triglyceride-rich lipoproteins is probably the result of either increased hepatic secretion of very low density lipoproteins (VLDLs), or decreased peripheral triglyceride-rich lipoprotein removal, or both [27–29]. Increased triglyceride production has been observed up to 3 years posttransplantation and this has been postulated to be a result of peripheral insulin resistance and secondary hyperinsulinemia [28]. Corticosteroids, thiazide diuretics, renal insufficiency, and weight gain could all participate in this phenomena. Whether cyclosporine affects peri-

KARGER
E-Mail karger@karger.ch
Fax +41 61 306 12 34
http://www.karger.ch

0378–0392/97/0236–0166 $12.00/0

William F. Keane, MD
Department of Medicine, Hennepin County Medical Center
701 Park Avenue South
Minneapolis, MN 55415 (USA)
Tel. +1 612 347 5809, fax +1 612 347 4321

pheral insulin resistance and hyperinsulinemia is unclear [30, 31]. Experimental evidence also suggests that a decrease in the catabolism of triglyceride-rich lipoproteins occurs in renal transplant patients [28, 29, 32]. The mechanism for this has not been convincingly established although a decrease in the activity of vascular and hepatic lipase enzymes has been reported occasionally [29, 32, 33]. Finally, reduced LDL receptor activity could also contribute to decreased VLDL removal. Both corticosteroids and cyclosporine could be associated with altered LDL receptor activity [34, 35].

Increased synthesis of VLDL with a normal conversion of VLDL to LDL could raise levels of LDL [27]. However, whether the conversion of VLDL to LDL is normal in renal transplant patients is unclear [29, 32, 33]. In addition, altered removal of LDL could result from changes in the composition of LDL (triglyceride enrichment), which could affect receptor-mediated uptake and degradation, or following down-regulation of the hepatic LDL receptor [36, 37]. Unfortunately, few studies have evaluated posttransplant LDL receptor number and/or function [35]. In general, posttransplant reverse cholesterol transport is similar to that seen in healthy subjects even though there may be some abnormalities of triglyceride concentration in these lipoproteins. It is not known why Lp(a) appears to be normal in some renal transplant patients and increased in others. However, while improvement in renal function after transplantation may be partly responsible for this observation [38, 39], other unidentified factors are probably also involved.

Consequences of Hyperlipidemia after Renal Transplantation

Cardiovascular disease is one of the most common causes of morbidity and mortality among long-term survivors of kidney transplantation [40–42]. There is increasing evidence that posttransplant lipoprotein abnormalities may contribute to the development of cardiovascular disease [40, 43, 44]. Interestingly, high-density lipoprotein and triglyceride levels appear to be more closely associated with cardiovascular disease than total or LDL cholesterol in these patients [24, 40]. The role of modified LDL in posttransplant cardiovascular disease is at present unknown. Similarly, whether Lp(a) has a role to play in this population remains to be established.

There is a growing body of evidence to suggest that posttransplant lipid abnormalities might influence chronic renal allograft loss [45–49]. Increased serum triglycerides appear to be the most consistent predictor of chronic renal allograft loss [45–49]. Whether these associations are causative or merely reflect a secondary effect of reduced renal function has not been clearly established. Few studies have examined the effect of antilipemic therapies on the rate of decline in renal function. However, in heart transplant patients, pravastatin reduced the incidence of severe acute rejection, 1-year graft survival, and coronary vasculopathy [50]. A similar study after kidney transplantation is still in progress and the results appear to be promising [51].

Management of Hyperlipidemia after Renal Transplantation

There are no interventional trials to evaluate clinically relevant end points in posttransplant patients with hyperlipidemia. However, it is reasonable to assume that the risks associated with hyperlipidemia and the benefits of correcting these abnormalities in renal transplant patients are at least comparable to those found in the general population. Using meta-analysis techniques, our group evaluated some 40 studies in renal transplant patients [52]. These studies were of short duration, included few patients and were mostly designed to evaluate safety and lipid-lowering efficacy. Nonetheless, some useful insights could be obtained from these analyses. Diet, 3-hydroxy-3-methylglutaryl coenzyme A (HMG CoA) reductase inhibitors, and fibric acid analogues consistently improved lipoprotein abnormalities [52]. Few studies have examined the use of niacin or probucol in these patients and additional studies are needed before these agents can be used routinely. The risk of rhabdomyolysis in renal transplant patients treated with HMG CoA reductase inhibitors appears to be small when the doses are reduced and concomitant therapy with other drugs associated with this complication is avoided. Increased serum creatinine levels in cyclosporine-treated transplant patients has been reported with bezafibrate, fenofibrate, and ciprofibrate, but not gemfibrozil [53]. The mechanism(s) and clinical implications of this increase in serum creatinine levels with the fibric acid analogues needs further study. Bile acid resins have also not been adequately studied in renal transplant patients. Concerns regarding altering the pharmacokinetics of cyclosporine have not been investigated.

In summary, antilipemic therapy is reasonable in high-risk renal transplant patients. Promotion of exercise and maintaining patients on the lowest possible dose of corticosteroids and cyclosporine compatible with prolonged allograft survival will help minimize hyperlipidemia. However, in the setting of levels of cholesterol that exceed current recommendations, therapy with an HMG CoA reductase inhibitor in conjunction with diet and weight reduction have been recommended [54]. It is reasonable to target LDL cholesterol to <130 mg/dl for primary prevention and <100 mg/dl for secondary prevention [54]. In patients with primarily hypertriglyceridemia (>200 mg/dl), a low-calorie, low-fat diet, and in some cases fish oil, may help to reduce triglyceride levels. A fibric acid derivative is effective in reducing triglycerides and may also reduce LDL cholesterol. However, the dose must be reduced in patients with reduced renal function. Combination antilipemic therapy in transplant patients has been associated with more frequent complications and its routine use at present cannot be recommended.

Acknowledgment

The author recognizes the assistance of Deanna Gunderson in the preparation of this manuscript.

References

1 Cattran DC, Steiner G, Wilson DR, Fenton SA: Hyperlipidemia after renal transplantation: Natural history and pathophysiology. Ann Intern Med 1979;91:554–559.

2 Vathsala A, Weinberg RB, Schoenberg L, Grevel J, Goldstein RA, VanBuren CT, Lewis RM, Kahan BD: Lipid abnormalities in cyclosporine-prednisone-treated renal transplant recipients. Transplantation 1989;48:37–43.

3 Ponticelli C, Barbi GL, Cantaluppi A, De Vecchi A, Annoni G, Donati C, Cecchettin M: Lipid disorders in renal transplant recipients. Nephron 1978; 20:189–195.

4 Kasiske BL, Umen AJ: Persistent hyperlipidemia in transplant patients. Medicine 1987;66:309–316.

5 Raine AEG, Carter R, Mann JI, Morris PJ: Adverse effect of cyclosporin on plasma cholesterol in renal transplant recipients. Nephrol Dial Transplant 1988;3:458–463.

6 Segoloni GP, Triolo G, Cassader M, Turello E, Ruiu G, Messina M, Pagano GF, Vercellone A: Dyslipidemia in renal transplantation: A 3-year follow-up. Transplant Proc 1993;25:2178–2179.

7 Cassader M, Ruiu G, Gambino R, Alemanno N, Triolo G, Pagano G: Lipoprotein-apolipoprotein changes in renal transplant recipients: A 2-year follow-up. Metabolism 1991;40:922–925.

8 Averna MR, Barbagallo CM, Sparacino V, Caputo F, Mancino C, Calabrese S, Notarbartolo A: Follow-up of lipid and apoprotein levels in renal transplant recipients. Nephron 1991;58:255–256.

9 Drüeke TB, Abdulmassih Z, Lacour B, Bader C, Chevalier A, Kreis H: Atherosclerosis and lipid disorders after renal transplantation. Kidney Int 1991; 39(suppl 31):S24–S28.

10 Jung K, Neumann R, Scholz D, Nugel E: Abnormalities in the composition of serum high density lipoprotein in renal transplant recipients. Clin Nephrol 1982;17:191–194.

11 Kasiske BL, Tortorice KL, Heim-Duthoy KL, Awni WM, Rao KV: The adverse impact of cyclosporine in serum lipids in renal transplant recipients. Am J Kidney Dis 1991;17:700–707.

12 Kobayashi N, Okubo M, Marumo F, Uchida H, Endo T, Nakamura H: De novo development of hypercholesterolemia and elevated high-density lipoprotein cholesterol: Apoprotein A-I ratio in patients with chronic renal failure following kidney transplantation. Nephron 1983;35:237–240.

13 Arnadottir M, Thysell H, Nilsson-Ehle P: Lipoprotein levels and post-heparin lipase activities in kidney transplant recipients: Ciclosporin- versus nonciclosporin-treated patients. Am J Nephrol 1991; 11:391–396.

14 Massy ZA, Bader CA, Chevalier A, Chauveau P, Zingraft J, Lambrey G, Drueke T, Jungers P, Kreis H, Lacour B: Serum lipoprotein (a) levels in chronic renal failure and renal transplant patients. J Nephrol 1994;7:229–236.

15 Webb AT, Reaveley DA, O'Donnell M, O'Connor B, Seed M, Brown EA: Does cyclosporin increase lipoprotein(a) concentrations in renal transplant recipients? Lancet 1993;341:268–270.

16 Gault MH, Longerich LL, Purchase L, Harnett J, Breckenridge C: Comparison of LP(a) concentrations and some potential effects in hemodialysis, CAPD, transplantation, and control groups, and review of the literature. Nephron 1995;70:155–170.

17 Azrolan N, Brown CD, Thomas L, Hayek T, Zhao ZH, Roberts KG, Scheiner C, Friedman EA: Cyclosporin A has divergent effects on plasma LDL cholesterol (LDL-C) and lipoprotein(a) [Lp(a)] levels in renal transplant recipients. Arterioscler Thromb 1994;14:1393–1398.

18 Brown JH, Anwar N, Short CD, Bhatnager D, Mackness MI, Hunt LP, Durrington PN: Serum lipoprotein (a) in renal transplant recipients receiving cyclosporin monotherapy. Nephrol Dial Transplant 1993;8:863–867.

19 Hilbrands LB, Demacher PNM, Hoitsma AJ: Cyclosporin and serum lipids in renal transplant recipients. Lancet 1993;341:765–766.

20 Dasgupta A, Powell D, Saldana S, Doria L: Elevated lipid peroxidation products and depleted transferrin levels in the plasma of kidney transplant recipients. Life Sci 1990;46:67–72.

21 Apanay DC, Neylan JF, Ragab MS, Sgoutas DS: Cyclosporine increases the oxidizability of low-density lipoproteins in renal transplant recipients. Transplantation 1994;58:663–669.

22 Sutherland WHF, Walker RJ, Ball MJ, Stapley SA, Robertson MC: Oxidation of low density lipoproteins from patients with renal failure or renal transplants. Kidney Int 1995;48:227–236.

23 Dorpel MA van den, Ghanem HAI, Zietse R: Effect of conversion from cyclosporin-A to azathioprine on LDL size and parameters of LDL oxidation in renal transplant recipients (abstract). J Am Soc Nephrol 1996;6:1119.

24 Abdulmassih Z, Chevalier A, Bader C, Drüeke T, Kreis H, Lacour B: Role of lipid disturbances in the atherosclerosis of renal transplant patients. Clin Transplant 1992;6:106–113.

25 Hricik DE, Mayes JT, Schulak JA: Independent effects of cyclosporine and prednisone on posttransplant hypercholesterolemia. Am J Kidney Dis 1991; 18:353–358.

26 Chan MK, Varghese Z, Persaud JW, Fernando ON, Moorhead JF: The role of multiple pharmacotherapy in the pathogenesis of hyperlipidemia after renal transplantation. Clin Nephrol 1981;15:309–313.

27 Chan MK, Varghese Z, Moorhead JF: Lipid abnormalities in uremia, dialysis, and transplantation. Kidney Int 1981;19:625–637.

28 Chan MK, Persaud JW, Varghese Z, Fernando ON, Moorhead JF : Fat clearances and hyperlipidaemia in renal allograft recipients – The role of insulin resistance. Clin Chim Acta 1981;114:61–67.

29 Savdie E, Gibson JC, Crawford GA, Simons LA, Mahony JF: Impaired plasma triglyceride clearance as a feature of both uremic and posttransplant triglyceridemia. Kidney Int 1980;18:774–782.

30 Jindal RM: Posttransplant diabetes mellitus – A review. Transplantation 1994;58:1289–1298.

31 Öst L, Tydén G, Fehrman I: Impaired glucose tolerance in cyclosporine-prednisolone-treated renal graft recipients. Transplantation 1988;46:370–372.

32 Föger B, Königsrainer A, Palos G, Brandstätter E, Ritsch A, König P, Miesenböck G, Lechleitner M, Margreiter R, Patsch JR: Effect of pancreas transplantation on lipoprotein lipase, postprandial lipemia, and HDL cholesterol. Transplantation 1994; 58:899–904.

33 Derfler K, Hayde M, Heinz G, Hirschl MM, Steger G, Hauser AC, Balcke P, Widhalm K: Decreased postheparin lipolytic activity in renal transplant recipients with cyclosporin A. Kidney Int 1991;40: 720–727.

34 de Groen PC: Cyclosporine, low-density lipoprotein, and cholesterol. Mayo Clin Proc 1988;63: 1012–1021.

35 Portman RJ, Scott RC, Rogers DD, Loose-Mitchell DS, Lemire JM, Weinberg RB: Decreased low-density lipoprotein receptor function and messenger mRNA levels in lymphocytes from uremic patients. Kidney Int 1992;42:1238–1246.

36 Princen HM, Meijer P, Wolthers BG, Vonk RJ, Kuipers F: Cyclosporin A blocks bile acid synthesis in cultured hepatocytes by specific inhibition of chenodeoxycholic acid synthesis. Biochem J 1991; 275:501–505.

37 Dahlback-Sjoberg H, Bjorkhem I, Princen HM: Selective inhibition of mitochondrial 27-hydroxylation of bile acid intermediates and 25-hydroxylation of vitamin D3 by cyclosporin A. Biochem J 1993;293:203–206.

38 Black IW, Wilcken DE: Decreases in apolipoprotein(a) after renal transplantation: Implications for lipoprotein(a) metabolism. Clin Chem 1992;38: 353–357.

39 Barbagallo CM, Averna MR, Sparacino V, Galione A, Caputo F, Scafidi V, Amato S, Mancino C, Cefalù AB, Notarbartolo A: Lipoprotein (a) levels in end-stage renal failure and renal transplantation. Nephron 1993;64:560–564.

40 Kasiske BL, Guijarro C, Massy ZA, Weiderkehr MR, Ma JZ: Cardiovascular disease after renal transplantation. J Am Soc Nephrol 1996;7:158–165.

41 Hill MN, Grossman RA, Feldman HI, Hurwitz S, Dafoe DC: Changes in causes of death after renal transplantation, 1966 to 1987. Am J Kidney Dis 1991;17:512–518.

42 Corboy J, Sutherland WH, Walker RJ, Robertson MC, Cox CM: Cholesteryl ester transfer in patients with renal failure or renal transplants. Kidney Int 1994;46:1147–1153.

43 Kasiske BL: Risk factors for accelerated atherosclerosis in renal transplant recipients. Am J Med 1988; 84:985–992.

44 Massy ZA, Millet P, Helenon O: Ultrasonographically assessed carotid plaques: Relation with cardiovascular risk factors in renal transplant recipients (abstract). J Am Soc Nephrol 1993;4:948.

45 Massy ZA, Guijarro C, Wiederkehr MR, Ma JZ, Kasiske BL: Chronic renal allograft rejection: Immunologic and nonimmunologic risk factors. Kidney Int 1996;49:518–524.

46 Markell MS, Sumrani N, Dibenedetto A, Friedman EA: Effect of early hyperlipidemia on graft and patient survival in cyclosporine-treated renal transplant patients. Am J Kidney Dis 1993;22:233–239.

47 Charlesworth JA, Edwards KD: Relation of hypertriglyceridaemia to kidney transplant dysfunction and rejection, and response to antilymphocyte globulin therapy. Prog Biochem Pharmacol 1974; 9:107–120.

48 Dimény E, Fellström B, Larsson E, Tufveson G, Lithell H: Chronic vascular rejection and hyperlipoproteinemia in renal transplant patients. Clin Transplant 1993;7:482–490.

49 Diményl E, Tufvenson G, Lithell H, Larsson E, Siegbahn A, Fellström B: The influence of pretransplant lipoprotein abnormalities on the early results of renal transplantation. Eur J Clin Invest 1993;23:572–579.

50 Kobashigawa JA, Katznelson S, Laks H, Johnson JA, Yeatman L, Wang XM, Chia D, Terasaki PI, Sabad A, Cogert GA: Effect of pravastatin on outcomes after cardiac transplantation. N Engl J Med 1995;333:621–627.

51 Katznelson S, Kobashigawa JA: Dual roles of HMG-CoA reductase inhibitors in solid organ transplantation: Lipid lowering and immunosuppression. Kidney Int 1995;48:S112–S115.

52 Massy ZA, Ma JZ, Louis TA, Kasiske BL: Lipid-lowering therapy in patients with renal disease. Kidney Int 1995;48:188–198.

53 Massy ZA, Guijarro C, Kasiske BL: Effect of fenofibrates in heart transplant patients. Transplantation 1995;59:451–452.

54 Expert Panel on Detection Evaluation, and Treatment of High Blood Cholesterol: Summary of the Second Report of the National Cholesterol Education Program (NCEP) Expert Panel on Detection, Evaluation, and Treatment of High Blood Cholesterol in Adults (Adult Treatment Panel II). JAMA 1993;269:3015–3023.

Mineral and Electrolyte Metabolism

Miner Electrolyte Metab 1997;23:170–173

Paul Jungers[a]
Philippe Chauveau[a]
Olivia Bandin[b], *Bernadette Chadefaux*[b]
Joëlle Aupetit[b]
Mathilde Labrunie[c]
Béatrice Descamps-Latscha[c]
Pierre Kamoun[b]

[a] Department of Nephrology,
[b] Department of Biochemistry, and
[c] INSERM U90, Necker Hospital, Paris, France

Hyperhomocysteinemia Is Associated with Atherosclerotic Occlusive Arterial Accidents in Predialysis Chronic Renal Failure Patients

Key Words
Accelerated atherosclerosis
Chronic renal failure
Homocysteine
Hyperhomocysteinemia
Occlusive arterial disease

Abstract
Hyperhomocysteinemia has been shown to constitute an independent risk factor for premature occlusive arterial disease. Moderate hyperhomocysteinemia is present in chronic uremic patients, who often develop premature atherosclerosis, but no direct evidence of an association between the occurrence of atherosclerotic cardiovascular accidents (CVAs) and hyperhomocysteinemia has yet been reported in such patients. We serially determined total plasma homocysteine (Hcy) levels in a cohort of 93 consecutive chronic renal failure, undialyzed patients (57 males, 36 females) with creatinine clearance (Ccr) <50 ml/min·1.73 m^2 and age ≥ 50 years at start of follow-up, together with serial assessment of Ccr and blood lipid parameters. From January 1989 to December 1995, 24 patients (group 1) experienced myocardial infarction (18 cases, 13 males) or cerebral infarction (6 cases, 3 males) while the remaining 69 (group 2) remained free of CVAs. Patients in groups 1 and 2 did not differ with respect to age (66 ± 1.8 vs. 65 ± 1.1 years, mean ± Se) or serum creatinine (227 ± 24 vs. 251 ± 36 μmol/l) at onset of a CVA (group 1) or at the end of follow-up (group 2). The mean Hcy level was significantly higher in group 1 (20.7 ± 1.6 vs. 12.8 ± 0.5 μmol/l, $p < 0.0001$), as was the proportion of patients with Hcy in excess of 14 μmol/l, the upper limit in healthy controls (83 vs. 30%, $p < 0.0001$). Logistic regression analysis identified Hcy as an independent risk factor for CVA, with an odds ratio of 11.4 (95% confidence interval 3.5–37.7), which remained significant after adjustment on other variables. We conclude that an elevated Hcy level is associated with a risk of occlusive arterial accidents in patients with chronic renal failure and that hyperhomocysteinemia contributes to the accelerated atherosclerosis complicating chronic uremia.

Introduction

Homocysteine is now a well-recognized risk factor favoring the development of premature atherosclerosis resulting in arterial occlusive accidents [1]. Massive hyperhomocysteinemia, as observed in patients with homocystinuria, the homozygous form of deficiency in the pyridoxine-dependent enzyme cysthathionine β-synthetase (CBS), is associated with a high incidence of premature coronary, cerebral or peripheral artery occlusive accidents [2]. Moderate hyperhomocysteinemia, as observed in patients heterozygous for CBS deficiency, was subsequently found to be associated with coronary, cerebrovascular, or peripheral artery occlusive accidents [3–7]. Moreover, moderate hyperhomocysteinemia constituted an independent risk factor for atherosclerotic arterial disease [4].

An elevated plasma homocysteine concentration is a consistent finding in chronic uremic patients. Hyperhomocysteinemia was first observed in patients with advanced chronic renal failure (CRF) [8]. It was subsequently shown that hyperhomocysteinemia is present from the early stage of CRF and increases with the progression of renal failure [9–11] in parallel with alterations in plasma amino acid levels [12]. As hyperhomocysteinemia is only partially corrected by dialysis, elevated homocysteine plasma levels have been repeatedly reported in patients treated by hemodialysis or peritoneal dialysis [10, 11, 13–15], and to a lesser degree in renal transplant recipients [16, 17].

Because chronic uremic patients exhibit an abnormally high incidence of occlusive arterial disease [18] and because atherosclerotic cardiovascular disease constitutes a major cause of mortality in chronic uremics [19], uremia-associated hyperhomocysteinemia, besides other risk factors such as hypertension, smoking, and dyslipemia, could be considered as a factor contributing to such accelerated atherosclerosis.

KARGER
E-Mail karger@karger.ch
Fax +41 61 306 12 34
http://www.karger.ch

0378–0392/97/0236–0170 $12.00/0

Prof. Paul Jungers
Department of Nephrology, Hôpital Necker
161 rue de Sèvres, F–75743 Paris Cedex 15 (France)
Tel. +33 1 44 49 54 57, fax +33 1 44 49 54 50

Previously, we found plasma homocysteine levels to be significantly higher in undialyzed CRF patients who had a past history of occlusive cardiovascular accidents (CVAs) than in their counterparts free of such accidents, but patients with CVAs were, on average, 9 years older than those free of CVAs, and occlusive accidents antedated the time of homocysteine determination by several years [10]. In the present study, we serially determined the plasma homocysteine concentration, together with other parameters involved in atherogenesis such as circulating lipids, in a cohort of nondialyzed CRF patients with various degrees of renal failure, and prospectively recorded all occlusive vascular events that occurred during a 7-year follow-up period. The contribution of hyperhomocysteinemia to atherosclerotic cardiovascular accidents in uremic patients can now be ascertained, since the plasma homocysteine level was determined within 1 year of CVA occurrence.

Patients and Methods

From January 1989 until December 1995, we prospectively studied a cohort of 93 adult patients (57 males, 36 females), all Caucasians, with progressive CRF followed as outpatients at our nephrology clinic. Their creatinine clearance (Ccr) was ≤50 ml/min/·1.73 m^2 but not less than 10 ml/min/·1.73 m^2 and they were aged 50 years or more at the start of the follow-up. Primary renal disease was chronic glomerulonephritis in 15, chronic interstitial nephritis in 35, polycystic kidney disease in 15, hypertensive nephrosclerosis in 25, and diabetic nephropathy in 3. All patients were ambulatory and in good nutritional status. None required dialysis therapy. They were recommended a protein intake between 0.7 and 1 g/kg body weight per day according to the degree of CRF. None received vitamin B supplementation at the start of follow-up.

The plasma total homocysteine concentration was serially determined once a year radioenzymatically. All occlusive arterial accidents of atherosclerotic origin that occurred for the first time during the observation period were recorded. Only cardiovascular accidents with a rapid onset, such as myocardial infarction (or angiographic evidence of coronary artery narrowing, or coronary angioplasty or surgery) and ischemic stroke were taken into account, in order to precisely identify the time of onset. Peripheral arterial disease, which was present in 7 patients (in combination with another manifestation of atherosclerotic disease in 4), was not taken into consideration because the time of onset could not be ascertained. When two types of accidents, coronary and cerebrovascular, occurred in a single patient, only data at the time of onset of the first accident were used.

Plasma homocysteine values determined within 1 year of the onset of coronary or cerebrovascular accidents were used in the group of patients who developed CVAs (CVA+ group), none of whom had previously received folic acid supplements. Values determined at the end of follow-up were used for patients who remained free of CVAs (CVA– group), 60% of whom received folate supplementation for 1 year or more.

Results are expressed as the mean ± SE. Statistical analysis used Student's t test, the χ^2 test, linear regression analysis, and logistic regression analysis using the Statistica® (Tulsa, Okla., USA) software.

Results

From 1989 to 1995, 24 patients (16 males, 8 females) experienced atherosclerotic occlusive arterial accidents, including myocardial infarction (or angiographic evidence of coronary stenosis) in 18 (13 males) and cerebral infarction in 6 (3 males), whereas the other 69 patients (41 males, 28 females) remained free of such accidents.

Table 1. Clinical characteristics, serum creatinine (Scr), creatinine clearance (Ccr) and plasma total homocysteine (Hcy) concentration in nondialyzed CRF patients with (CVA+) and without (CVA–) atherosclerotic cardiovascular accidents

Parameter	CVA+ (n=24)	CVA– (n=69)	p
Males/females	16/8	41/28	NS
Age[a], years	66 ± 1.8	65 ± 1.1	NS
Scr[a], µmol/l	227 ± 24	251 ± 36	NS
Ccr[a], ml/min·1.73 m^2	34 ± 2	31 ± 3	NS
Hcy[a], µmol/l	20.7 ± 1.6	12.8 ± 0.5	<0.0001
Hcy >14 µmol/l	20 (83%)	21 (30%)	<0.0001

NS = Not significant.
[a] Mean ± SE.

The mean age and Ccr value at the onset of atherosclerotic accidents in the CVA+ group and at the end of follow-up in the CVA– group did not differ (table 1). In contrast, the mean homocysteine concentration was significantly higher (20.7 ± 1.6 vs. 12.8 ± 0.5 µmol/l, $p < 0.0001$) in the CVA+ than in the CVA– group. The proportion of patients with a plasma homocysteine concentration in excess of 14 µmol/l (the upper limit of normal in 45 healthy controls) was as high as 83% in the CVA+ group and only 30% in the CVA– group, a highly significant difference ($p < 0.0001$, fig. 1).

Logistic regression analysis identified the plasma homocysteine concentration as an independent risk factor for the occurrence of atherosclerotic cardiovascular accidents, with an odds ratio of 11.4 (95% confidence interval 3.5–37.7), which remained significant after adjustment on other variables.

In addition, regression analysis showed, as expected, a negative relationship between homocysteine concentration and Ccr ($r = -0.35$, $p < 0.001$), and a significant correlation of homocysteine concentration with tobacco smoking ($r = 0.29$, $p = 0.005$), triglyceride level ($r = 0.50$, $p < 0.001$) and lipoprotein(a) concentration ($r = 0.30$, $p = 0.003$).

Discussion

Our present study affords direct evidence that hyperhomocysteinemia is associated with an increased risk of atherosclerotic CVAs in nondialyzed, chronic uremic patients. Because our patients in the CVA+ group were similar in age and degree of renal dysfunction at the onset of myocardial infarction or ischemic stroke to patients devoid of such accidents at latest follow-up, the difference between the two groups with respect to mean plasma homocysteine concentration and the proportion with a supranormal homocysteine level cannot be attributed to older age or a more advanced degree of renal failure in patients who developed atherosclerotic accidents, as might have been the case in our preliminary studies [9, 10].

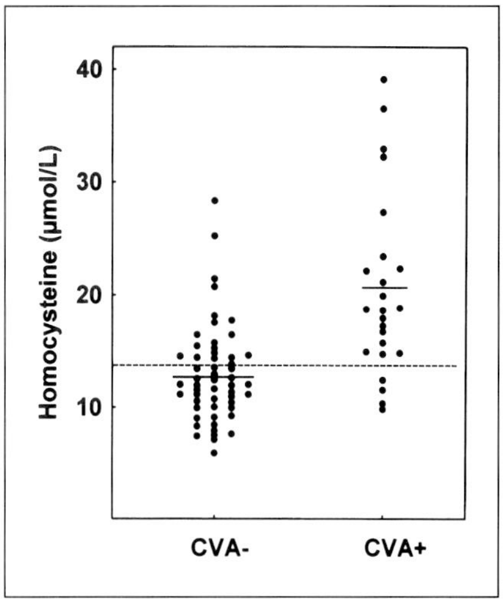

Fig. 1. Plasma total homocysteine concentration in nondialyzed uremic patients with (CVA +) or without (CVA −) atherosclerotic cardiovascular accidents. The dashed line indicates the upper limit of plasma homocysteine concentration in healthy controls.

Moreover, logistic regression analysis identified hyperhomocysteinemia as an independent risk factor for atherosclerotic accidents, a factor which remained significant after adjustment on all other variables (including blood pressure, tobacco smoking, apolipoprotein B, triglyceride, fibrinogen, and lipoprotein(a) level).

In keeping with our present results in predialysis CRF patients, hyperhomocysteinemia was also recently shown to be a significant, independent risk factor for atherosclerotic arterial disease in renal transplant recipients [17] and in patients on regular hemodialysis treatment [15]. The positive correlation we observed between homocysteine concentration, tobacco smoking, and lipoprotein(a) may be relevant in view of the fact that tobacco smoking may increase homocysteine accumulation and homocysteine enhances the binding of lipoprotein(a) to fibrin, a mechanism which possibly contributes to the proatherogenic and prothrombogenic properties of homocysteine [20].

A beneficial effect of folic acid pharmacologic supplementation in lowering homocysteine concentration has been observed in nondialyzed uremic patients [21, 22], and was also reported in patients on hemodialysis or peritoneal dialysis [10, 23–25]. Nearly 60% of our CVA − patients had received folic acid supplements (5 mg/day) for 1–3 years at the time of latest follow-up. This may explain why our patients who remained free of atherosclerotic accidents had a markedly lower homocysteine concentration than those who suffered such accidents despite a similar age and degree of renal failure. Therefore, even if folate supplements are not routinely needed in uremics on a purely hematologic basis [26, 27], concordant evidence of the homocysteine-lowering effect of folic acid supplementation in uremic patients and nonuremic subjects [28, 29] strongly suggests that folate supplementation should be offered to chronic uremic patients with moderate to severe renal failure, as well as to patients on regular dialysis treatment and following renal transplantation, as an innocuous and efficient means of reducing hyperhomocysteinemia.

In conclusion, the data presented here provide direct evidence that uremia-associated hyperhomocysteinemia is an independent risk factor for atherosclerotic CVAs in nondialyzed CRF patients. Because chronic uremic patients are exposed to long-lasting homocysteine accumulation, therapeutic measures aimed at reducing hyperhomocysteinemia, such as folic acid supplementation, should be considered in the long-term management of chronic uremics in order to reduce the risk of accelerated atherosclerosis and occlusive CVAs.

References

1 Malinow MR: Homocyst(e)ine and arterial occlusive diseases. J Intern Med 1994;236:603–617.

2 Mudd SH, Levy HL, Skovby F: Disorders of transsulfuration; in Scriver CR, Beaudet AL, Sly WS, Valle DS (eds): The Metabolic and Molecular Bases of inherited Disease, ed 7. New York, McGraw-Hill, 1995, vol 2, pp 1279–1327.

3 Kang S, Wong PWK, Cook HY, Norusis M, Messer JV: Protein-bound homocyst(e)ine: A posible risk factor for coronary artery disease. J Clin Invest 1986;77:1482–1486.

4 Clarke R, Daly L, Robinson K, Naughten E, Cahalane S, Fowler B, Graham I: Hyperhomocysteinemia: An independent risk factor for vascular disease. N Engl J Med 1991;324:1149–1155.

5 Stampfer MJ, Malinow MR, Willett WC: A prospective study of plasma homocyst(e)ine and risk of myocardial infarction in US physicians. JAMA 1992;268:877–881.

6 Brattstrom L, Israelsson B, Norrving B, Bergqvist D, Thörne J, Hultberg B, Hamfelt A: Impaired homocysteine metabolism in early-onset cerebral and peripheral occlusive arterial disease. Atherosclerosis 1990;81:51–60.

7 Selhub J, Jacques PF, Bostom AG, D'Agostino RB, Wilson PWF, Belanger AJ, O'Leary DH, Wolf PA, Schaeffer EJ, Rosenberg IH: Association between plasma homocysteine concentrations and extracranial carotid artery stenosis. N Engl J Med 1995; 332:286–291.

8 Wilcken DEL, Gupta VJ: Sulphur-containing amino acids in chronic renal failure with particular reference to homocysteine and cysteine-homocysteine mixed disulphide. Eur J Clin Invest 1979; 9:301–307.

9 Chauveau P, Chadefaux B, Coudé M, Aupetit J, Hannedouche T, Kamoun P, Jungers P: Increased plasma homocysteine concentration in patients with chronic renal failure. Miner Electrolyte Metab 1992;18:196–198.

10 Chauveau P, Chadefaux B, Coudé M, Aupetit J, Hannedouche T, Kamoun P, Jungers P: Hyperhomocysteinemia, a risk factor for atherosclerosis in chronic uremic patients. Kidney Int 1993;43(suppl 41):S72–S77.

11 Hultberg B, Anderson A, Sterner G: Plasma homocysteine in renal failure. Clin Nephrol 1993;40:230–234.

12 Jungers P, Chauveau P, Ceballos I, Bardet J, Parvy P, Hannedouche T, Kamoun P: Plasma free aminoacid alterations from early to end-stage chronic renal failure. J Nephrol 1994;7:48–54.

13 Wilcken DEL, Gupta VJ, Reddy SG: Accumulation of sulphur-containing amino acids including cysteine-homocysteine in patients on maintenance haemodialysis. Clin Sci 1980;58:427–430.
14 Kang SS, Wong PWK, Bidani A, Milanez S: Plasma protein-bound homocyst(e)ine in patients requiring chronic haemodialysis. Clin Sci 1983;65:335–336.
15 Bachmann J, Tepel M, Raidt H, Riezler R, Graefe U, Langer K, Zidek W: Hyperhomocysteinemia and the risk for vascular disease in hemodialysis patients. J Am Soc Nephrol 1995;6:121–125.
16 Wilcken DEL, Gupta VJ, Betts AK: Homocysteine in the plasma of renal transplant recipients: Effects of cofactors for methionine metabolism. Clin Sci 1981;61:743–749.
17 Massy ZA, Chadefaux-Vekemans B, Chevalier A, Bader CA, Druëke TB, Legendre C, Lacour B, Kamoun P, Kreis H: Hyperhomocysteinemia: A significant risk factor for cardiovascular disease in renal transplant recipients. Nephrol Dial Transplant 1994;9:1103–1108.
18 Ibels LS, Stesart JH, Mahony JF, Neale FC, Sheil AGR: Occlusive arterial disease in uraemic and haemodialysis patients and renal transplant recipients: A study of the incidence of arterial disease and of the prevalence of risk factors implicated in the pathogenesis of arteriosclerosis. Q J Med 1977; 46:197–214.
19 Parfrey PS: Cardiac and cerebrovascular disease in chronic uremia. Am J Kidney Dis 1993;21:77–80.
20 Harpel PC, Chang VT, Borth W: Homocysteine and other sulfhydryl compounds enhance the binding of lipoprotein(a) to fibrin: A potential biochemical link between thrombosis, atherogenesis, and sulfhydryl compound metabolism. Proc Natl Acad Sci USA 1992;89:10193–10197.
21 Wilcken DEL, Dudman NPB, Tyrrell PA, Robertson MR: Folic acid lowers elevated plasma homocysteine in chronic renal insufficiency: Possible implications for prevention of vascular disease. Metabolism 1988;37:697–701.
22 Chauveau P, Chadefaux B, Coudé M, Aupetit J, Kamoun P, Jungers P: Long-term folic acid (but not pyridoxine) supplementation lowers elevated plasma homocysteine level in chronic renal failure. Miner Electrolyte Metab 1996;22:106–109.
23 Arnadottir M, Brattstrom L, Simonsen O, Thysell H, Hultberg B, Anderson A, Nilsson-Ehle P: The effect of high-dose pyridoxine and folic acid supplementation on serum lipid and plasma homocysteine concentrations in dialysis patients. Clin Nephrol 1993;40:236–240.
24 Bostom AG, Shemin D, Lapane KL, Hume AL, Yoburn D, Nadeau MR, Bendich A, Selhub J, Rosenberg IH: High dose B-vitamin treatment of hyperhomocysteinemia in dialysis patients. Kidney Int 1996;49:147–152.
25 Janssen MJ, Van Guldener C, De Jong GMT, Van den Berg M, Stehouver CDA, Donker AJM: Folic acid treatment of hyperhomocysteinemia in dialysis patients. Miner Electrolyte Metab 1996;22:110–114.
26 Swainson C, Winney R: Do dialysis patients need extra folate? Lancet 1983;i:239.
27 Westhuyzen J, Matherson K, Tracey R, Fleming SJ: Effect of withdrawal of folic acid supplementation in maintenance hemodialysis patients. Clin Nephrol 1993;40:96–99.
28 Brattstrom L, Israelsson B, Jeppson H, Hultberg B: Folic acid, an innocuous means of reducing plasma homocysteine. Scand J Clin Lab Invest 1988;48: 215–221.
29 Stampfer MJ, Malinow MR: Can lowering homocysteine levels reduce cardiovascular risk? N Engl J Med 1995;332:328–329.

Mineral and Electrolyte Metabolism

Miner Electrolyte Metab 1997;23:174–178

Alessandra F. Perna
Natale G. De Santo
Diego Ingrosso

Institute of Biochemistry of Macromolecules and Division of Nephrology/Department of Pediatrics, School of Medicine, Second University of Naples, Italy

Adverse Effects of Hyperhomocysteinemia and Their Management by Folic Acid

Key Words
Adenosylhomocysteine
Adenosylmethionine
Cardiovascular risk factors
Erythrocytes
Homocysteine
Methylation reactions
Uremia

Abstract
A moderate increase in plasma homocysteine is an independent risk factor for cardiovascular disease. Plasma homocysteine is frequently elevated in chronic renal failure and in uremic patients, and the major causes of death in these patients are cardiovascular accidents. Homocysteine metabolism and mechanisms of toxicity are reviewed. Homocysteine elevation in blood leads to the intracellular increase of its precursor, adenosylhomocysteine, a powerful inhibitor of adenosylmethionine-dependent transmethylations. In vitro evidence shows that this increase is reversible upon homocysteine removal. Membrane protein methylation levels are consistently reduced in erythrocytes of both chronic renal failure and hemodialysis patients. This widespread enzymatic methylation is a key step for the repair of molecular damage resulting from the spontaneous deamidation and isomerization reactions of susceptible residues in proteins. In agreement with these findings is the observation that the concentration of a stable side product, *D*-Asx, of the repair process is significantly lower in erythrocyte membrane proteins from hemodialysis patients than from controls, showing that the repair of damaged membrane proteins is actually defective. It has been shown that treatment with folates dramatically lowers plasma homocysteine, presumably by improving remethylation to methionine. This indicates that folates and/or their active derivative, i.e., methyltetrahydrofolate, could be effective in ameliorating transmethylations as well.

Homocysteine Metabolism

Homocysteine is a sulfur amino acid not normally utilized as a precursor amino acid for protein biosynthesis [1], but which can bind to proteins posttranslationally through the formation of disulfide bonds or by homocysteinylation [2]. A substantial increase in its concentration was originally described in the plasma (as well as in the urine, mainly in its oxidized form, i.e., homocystine) of homocystinuric patients. Homocystinuria is a rare hereditary disease but, nonetheless, second only to phenylketonuria among the enzymatic defects of amino acid metabolism. This syndrome is strongly associated with an increased risk of premature cerebral, coronary, and peripheral vascular disease [3].

Homocysteine is an intermediate in the metabolism of the essential amino acid methionine (fig. 1). Methionine from the diet or protein degradation can be incorporated into proteins, or it is converted into the nucleoside S-adeno sylmethionine (AdoMet) by methionine adenosyl transferase, which consumes ATP as well. AdoMet donates the methyl group to many possible acceptors, being the universal methyl donor of transmethylation reactions. The aminopropyl moiety of decarboxylated AdoMet is required, instead, for polyamine biosynthesis.

The demethylated product of AdoMet, the thioether S-adenosylhomocysteine (AdoHcy), is the inhibitor of AdoMet-dependent methyl transfer reactions, and the precursor of homocysteine, through a reversible enzymatic hydrolysis yielding homocysteine and adenosine [4–7]. Homocysteine, in turn, is metabolized through two major pathways. The first is via cystathionine β-synthase, a vitamin-B_6-dependent pathway yielding cysteine as the final product (transsulfuration). This enzyme binds serine and homocysteine to produce cystathionine, and then a cystathioninase, which also requires vitamin B_6, cleaves cystathionine to 2-ketobutyrate and cysteine. Cysteine is eventually converted to glutathione, taurine, and other sulfur-containing metabolites.

The second pathway is via methyltetrahydrofolate (MTHF)-homocysteine methyltransferase (methionine synthase), leading to methionine synthesis (remethylation). This latter metabolism requires folate in its active form and vitamin B_{12}. The active form of folate is represented by MTHF, which donates the methyl group in the remethylating reaction catalyzed by methionine synthase.

MTHF is formed from methylenetetrahydrofolate by enzymatic reduction catalyzed by methylenetetrahydrofolate reductase. Betaine can act as an alternative methyl donor to MTHF in methionine biosynthesis, particularly in the liver [4–7].

KARGER
E-Mail karger@karger.ch
Fax + 41 61 306 12 34
http://www.karger.ch

0378–0392/97/0236–0174 $12.00/0

Alessandra F. Perna, MD, PhD
Istituto di Biochimica delle Macromolecole
Seconda Università degli Studi di Napoli
via Costantinopoli 16, I–80131 Napoli (Italy)
Tel. +39 81 441751, fax +39 81 441688

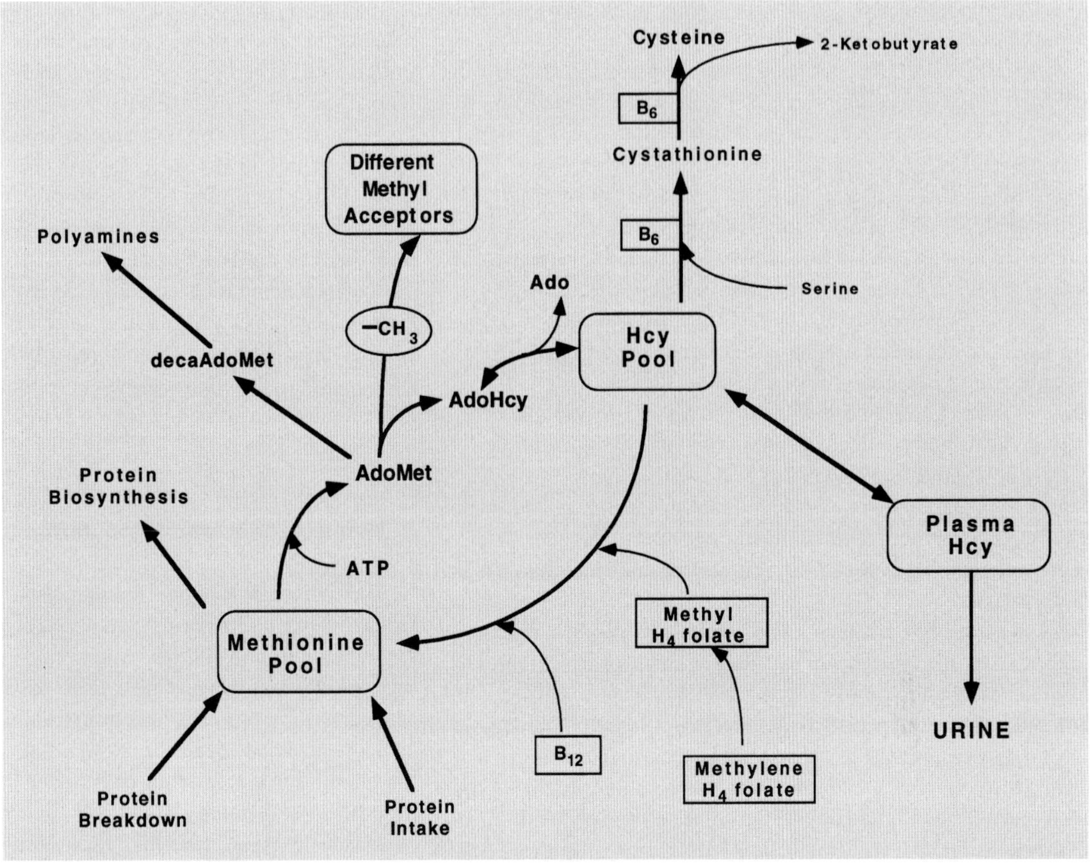

Fig. 1. Homocysteine metabolism. Hcy = Homocysteine; B_6 = vitamin B_6; B_{12} = vitamin B_{12}; Ado = adenosine; AdoHcy = adenosylhomocysteine; AdoMet = S-adenosylmethionine.

Thus, three vitamins are required in homocysteine metabolism: vitamin B_6, vitamin B_{12}, and folic acid in its active form.

Homocysteine metabolism is tightly regulated by Ado-Met. The latter is in fact not only the methyl donor of transmethylation reactions, but also functions as a positive modulator of cystathionine β-synthase, and a negative modulator for methylenetetrahydrofolate reductase. Considering the K_m for homocysteine of the relevant enzymes, high for cystathionine β-synthase, and low for methionine synthase, it can be argued that when methionine is high, after a meal for example, the transsulfuration pathway is activated. In contrast, during methionine deficiency, the remethylation pathway recuperates homocysteine to yield methionine.

Dietary folates are absorbed by the intestine only after the additional glutamate residues have been cleaved from the pteroyl polyglutamate molecule by a conjugase. The only circulating form is the monoglutamate form. Intracellular conjugases add glutamate residues once again to the pteroylmonoglutamate moiety.

In the typical form of homocystinuria, the defect lies in very low expression levels of cystathionine β-synthase. Most of these patients respond to treatment with high amounts of vitamin B_6, which activates the metabolic pathway, and the little available enzyme. Other forms of hyperhomocysteinemia have been linked to a deficiency of the enzyme methylenetetrahydrofolate reductase, or to a thermolabile form of the reductase, which has been recently described [8], or to inherited disorders of vitamin B_{12} [3]. A relative or an absolute deficiency of folates can also induce homocysteine increase [3].

Homocysteine as a Cardiovascular Risk Factor

Homocystinuria, which gives rise to very high blood levels of homocysteine (in untreated patients up to 500 μM), is associated with premature vascular disease, and in 1976 Wilcken and Wilcken [9] recognized that altered methionine metabolism is a risk factor for premature coronary artery disease. However, the large prospective or cross-sectional population studies which point to moderate homocysteine increase as a risk factor are more recent [10–14].

Clarke et al. [10] studied, in a case-control study, 123 patients with various vascular disorders including cerebrovascular, peripheral, and coronary vascular disease. After adjustments for the effects of conventional risk factors, the odds ratio for vascular disease in hyperhomocysteinemic patients was 3.2 compared to normals. The authors concluded that hyperhomocysteinemia is an independent risk factor for vascular disease, in most cases due to cystathionine β-synthase deficiency.

Stampfer et al. [11] carried out a prospective study on 14,916 male physicians followed up for 5 years. Samples from 271 men who subsequently developed myocardial infarction were analyzed for homocysteine levels together with paired controls. Homocysteine levels were 11.1 ± 4 (SD) μM in cases vs. 10.5 ± 2.8 μM in controls. Men with homocysteine levels situated in the top 5% (15.8 ± 2.8 μM) had a threefold increased risk of myocardial infarction. The increased risk was virtually unchanged after considering other coronary risk factors. Conclusions were that moderately high levels of plasma homocysteine are associated with subsequent risk of myocardial infarction independent of other coronary risk factors. Because high homocysteine levels can be responsive to treatment with vitamin supplements, homocysteine may be an independent, modifiable, risk factor.

Taken together, the epidemiological evidence supports a graded response, as in the case of cholesterol, rather than a threshold effect. The relationship between high homocysteine and vascular disease is as strong as that between vascular disease and smoking or high blood pressure.

A causative link between homocysteine and vascular disease is supported by the observation that the relation is present regardless of the underlying metabolic defect, and that baboons infused with homocysteine develop vascular lesions [15].

It has been postulated that several thousand cardiovascular deaths could be prevented by lowering plasma homocysteine (it has been called the prevention opportunity of the decade), either by identifying hyperhomocysteinemic subjects and treating them with folic acid and its active derivatives, or by supplementing breads, cereals, and other dietary components with low doses of folic acid [16].

Regarding homocysteine toxicity, the high reactivity of homocysteine has been considered an important mechanism in vascular injury, which will ultimately induce atherosclerotic narrowing. As possible mechanisms, it has been reported that homocysteine increases the binding of lipoprotein(a) to fibrin, induces endothelial activation of coagulation factor V, reduces protein C activation by arterial and venous endothelial cells, and induces endothelial cell injury in vitro. Homocysteine facilitates peroxidation of lipids by oxygen and iron [17–22].

Homocysteine in Uremia

Hyperhomocysteinemia is almost constantly associated with chronic renal failure, and these patients have an increased risk of vascular disease, which accounts for 40% of deaths [23–28].

While normal levels range between 6 and 12 μM, end-stage renal disease subjects have values between 25 and 40 μM or more, a very high range considering that high-risk cardiovascular subjects are those with values above 15 μM [11].

The basic observation by Wilcken and Wilcken [9] has been firmly established by other studies, which show hyperhomocysteinemia in patients with various degrees of renal failure, as well as in those on hemodialysis and peritoneal dialysis [24–28]. A very recent large study by Robinson and coworkers [29, 30] shows that, in 176 patients on maintenance hemodialysis or peritoneal dialysis, the prevalence of hyperhomocysteinemia is 85%, and patients with more than 27 μM homocysteine have an independent odds ratio of 2.9 for vascular complications. B vitamin levels, in particular folic acid, were lower in patients with vascular complications than in those without. The conclusion of the study is that these patients may benefit from supplementation with much higher doses of B vitamins than those currently recommended.

Renal excretion is not considered to be a major cause of hyperhomocysteinemia in kidney failure, since the fractional clearance of homocysteine increases as renal function worsens, and urinary clearance is low anyway. Reduced renal metabolism is a mechanism for hyperhomocysteinemia, and some extrarenal disturbances have also been invoked [30].

Homocysteine and Methylation Reactions

We have been studying homocysteine metabolism in uremia. Homocysteine under normal conditions is rapidly removed by both the transsulfuration and the remethylation pathways, and does not accumulate significantly. Under these conditions, the AdoHcy hydrolase reaction proceeds in the direction of AdoHcy hydrolysis to homocysteine and adenosine. However, this is a reversible reaction (thermodynamics actually favor AdoHcy biosynthesis) and thus, if the products accumulate, AdoHcy, a powerful competitive inhibitor of transmethylations, can build up within a cell until a new equilibrium is reached. In the steady-state we expect that patients will show a higher intracellular concentration of the transmethylation inhibitor AdoHcy than the methyl donor AdoMet, and this will be reflected in a substantial, chronic inhibition of transmethylation reactions.

We measured homocysteine and AdoHcy in the plasma and red blood cells of hemodialysis patients and found a significant increase in homocysteine and AdoHcy, which are positively correlated. AdoMet, the nucleoside which by donating the methyl group forms AdoHcy, adenosine, and AdoHcy hydrolase specific activity were not significantly different between the two groups [31].

The concentration ratio between AdoMet and AdoHcy is the most important indicator of the presence and degree of methylation inhibition due to the thioether. In chronic renal failure patients, and in hemodialysis patients, this ratio is significantly reduced (fig. 2), compared to normal controls, indicating a strong inhibition [32].

Methylation Reactions and Protein Repair in Renal Failure

Proteins are subject to spontaneous molecular damage (so-called 'protein fatigue'), including spontaneous deamidation of asparaginyl residues or isomerization of normal aspartyl residues [33], which can significantly impair protein structure and function.

The only AdoMet-dependent reaction still active in mature red blood cells is that catalyzed by the *L*-isoaspartyl protein O-methyltransferase (or PCMT, EC 2.1.1.77). This enzyme methyl esterifies

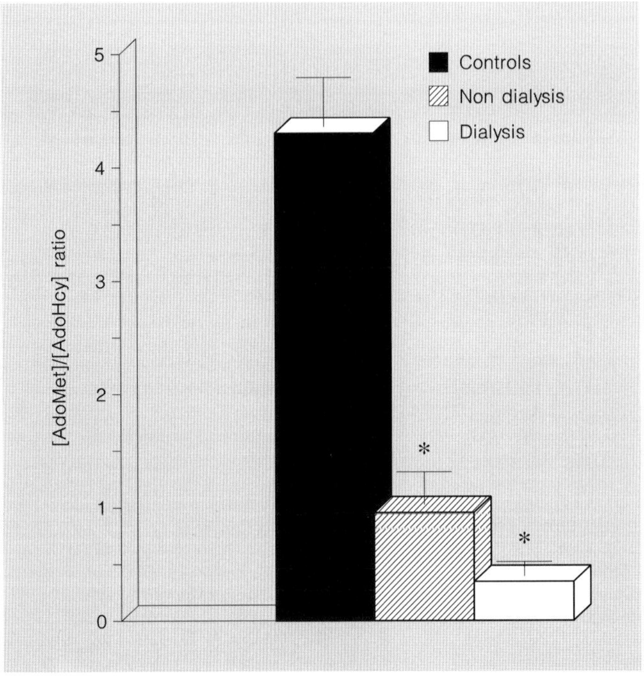

Fig. 2. Erythrocyte AdoMet/AdoHcy concentration ratio in controls, chronic renal failure patients on conservative treatment (nondialysis), and hemodialysis patients.

the free carboxyl groups of damaged aspartyl residues in protein and peptide substrates. This methylation is the key step in a mechanism devoted to the repair of abnormal *L*-isoaspartyl residues in fatigue-damaged proteins. The molecular mechanisms for the occurence of *L*-isoaspartyl sites and their repair have been described in detail [see ref. 33 for a review].

We studied erythrocyte membrane protein methyl esterification, as a typical model transmethylation reaction, in chronic renal failure patients on hemodialysis, compared to controls, and observed a significant reduction of total methyl esterification in both patient groups. The results are identical either by normalizing methylation levels per milligram protein, or per cell number [32].

Characterization of methylated protein species through SDS/PAGE (polyacrylamide gel electrophoresis) showed that the decrease was present for all methyl-accepting substrates, but strikingly evident for the cytoskeletal component ankyrin (band 2.1), which is involved in the maintenance of membrane structural integrity and functional stability [32].

We determined the amount of a stable side-product, *D*-Asx, of the repair process in erythrocyte membrane proteins from hemodialysis patients and found that it was significantly lower than in controls. The results fit a mathematical computer model in which, taking into account all possible reactions generating altered aspartyl residues, the ability of these cells to repair altered *L*-isoaspartyl sites in proteins is dramatically impaired. The calculations predict a 12- to 600-fold increase (depending on the model used) in *L*-isoaspartyl residues, greatly overshadowing the small decrease in *D*-Asx resulting from PCMT inhibition [34].

The accumulation of structural damage in erythrocyte membrane proteins, which is not adequately repaired in uremia, is therefore consequent to an increase in homocysteine and, subsequently, AdoHcy.

The erythrocyte has been a useful model for studying PCMT-dependent protein repair; nevertheless, this is a ubiquitous enzyme and this methylation reaction potentially involves all proteins containing deamidation- and isomerization-susceptible residues, such as those involved in atherogenesis.

In addition, AdoHcy inhibits all AdoMet-dependent transmethylation reactions that in different cells and tissues involve a variety of methyl acceptors, such as proteins, nucleic acids, phospholipids, neurotransmitters like norepinephrine, and other small molecules, with different biological functions. For example, DNA methylation represents a strong modulator of cell growth and oncogene expression [35]. Therefore a reduction of the AdoMet/AdoHcy ratio can potentially interfere with a number of different processes in which AdoMet acts as the methyl donor.

Management of Hyperhomocysteinemia by Folic Acid

Regarding the management of hyperhomocysteinemia by folic acid in uremia, the rationale lies in the reinforcement of the remethylation reaction of homocysteine to metionine. The first study in chronic renal failure patients was by Wilcken et al. [36], who administered 5 mg of folic acid daily for 15 days, obtaining a significant reduction in plasma homocysteine. Other studies confirmed these results in hemodialysis patients and peritoneal dialysis patients [see ref. 30 for a review].

Recently, Bostom et al. [37] treated 15 hemodialysis patients with 15 mg folic acid, plus 100 mg B_6 and 1 mg B_{12}. Homocysteine was reduced in all patients, and in 3 out of 12 with more than 20 μM, homocysteine was reduced to normal levels, that is less than 15 μM.

In the study by Robinson et al. [29], homocysteine levels correlated negatively with folate levels, and these authors suggest that the homocysteine concentration may be correctable only by attaining a very high circulating concentration of folate.

All these studies lead to the conclusion that in renal failure patients, a relative deficiency of folates is present, in the face of their increased plasma and/or erythrocyte concentration. The mechanism of such resistance to folate needs to be clarified. However, inhibitor(s) of folate conjugase have been reported to occur in hemodialysis patients [38]. This enzyme activity is crucial in the conversion of poly- to monoglutamate forms of folate, and is therefore required for normal uptake and suitability by the cells [39]. In addition, the oral methionine load test is altered in these patients, indicating that transsulfuration may also be impaired [40].

The active derivative of folic acid, MTHF, bypasses the reduction step of folates, providing the metabolically active species to the cell. It has been recently shown that MTHF effectively reduces plasma homocysteine in hemodialysis patients, and ameliorates the ratio between AdoMet and AdoHcy, thus inducing a beneficial effect on AdoMet-dependent transmethylation reactions [41].

References

1 Jakuboski H, Goldman E: Synthesis of homocysteine thiolactone by methionyl-tRNA synthetase in cultured mammalian cells. FEBS Lett 1993;317:237–240.

2 McCully KS: Chemical pathology of homocysteine. I. Atherogenesis. Ann Clin Lab Sci 1993;23:477–493.

3 Mudd SH, Levy HL, Skovby F: Disorders of transsulfuration; in Scriver CR, Beaudet AL, Sly WS, Valle D (eds): The Metabolic Basis of Inherited Disease. New York, Mc Graw Hill, 1989, p 693.

4 Selhub J, Miller JW: The pathogenesis of homocysteinemia: Interruption of the coordinate regulation by S-adenosylmethionine of the remethylation and transsulfuration of homocysteine. Am J Clin Nutr 1992;55:131–138.

5 Zappia V, Zydek-Cwick CR, Schlenk F: The specificity of S-adenosylmethionine derivatives in methyl transfer reactions. J Biol Chem 1969;244:4499–4505.

6 Cantoni GL, Chiang PK: The role of S-adenosylhomocysteine and S-adenosylhomocysteine hydrolase in the control of biologic methylations; in Cavallini D, Gaull GE, Zappia V (eds): Natural Sulfur Compounds: Novel Biochemical and Structural Aspects. New York, Plenum, 1980, p 67.

7 Barber JR, Clarke S: Inhibition of protein carboxyl methylation by S-adenosyl-*L*-homocysteine in intact erythrocytes: Physiological consequences. J Biol Chem 1984;259:7115–7122.

8 Kang SS, Wong PWK, Susmano A, Sora J, Norusis M, Ruggie N: Thermolabile methylenetetrahydrofolate reductase: An inherited risk factor for coronary artery disease. Am J Hum Genet 1991;48:536–545.

9 Wilcken DEL, Wilcken B: The pathogenesis of coronary artery disease: A possible role for methionine metabolism. J Clin Invest 1976;57:1079–1082.

10 Clarke R, Daly L, Robinson K, Naughten E, Cahalane S, Fowler B, Graham I: Hyperhomocysteinemia: An independent risk factor for vascular disease. N Engl J Med 1991;324:1149–1155.

11 Stampfer MJ, Malinow MR, Willett WC, Newcomer LM, Upson B, Ullmann D, Tishler PV, Hennekens CH: A prospective study of plasma homocyst(e)ine and risk of myocardial infarction in US physicians. JAMA 1992;268:877–881.

12 Malinow MR, Nieto FJ, Szklo M, Chambless LE, Bond G: Carotid artery intimal-medial wall thickening and plasma homocyst(e)ine in asymptomatic adults: The atherosclerosis Risk in Communities Study. Circulation 1993;87:1107–1113.

13 Nygard O, Vollset SE, Refsum H, Stensvold I, Tverdal A, Nordrehaug JE, Ueland PM, Kvale G: Total plasma homocysteine and cardiovascular risk profile: The Hordaland Homocysteine Study. JAMA 1995;274:1526–1533.

14 Robinson K, Mayer EL, Miller DP, Green R, Lente F van, Gupta A, Kottke-Marchant K, Savon SR, Selhub J, Nissen SE, Kutner M, Topol EJ, Jacobsen DW: Hyperhomocysteinemia and low pyridoxal phosphate: Common and independent reversible risk factors for coronary artery disease. Circulation 1995;92:2825–2830.

15 Harker LA, Ross R, Slichter SJ, Scott CR: Homocystine-induced arteriosclerosis: The role of endothelial cell injury and platelet response to its genesis. J Clin Invest 1976;58:731–741.

16 Oakley GP: Folic acid fortification – Prevention opportunity of the decade? Irish J Med Sci 1995;164:S14.

17 Tsai C, Perrella MA, Yoshizumi M, Hsien CM, Habe E, Schlagel R, Lee ME: Promotion of vascular smooth muscle cell growth by homocysteine: A link to atherosclerosis. Proc Natl Acad Sci USA 1994;91:6369–6373.

18 Harpel PC, Chang VT, Bort W: Homocysteine and other sulfhydryl compounds enhance the binding of lipoprotein(a) to fibrin: A potential biochemical link between thrombosis, atherogenesis, and sulfhydryl compound metabolism. Proc Natl Acad Sci USA 1992;89:10193–10197.

19 Preibisch G, Kuffner C, Elstner EF: Biochemical model reactions on the prooxidative activity of homocysteine. J Biosci 1993;48:58–62.

20 Hajjar KA: Homocysteine-induced modulation of tissue plasminogen activator binding to its endothelial cell membrane receptor. J Clin Invest 1993;91:2873–2879.

21 Rodgers GM, Kane WH: Activation of endogenous factor V by a homocysteine-induced vascular endothelial cell activator. J Clin Invest 1986;77:1909–1916.

22 Hayashi T, Honda G, Suzuki K: An atherogenic stimulus homocysteine inhibits cofactor activity of thrombomodulin and enhances thrombomodulin expression in human umbilical vein endothelial cells. Blood 1992;79:2930–2936.

23 Parfrey PS: Cardiac and cerebrovascular disease in chronic uremia. Am J Kidney Dis 1993;21:77–80.

24 Wilcken DEL, Gupta VJ, Reddy SG: Accumulation of sulphur-containing amino acids including cysteine-homocysteine in patients on maintenance haemodialysis. Clin Sci 1980;58:427–430.

25 Laidlaw SA, Smolin LA, Davidson WD, Kopple JD: Sulfur amino acids in maintenance hemodialysis patients. Kidney Int 1987;32:S191–S196.

26 Chauveau P, Chadefaux B, Coudé M, Aupetit J, Hannedouche T, Kamoun P, Jungers P: Increased plasma homocysteine concentration in patients with chronic renal failure. Miner Electrolyte Metab 1992;18:196–198.

27 Chauveau P, Chadefaux B, Coudé M, Aupetit J, Hannedouche T, Kamoun P, Jungers P: Hyperhomocysteinemia, a risk factor for atherosclerosis in chronic uremic patients. Kidney Int 1993;43:S72–S77.

28 Bachmann J, Tepel M, Raidt H, Riezler R, Graefe U, Langer K, Zidek W: Hyperhomocysteinemia and the risk for vascular disease in hemodialysis patients. J Am Soc Nephrol 1995;6:121–125.

29 Robinson K, Gupta A, Dennis V, Arheart D, Chaudhary D, Green R, Vigo P, Mayer EL, Selhub J, Kutner M, Jacobsen DW: Hyperhomocysteinemia confers an independent increased risk of atherosclerosis in end-stage renal disease and is closely linked to plasma folate and piridoxine concentrations. Circulation, 1996;94:2743–2748.

30 Dennis VW, Robinson K: Homocysteinemia and vascular disease in end stage renal disease. Kidney Int 1996;50:S11–S17.

31 Perna AF, Ingrosso D, De Santo NG, Galletti P, Zappia V: Mechanism of erythrocyte accumulation of methylation inhibitor S-adenosylhomocysteine in uremia. Kidney Int 1995;47:247–253.

32 Perna AF, Ingrosso D, Galletti P, Zappia V, De Santo NG: Enzymatic methyl esterification of erythrocyte membrane proteins is impaired in chronic renal failure: Evidence for high levels of the natural inhibitor S-adenosylhomocysteine. J Clin Invest 1993;91:2497–2503.

33 Galletti P, Ingrosso D, Manna C, Clemente G, Zappia V: Protein damage and methylation mediated repair in the erythrocyte. Biochem J 1995;306:313–325.

34 Perna AF, D'Aniello A, Lowenson JD, Clarke S, De Santo NG, Ingrosso D: *D*-aspartate content of erythrocyte membrane proteins is decreased in uremia: Implications for the repair of damaged proteins. J Am Soc Nephrol 1997;8:95–104.

35 Perna AF, Ingrosso D, Galletti P, Zappia V, De Santo NG: Membrane protein damage and methylation reactions in chronic renal failure. Kidney Int 1996;50:358–366.

36 Wilcken DEL, Dudman NPB, Tyrrell PA, Robertson MR: Folic acid lowers elevated plasma homocysteine in chronic renal insufficiency: Possible implications for prevention of vascular disease. Metabolism 1988;37:697–701.

37 Bostom AG, Shemin D, Lapane KL, Hume AL, Yoburn D, Nadeau MR, Bendich A, Selhub J, Rosenberg IH: High dose B-vitamin treatment of hyperhomocysteinemia in dialysis patients. Kidney Int 1996;49:147–152.

38 Livant EJ, Tamura T, Johnston KE, Vaughn WH, Bergman SM, Forehand J, Walthaw J: Plasma folate conjugase activities and folate concentration in patients receiving hemodialysis. J Nutr Biochem 1994;5:504–508.

39 Krudieck CL, Eto I: Folate in tissues and cells: Support for a 'two tier' hypothesis of one carbon metabolism; in Cooper BA, Whitehead MV (eds): Chemistry and Biology of Pteridines. Berlin, Walter de Gruyter, 1986, pp 447–466.

40 Hultberg B, Andersson A, Sterner G: Plasma homocysteine in renal failure. Clin Nephrol 1993;40:230–234.

41 Perna AF, Ingrosso D, De Santo NG, Galletti P, Brunone M, Zappia V: Metabolic consequences of folate-induced reduction of hyperhomocysteinemia in uremia. J Am Soc Nephrol, in press.

Mineral and Electrolyte Metabolism

Miner Electrolyte Metab 1997;23:179–184

Indoxyl Sulfate and Progression of Renal Failure: Effects of a Low-Protein Diet and Oral Sorbent on Indoxyl Sulfate Production in Uremic Rats and Undialyzed Uremic Patients

Toshimitsu Niwa[a]
Saori Tsukushi[b]
Michihiko Ise[c]
Takashi Miyazaki[a]
Yoshiharu Tsubakihara[d]
Akira Owada[e]
Tatsuo Shiigai[e]

[a] Nagoya University, Daiko Medical Center, Higashi-ku, Nagoya,
[b] Tokai-Chuou Hospital, Kagamihara, Gifu,
[c] Kureha Chemical Co., Tokyo,
[d] Osaka Prefectural Hospital, Sumiyoshi, Osaka, and
[e] Toride Kyodo General Hospital, Toride, Ibaraki, Japan

Key Words
AST-120
Chronic renal failure
Indoxyl sulfate
Low-protein diet
Oral sorbent

Abstract
We have previously demonstrated that indoxyl sulfate is a stimulating factor for the progression of glomerular sclerosis in uremic rats. In this study we determined if a low-protein diet or oral sorbent (AST-120) could reduce the serum and urine levels of indoxyl sulfate in 5/6-nephrectomized uremic rats and undialyzed uremic patients. The uremic rats were treated by fasting or AST-120 for 2 days. The serum and urine levels of indoxyl sulfate dramatically decreased 1–2 days after fasting or AST-120 treatment. We then measured the serum and urine levels of indoxyl sulfate and calculated protein intake from urinary amounts of urea nitrogen using Maroni's equation in 80 undialyzed uremic patients with creatinine clearance less than 30 ml/min. The serum and urine levels of indoxyl sulfate were significantly lower in the patients on a low-protein diet than in those in the normal-protein diet group. Administration of AST-120 significantly decreased serum and urine levels of indoxyl sulfate in 22 undialyzed uremic patients. In conclusion, a low-protein diet or AST-120 reduced the serum and urine levels of indoxyl sulfate, a stimulating factor for glomerular sclerosis, in both uremic rats and undialyzed uremic patients.

Introduction

Circulating uremic toxins are thought to be involved in the progression of glomerular sclerosis. Motojima et al. [1] reported that peritoneal dialysis or the administration of oral sorbent (AST-120; Kremezin, Kureha-Chemical Co., Tokyo, Japan) attenuated the progression of glomerular sclerosis in subtotally nephrectomized rats. Their results support the notion that nephrectomy increases circulating levels of substances stimulating sclerosis. Hypertrophy of remnant nephrons following nephrectomy involves circulating substances [2–4], and glomerular hypertrophy and the development of glomerular sclerosis are closely related [5–10]. We have demonstrated that indoxyl sulfate is one of the circulating substances that stimulate glomerular sclerosis [11, 12]. The oral administration of indoxyl sulfate to uremic rats stimulated the progression of glomerular sclerosis accompanied by a reduction in renal function [11]. The oral administration of indole, a precursor of indoxyl sulfate, to uremic rats also stimulated the progression of glomerular sclerosis through the production of indoxyl sulfate [12]. Since only indoxyl sulfate but not indole could be detected in the urine of the rats given indole orally, indole itself was not responsible for the progression of renal failure [12]. Because indoxyl sulfate accumulating in uremic blood can be removed by peritoneal dialysis [13] and its serum level can be reduced by administration of AST-120 [14–16], the findings of Motojima et al. [1] can be at least partially explained by the removal of indoxyl sulfate.

Protein restriction delays the progression of renal disease in laboratory animals [17, 18]. Most studies in humans have suggested that a restriction of dietary protein is beneficial in the progression of chronic renal failure [19–26]. AST-120 also delays the progression of chronic renal failure in both uremic rats [27–29] and undialyzed uremic patients [30]. AST-120 consists of fine spherical particles approximately 0.2–0.4 mm in diameter, composed of porous microcrystalline carbon with an oxygen complex including a surface oxide.

KARGER
E-Mail karger@karger.ch
Fax +41 61 306 12 34
http://www.karger.ch

0378–0392/97/0236–0179 $12.00/0

Toshimitsu Niwa, MD
Nagoya University Branch Hospital, 1-1-20, Daiko-minami
Higashi-ku, Nagoya 461 (Japan)
Tel. +81 52 723 1111, fax +81 52 722 0973

AST-120 is considered to retard the progression of renal failure by intestinal absorption of substances stimulating the progression of renal failure, or their precursors.

We hypothesize that the overload of dietary protein metabolites such as indoxyl sulfate on the remnant nephrons stimulates the progression of glomerular sclerosis and chronic renal failure. Thus, the protective effects of low-protein diet and AST-120 on the progression of chronic renal failure may be explained by alleviating their overload through reduced production of dietary protein metabolites, such as indoxyl sulfate, that stimulate the progression of glomerular sclerosis. In fact, we demonstrated in uremic rats that both a low-protein diet and AST-120 reduced the serum and urine levels of indoxyl sulfate and protected against the progression of renal dysfunction and glomerular sclerosis [11].

In this study, we examined the short-term effects of fasting and AST-120 on the serum and urine levels of indoxyl sulfate in uremic rats, and the effects of a low-protein diet and AST-120 on their levels in undialyzed uremic patients.

Materials and Methods

Effects of Fasting and AST-120 in Uremic Rats

The effects of fasting and AST-120 were studied using 10 male Sprague-Dawley rats (Clea Japan, Tokyo, Japan) weighing 327–401 g. The rats (8 weeks old) were anesthetized with an intraperitoneal injection of sodium pentobarbital (45 mg/kg), and two of three branches of the left renal artery were ligated through a lateral incision. One week later, the hilum of the right kidney was totally ligated via a lateral incision performed under ether anesthesia. Thus, five-sixths of the normal kidney mass was removed, which induced uremia in most of the rats. Five weeks after the second operation, the serum levels of creatinine and blood urea nitrogen (BUN) were determined. Animals with an elevated serum creatinine (1.1–2.0 mg/dl) were chosen for the study. Before the study, all the rats ate a normal-protein diet containing 24.8% protein (Clea Japan) at about 20 g/day. They were randomly divided into two experimental groups as follows: (1) the fasting group ($n = 5$), in which the rats fasted for 2 days but drank water freely, followed by 20 g/day of a normal-protein diet containing 24.8% protein for 2 days; (2) the AST-120 group on a normal-protein diet ($n = 5$), in which the rats were fed with 20 g/day of the normal-protein diet containing 24.8% protein and 5% AST-120 for 2 days, followed by 20 g of the normal-protein diet containing 24.8% protein for 2 days. The serum and urine levels of indoxyl sulfate, serum creatinine, BUN, and urine volume were measured daily.

Patients

The serum and urine levels of indoxyl sulfate were measured in 80 randomly selected undialyzed patients with chronic renal failure [55 males and 25 females, 57.7 ± 13.1 (mean $\pm$ SD) years old] showing creatinine clearance (Ccr; 24 h) less than 30 ml/min. The causes of chronic renal failure were chronic glomerulonephritis ($n = 65$), diabetic nephropathy ($n = 4$), polycystic kidney ($n = 5$), nephrosclerosis ($n = 3$), pyelonephritis ($n = 2$), and rheumatoid arthritis ($n = 1$). BUN, serum creatinine, and Ccr levels in the 80 patients were 52.5 ± 20.0 mg/dl, 5.7 ± 2.0 mg/dl, and 13.6 ± 6.4 ml/min, respectively. We classified the uremic patients into the following four groups according to dietary protein intake: (1) 0.25–0.50 g/kg per day, (2) 0.50–0.75 g/kg per day, (3) 0.75–1.00 g/kg per day, and (4) 1.00–1.25 g/kg per day, and retrospectively studied the effects of a low-protein diet on the urine and serum levels of indoxyl sulfate.

Twenty-two undialyzed patients with chronic renal failure (15 males and 7 females, 52.9 ± 14.8 years old) who formed another group of patients, were given AST-120 (6 g/day) orally for 1 month. Serum and urine samples were obtained from the patients before and 1 month after AST-120 treatment.

As a control group, serum and urine levels were also obtained from 10 healthy subjects (5 males and 5 females, 27.2 ± 4.4 years old). The samples were stored at $-20\,^{\circ}C$ until analysis.

Protein intake was calculated from Maroni's equation [31]:

$$\text{protein intake (g/day)} = (\text{UUN} + 0.031 \times \text{BW}) \times 6.25$$

where UUN is urinary urea nitrogen (g/day) and BW is body weight (kg).

Analysis of Indoxyl Sulfate

Serum and urine levels of indoxyl sulfate were measured using a newly developed high-performance liquid chromatography (HPLC) [32], which is more useful for the analysis of a variety of compounds and is more durable than that reported previously [33]. The identification of indoxyl sulfate was confirmed by liquid chromatography/electrospray ionization-mass spectrometry, which was performed by directly connecting the HPLC with a mass spectometer (Hitachi MS-1000S). Selected ion monitoring of m/z 212, $(M-H)^-$, was used for the identification of indoxyl sulfate in the samples.

Statistical Analysis

Results are expressed as the mean $\pm$ SE. Differences between two groups were compared using unpaired or paired t tests as appropriate. For multiple comparisons, analysis of variance (ANOVA) was used. Results were considered statistically significant when the p value was less than 0.05.

Results

Effects of Fasting and AST-120 in Uremic Rats

Figure 1 shows the short-term effects of fasting or AST-120 administration on the serum and urine levels of indoxyl sulfate. Both fasting and oral administration of AST-120 markedly decreased the serum and urine levels of indoxyl sulfate even at days 1–2, and after cessation of fasting or AST-120 treatment, their levels partially recovered. The decrease in the serum ($p < 0.05$ at days 1 and 2) and urine ($p < 0.01$ at days 2 and 3) levels of indoxyl sulfate was more prominent in the AST-120 group than in the fasting group.

Figure 2 shows the effects of fasting or AST-120 on the levels of BUN and serum creatinine. Both fasting and oral administration of AST-120 significantly decreased BUN, although the decrease in BUN was not as prominent as the decrease in serum and urine levels of indoxyl sulfate. Fasting significantly increased serum creatinine at day 1, but AST-120 did not change serum creatinine during the study.

Effects of a Low-Protein Diet and AST-120 in Uremic Patients

The serum and urine levels of indoxyl sulfate in healthy subjects were 0.15 ± 0.02 mg/dl and 62.1 ± 4.1 mg/day, respectively. Serum levels of indoxyl sulfate in undialyzed uremic patients were significantly reduced as the protein intake decreased to less than 0.75 g/kg per day (fig. 3). Urinary excretion of indoxyl sulfate was also significantly reduced as the protein intake decreased to less than 0.75 g/kg per day (fig. 4). Table 1 shows the effects of a low-protein diet on the other clinical parameters in four groups of undialyzed uremic patients. There were no significant differences in renal function (serum creatinine and Ccr) among the four patient groups. Serum and urine ratios of indoxyl sulfate to creatinine were also

Fig. 1. Effects of fasting or AST-120 administration on the serum and urine levels of indoxyl sulfate in 5/6-nephrectomized uremic rats (* $p<0.05$, ** $p<0.01$ as compared with day 0 by paired t test).

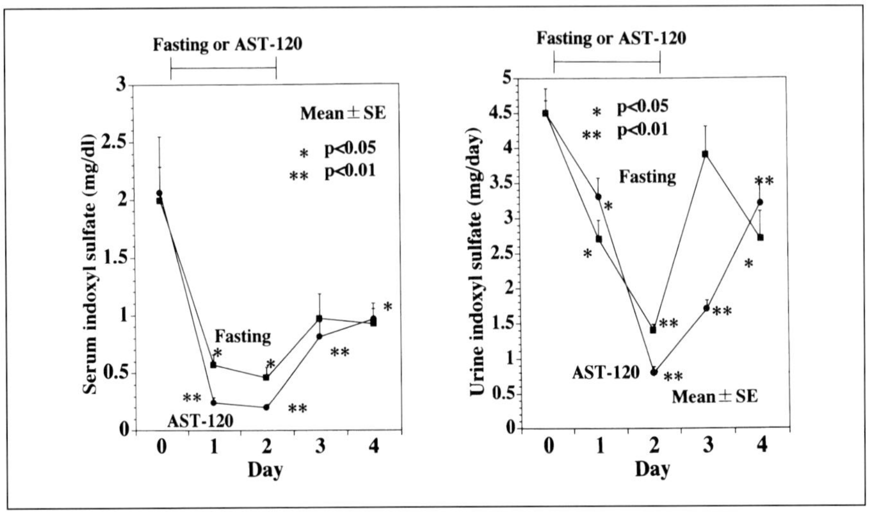

Table 1. Effects of a low-protein diet on indoxyl sulfate and the other clinical parameters in patients with chronic renal failure

	Protein intake (g/kg per day)				ANOVA
	0.25–0.50 (n=11)	0.50–0.75 (n=29)	0.75–1.00 (n=28)	1.00–1.25 (n=12)	
Protein intake, g/kg per day	0.46±0.02	0.63±0.01	0.87±0.01	1.10±0.02	**
Serum indoxyl sulfate, mg/dl	0.91±0.14	1.59±0.18	2.81±0.44	2.36±0.63	*
Urine indoxyl sulfate, mg/day	34.2±8.9	49.6±6.3	77.9±7.8	77.7±16.7	**
Indoxyl sulfate clearance, ml/min	1.92±0.43	2.62±0.40	3.04±0.54	3.09±0.69	NS
Serum indoxyl sulfate/serum creatinine	0.16±0.02	0.28±0.03	0.44±0.05	0.47±0.11	**
Urine indoxyl sulfate/urine creatinine	0.040±0.012	0.053±0.007	0.092±0.011	0.076±0.018	**
BUN, mg/dl	36.4±2.6	48.1±3.0	60.1±4.0	59.9±6.6	**
Serum creatinine, mg/dl	5.8±0.4	5.9±0.4	6.0±0.42	4.8±0.5	NS
Total protein, g/dl	6.3±0.2	6.6±0.1	6.8±0.1	6.5±0.2	NS
Albumin, g/dl	3.6±0.2	3.8±0.1	3.9±0.1	3.9±0.1	NS
Total cholesterol, mg/dl	162±10	179±6	193±9	181±10	NS
Hematocrit, %	29.1±1.9	31.0±1.1	29.0±1.4	31.0±2.1	NS
Ccr, ml/min	11.9±1.2	13.1±1.0	13.1±1.3	17.3±2.2	NS
Urine volume, ml/day	1,377±105	1,551±82	1,870±73	2,062±201	**
Urine protein, g/day	0.23±0.07	0.88±0.21	1.76±0.31	2.68±0.62	*

All results are expressed as the mean±SE. NS=Not significant; * $p<0.05$; ** $p<0.01$ by ANOVA.

significantly reduced as the protein intake decreased. BUN, urine volume, and urine protein were significantly reduced in the low-protein diet groups. Serum indoxyl sulfate was significantly correlated with serum creatinine ($r=0.50$, $p<0.01$) and BUN ($r=0.61$, $p<0.01$). Urinary excretion of indoxyl sulfate did not show any significant correlation with urinary protein ($r=0.21$), although protein intake significantly correlated with urinary protein ($r=0.50$, $p<0.01$).

Table 2 shows the effects of oral administration of AST-120 on the serum and urine levels of indoxyl sulfate in patients with chronic renal failure. Oral administration of AST-120 significantly reduced the serum and urine levels of indoxyl sulfate without any change in the protein intake.

Discussion

We first demonstrated that the serum and urine levels of indoxyl sulfate could be rapidly reduced by fasting or oral sorbent in uremic rats, and are also decreased in undialyzed uremic patients on a low-protein diet or taking oral sorbent. Notably, serum indoxyl sulfate was more intensely affected by dietary protein than BUN. These results suggest that the protective effects of a low-protein diet and of the oral sorbent on the progression of renal failure may be explained at least partially by an alleviation of the overload of indoxyl sulfate on the remnant nephrons, because indoxyl sulfate has been demonstrated to stimulate the progression of glomerular sclerosis in uremic rats [11, 12].

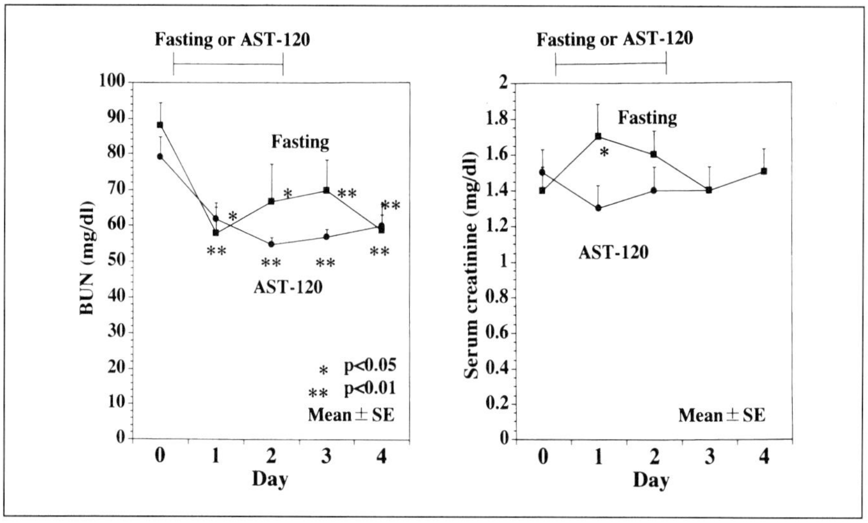

Fig. 2. Effects of fasting or AST-120 administration on the levels of BUN and serum creatinine in 5/6-nephrectomized uremic rats (*$p<0.05$, **$p<0.01$ as compared with day 0 by paired t test).

Table 2. Effects of AST-120 on indoxyl sulfate and other parameters in patients with chronic renal failure

	Before AST-120 (n = 22)	1 month after AST-120 (n = 22)
Protein intake, g/kg per day	0.75 ± 0.06	0.81 ± 0.06
Serum indoxyl sulfate, mg/dl	2.52 ± 0.43	1.55 ± 0.33**
Urine indoxyl sulfate, mg/day	75.3 ± 11.0	40.8 ± 7.2**
Indoxyl sulfate clearance, ml/min	3.8 ± 0.9	4.3 ± 1.7
Serum indoxyl sulfate/serum creatinine	0.375 ± 0.062	0.234 ± 0.050**
Urine indoxyl sulfate/urine creatinine	0.081 ± 0.012	0.043 ± 0.009**
BUN, mg/dl	54.6 ± 4.7	57.4 ± 5.0
Serum creatinine, mg/dl	7.4 ± 1.0	7.6 ± 1.1
Total protein, g/dl	6.7 ± 0.1	6.6 ± 0.2
Albumin, g/dl	3.9 ± 0.1	3.8 ± 0.1
Total cholesterol, mg/dl	188 ± 6	179 ± 7*
Hematocrit, %	32.4 ± 1.5	31.3 ± 1.7
Ccr, ml/min	12.7 ± 1.7	13.2 ± 1.7
Urine volume, ml/day	1,719 ± 142	1,831 ± 137
Urine protein, g/day	2.11 ± 0.36	1.78 ± 0.32

All results are expressed as the mean ± SE. *$p<0.05$, **$p<0.01$ compared with before AST-120 by paired t test.

Indoxyl sulfate is a uremic toxin, since it inhibits the drug binding of albumin [33], erythroid colony formation [34], lymphocyte blast formation [34], and thyroxine hepatocyte transport [35], and it stimulates the progression of glomerular sclerosis [11, 12]. The serum concentration of indoxyl sulfate was markedly increased in uremic patients [11, 13, 16, 33]. This substance is synthesized in the liver from indole, which is synthesized from tryptophan by tryptophase by intestinal bacteria such as *Escherichia coli.* Indole is absorbed into the blood from the intestine, and is metabolized mainly to indoxyl sulfate and partially to other metabolites such as indoxyl-β-*D*-glucuronide [32], which is present in serum and urine at much lower concentrations than indoxyl sulfate. Indoxyl sulfate and indoxyl-β-*D*-glucuronide are normally excreted into urine. However, uremic patients cannot efficiently excrete these metabolites into urine because of reduced renal clearance, leading to their accumulation in uremic serum. AST-120 binds indole in the intestine, and stimulates its excretion into feces, thus inhibiting indole absorption from the intestine. Consequently, AST-120 decreases the serum and urinary levels of indoxyl sulfate and indoxyl-β-*D*-glucuronide. We reported that AST-120 reduced the serum levels of indoxyl sulfate in uremic rats [14, 15] and hemodialyzed uremic patients [16], and also reduced the serum and urine levels of indoxyl-β-*D*-glucuronide in uremic patients [32]. In this study we demonstrated that AST-120 reduced the serum and urine levels of indoxyl sulfate in undialyzed uremic patients. However, AST-120 does not specifically bind indole, because it also binds intestinal phenols, especially *p*-cresol, lowering serum and urine levels of *p*-cresol as demonstrated in uremic rats [36].

We reported that administration of tryptophan directly into the large bowel increased serum and urine levels of indoxyl sulfate, but its oral administration did not [12]. Oral tryptophan may be rapidly absorbed from the small bowel so that it cannot reach the large bowel, and thus cannot be further metabolized into indole by intestinal bacteria mainly present in the large intestine. However, a small portion of tryptophan given directly into the large bowel of uremic rats is metabolized to indole by the intestinal bacteria, and is further

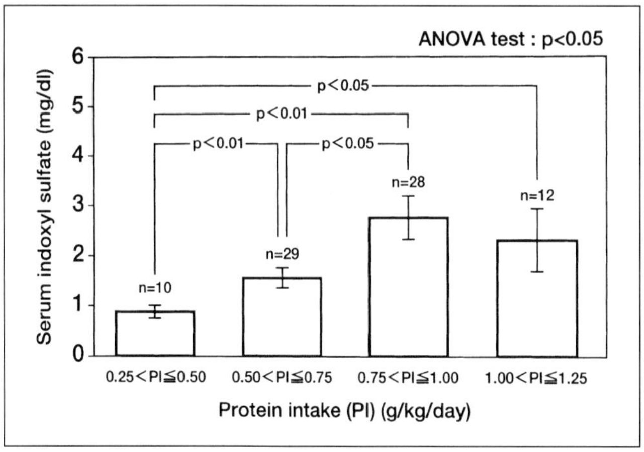

Fig. 3. Effect of a low-protein diet on serum levels of indoxyl sulfate in four undialyzed uremic patient groups with different dietary protein intake.

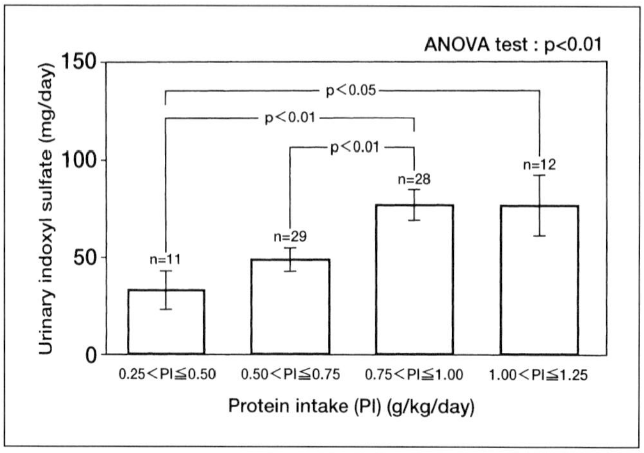

Fig. 4. Effect of a low-protein diet on urinary excretion of indoxyl sulfate in four undialyzed uremic patient groups with different dietary protein intake.

metabolized to indoxyl sulfate in the liver. In contrast to tryptophan, oral administration of indole increased urinary excretion of indoxyl sulfate. The present findings demonstrating that the serum and urine levels of indoxyl sulfate were rapidly reduced by fasting, and also fell in the undialyzed uremic patients on a low-protein diet, suggets that a small proportion of dietary proteins enters the large bowel, where tryptophan released by further digestion is metabolized to indole by intestinal bacteria, and subsequently metabolized to indoxyl sulfate in the liver.

High-protein intake accelerates the development of glomerular sclerosis, leading to a more rapid loss of renal function in rats [37]. The high-protein diet induces renal vasodilation, resulting in glomerular hyperfiltration and glomerular hypertension, which are thought to play an important role in the development of glomerular sclerosis. We suggest that increased production of indoxyl sulfate by a high-protein diet may at least partially explain the acceleration of progressive glomerular sclerosis by such a diet.

We propose that endogenous protein metabolites such as indoxyl sulfate play a significant role in the progression of chronic renal failure, namely the 'protein metabolite hypothesis' [38]. The initial insult leads to a loss of functioning nephrons via disease-specific pathophysiological processes. The progressive fall in glomerular filtration leads to increased levels of circulating endogenous protein metabolites such as indoxyl sulfate [11, 13, 16, 33] and their overload in the remnant nephrons. Indoxyl sulfate promotes progressive glomerular sclerosis [11, 12] leading to a further loss of nephrons. Thus, the vicious cycle of progressive renal injury is complete. Our present results indicate that a low-protein diet or oral sorbent can reduce the serum and urine levels of indoxyl sulfate. If the overload of indoxyl sulfate is alleviated, e.g., by restriction of dietary protein intake and/or by the administration of the oral sorbent, AST-120, this chain of events might be interrupted. In fact, many clinical and experimental studies demonstrate that dietary protein restriction [17–26] and AST-120 [27–30] delay the progression of chronic renal failure. However, we consider that indoxyl sulfate is not the sole stimulating factor for progression of chronic renal failure, but a representative of the endogenous protein metabolites stimulating that progression.

References

1 Motojima M, Nishijima F, Ikoma M, Kawamura T, Yoshioka T, Fogo AB, Sakai T, Ichikawa I: Role for 'uremic toxin' in the progressive loss of intact nephrons in chronic renal failure. Kidney Int 1991; 40:461–469.

2 Kurnick NB, Lindsey PA: Compensatory renal hypertrophy in parabiotic mice. Lab Invest 1968;10: 45–48.

3 Obertop H, Malt RA: Lost mass and excretion as stimuli to parabiotic compensatory renal hypertrophy. Am J Physiol 1977;232:F405–F408.

4 Harris RH, Best CF: Circulatory retention of urinary factors as a stimulus to renal growth. Kidney Int 1977;12:305–312.

5 Yoshida Y, Fogo A, Ichikawa I: Glomerular hemodynamic changes vs hypertrophy in experimental glomerular sclerosis. Kidney Int 1989;35:654–660.

6 Yoshida Y, Kawamura T, Ikoma M, Fogo A, Ichikawa I: Effects of antihypertensive drugs on glomerular morphology. Kidney Int 1989;36:626–635.

7 Doi T, Striker LJ, Quaife C, Conti FG, Palmiter R, Behringer R, Brinster R, Striker GE: Progressive glomerular sclerosis develops in transgenic mice chronically expressing growth hormone and growth-hormone-releasing factor but not in those expressing insulin-like growth factor-1. Am J Pathol 1988;131:398–403.

8 MacKay K, Striker LJ, Stauffer JW, Agodoa LY, Striker GE: Relationship of glomerular hypertrophy and sclerosis: Studies in SV40 transgenic mice. Kidney Int 1990;37:741–748.

9 Klahr S, Schreiner G, Ichikawa I: The progression of renal disease. N Engl J Med 1988;318:1657–1666.

10 Fogo A, Ichikawa I: Evidence for the central role of glomerular growth promoters in the development of sclerosis. Semin Nephrol 1989;9:329–342.

11 Niwa T, Ise M: Indoxyl sulfate, a circulating uremic toxin, stimulates the progression of glomerular sclerosis. J Lab Clin Med 1994;124:96–104.

12 Niwa T, Ise M, Miyazaki T: Progression of glomerular sclerosis in experimental uremic rats by administration of indole, a precursor of indoxyl sulfate. Am J Nephrol 1994;14:207–212.

13 Niwa T, Yazawa T, Kodama T, Uehara Y, Maeda K, Yamada K: Efficient removal of albumin-bound furancarboxylic acid, an inhibitor of erythropoiesis, by continuous ambulatory peritoneal dialysis. Nephron 1990;56:241–245.

14 Niwa T, Yazawa T, Ise M, Sugano M, Kodama T, Uehara Y, Maeda K: Inhibitory effect of oral sorbent on accumulation of albumin-bound indoxyl sulfate in serum of experimental uremic rats. Nephron 1991;57:84–88.

15 Niwa T, Miyazaki T, Hashimoto N, Hayashi H, Ise M, Uehara Y, Maeda K: Suppressed serum and urine levels of indoxyl sulfate by oral sorbent in experimental uremic rats. Am J Nephrol 1992;12: 201–206.

16 Niwa T, Emoto Y, Maeda K, Uehara Y, Yamada N, Shibata M: Oral sorbent suppresses accumulation of albumin-bound indoxyl sulfate in serum of haemodialysis patients. Nephrol Dial Transplant 1991;6:105–109.

17 Klahr S, Buerkert J, Purkerson ML: Role of dietary factors in the progression of chronic renal disease. Kidney Int 1983;24:579–587.

18 Brenner BM: Hemodynamically mediated glomerular injury and the progressive nature of kidney disease. Kidney Int 1983;23:647–655.

19 Mitch WE, Walser M, Steinman TI, Hill S, Zeger S, Tungsanga K: The effect of a keto acid-amino acid supplement to a restricted diet on the progression of chronic renal failure. N Engl J Med 1984; 311:623–629.

20 Rosman JB, ter Wee PM, Meijer S, Piers-Becht TPM, Sluiter WJ, Donker AJM: Prospective randomized trial of early dietary protein restriction in chronic failure. Lancet 1984;ii:1291–1296.

21 Hunsicker LG: Studies of therapy of progressive renal failure in humans. Semin Nephrol 1989;9:380–394.

22 Ihle BU, Becker GJ, Whitworth JA, Charlwood RA, Kincaid-Smith PS: The effect of protein restriction on the progression of renal insufficiency. N Engl J Med 1989;321:1773–1777.

23 Levey AS, Gassman JJ, Hall PM, Walker WG: Assessing the progression of renal disease in clinical studies: Effects of duration of follow-up and regression to the mean: Modification of Diet in Renal Disease (MDRD) Study Group. J Am Soc Nephrol 1991;1:1087–1094.

24 Fouque D, Laville M, Boissel JP, Chifflet R, Labeeuw M, Zech PY: Controlled low protein diets in chronic renal insufficiency: Meta-analysis. BMJ 1992;304:216–220.

25 Walser M, Hill SB, Ward L, Magder L: A crossover comparison of progression of chronic renal failure: Ketoacids vesus amino acids. Kidney Int 1993;43: 933–939.

26 Klahr S, Levey AS, Beck GJ, Caggiula AW, Hunsicker L, Kusek JW, Striker G: The effect of dietary protein restriction and blood-pressure control on the progression of chronic renal disease. N Engl J Med 1994;330:877–884.

27 Kanai F, Takahama T, Yamazaki Z, Idezuki Y, Koide K: Effect of oral sorbent on experimental uremic rats. Jpn J Nephrol 1986;28:1249–1259.

28 Sakai T, Kumano K, Takara S, Kuwao S, Ise M, Sugano M, Uehara Y: Effect of oral-sorbent (AST-120) to chronic renal failure (CRF) in rats. Jpn J Nephrol 1989;31:359–365.

29 Yoshida Y, Sakai T, Ise M: Effects of oral adsorbent in the rat model of chronic renal failure. Nephron 1992;62:305–314.

30 Sanaka T, Sugino N, Teraoka S, Ota K: Therapeutic effects of oral sorbent in undialyzed uremia. Am J Kidney Dis 1988;12:97–103.

31 Maroni BJ, Steinman TI, Mitch WE: A method for estimating nitrogen intake of patients with chronic renal failure. Kidney Int 1985;27:58–65.

32 Niwa T, Miyazaki T, Tsukushi S, Maeda K, Tsubakihara Y, Owada A, Shiigai T: Accumulation of indoxyl-β-*D*-glucuronide in uremic serum: Suppression of its production by oral sorbent and efficient removal by hemodialysis. Nephron 1996;74:72–78.

33 Niwa T, Takeda N, Tatematsu A, Maeda K: Accumulation of indoxyl sulfate, an inhibitor of drug-binding, in uremic serum as demonstrated by internal-surface reversed-phase liquid chromatography. Clin Chem 1988;34:2264–2267.

34 Kawashima Y: Study on the uremic protein binding inhibitors as uremic toxin: Toxic effect on erythroid colony formation, lymphocyte blast formation and renal function. Jpn J Nephrol 1989;31:1151–1161.

35 Lim CF, Bernard BF, Jong M de, Docter R, Krenning EP, Hennemann G: A furan fatty acid and indoxyl sulfate are the putative inhibitors of thyroxine hepatocyte transport in uremia. J Clin Endocrinol Metab 1993;76:318–324.

36 Niwa T, Ise M, Miyazaki T, Maeda K: Suppressive effect of oral sorbent on accumulation of *p*-cresol in the serum of experimental uremic rats. Nephron 1993;65:82–87.

37 Brenner BM, Meyer TW, Hostetter TH: Dietary protein intake and the progressive nature of kidney disease: The role of hemodynamically mediated glomerular injury in the pathogenesis of progressive glomerular sclerosis in aging, renal ablation, and intrinsic renal disease. N Engl J Med 1992;307: 652–659.

38 Niwa T: Organic acids and the uremic syndrome: Protein metabolite hypothesis in the progression of chronic renal failure. Semin Nephrol 1996;16: 167–182.

Mineral and Electrolyte Metabolism

Miner Electrolyte Metab 1997;23:185–188

Protein Turnover in the Kidney and the Whole Body in Humans

Giacomo Garibotto [a]
Paolo Tessari [b]
Cristina Robaudo [a]
Michela Zanetti [b]
Stefano Saffioti [a]
Monica Vettore [b]
Sandro Inchiostro [b]
Paolo Sacco [a]
Giacomo Deferrari [a]
Alberto Tizianello [a]

[a] Division of Nephrology, Department of Internal Medicine, University of Genoa and
[b] Department of Metabolic Diseases, University of Padua, Italy

Key Words
Amino acids
Kidney
Leucine
Protein degradation
Protein synthesis

Abstract
For a better understanding of protein synthesis and degradation in the human kidney, the arteriovenous difference technique across the kidney, splanchnic organs, and leg muscle was combined with labeled leucine and phenylalanine isotope dilution models. Results indicate that in the postabsorptive state, the protein balance across the human kidney is negative because the rate of leucine release from protein degradation is greater than the amount used for protein synthesis. In the splanchnic bed, net protein balance is neutral since the amount of leucine deriving from protein degradation is similar to the amount utilized for protein synthesis. In the leg muscle, protein degradation exceeds protein synthesis. The kidney exhibits the highest leucine metabolic activity when expressed in terms of total organ leucine content. The estimated fractional protein synthesis rate in the human kidney is about 40% per day (vs. about 2% in muscle and 12% in the splanchnic bed). The human kidney presents high rates of protein turnover and accounts for a significant fraction of whole-body protein degradation, protein synthesis, and leucine oxidation.

Introduction

Under normal physiological conditions, kidney protein content reflects the balance between synthesis and degradation resulting from protein turnover. Understanding the processes leading to maintenance of the kidney protein content requires knowledge of the mechanisms by which protein mass may change. Better understanding of the mechanisms regulating protein turnover in the human kidney in physiological and pathological situations may be practically important. Conditions such as compensatory renal hypertrophy, diabetes mellitus, and decrease in functioning nephrons are associated with changes in kidney protein content and function [1]. However, protein turnover in the human kidney has never been fully evaluated, and the role played by the kidney in whole-body protein turnover is not understood. To gain more knowledge about protein synthesis and degradation in the human kidney, we combined the arteriovenous difference technique across the kidney, as well as splanchnic organs and leg muscle, with labeled leucine and phenylalanine isotope dilution models [2].

Net Uptake and Release of Leucine, Ketoisocaproate and Phenylalanine by the Kidney and Splanchnic Organs

Figure 1 shows the renal metabolism of phenylalanine. In the postabsorptive state, the balance of this amino acid across the kidney is positive owing less to the release than to the uptake of phenylalanine, which is about 100% higher than the release. It is interesting to note that the amount of phenylalanine taken up by the kidney is met by an almost identical release of tyrosine, thus suggesting a nearly complete metabolic conversion. The human kidney also behaves in a unique fashion regarding leucine and ketoisocaproate metabolism, since it takes up the latter and releases leucine [2]. Nearly opposite behaviour is shown by the splanchnic organs. In fact, they release ketoisocaproate in very similar amounts to those taken up by the kidney, while the leucine balance is close to zero [2]. Therefore splanchnic organs cooperate with the kidney providing it with ketoisocaproate which is likely utilized for producing energy. In vitro studies have shown that leucine is preferentially

KARGER
E-Mail karger@karger.ch
Fax +41 61 306 12 34
http://www.karger.ch

0378–0392/97/0236–0185 $12.00/0

Giacomo Garibotto, MD
Department of Internal Medicine-Nephrology Division
Viale Benedetto XV, 6, I–16132 Genoa (Italy)
Tel. +39 10 3537970, fax +39 10 352324

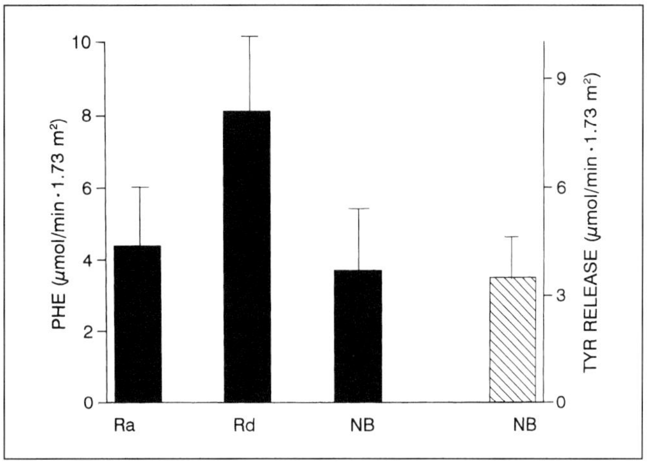

Fig. 1. Renal metabolism of phenylalanine (PHE) and tyrosine (TYR). Ra = Rate of appearance; Rd = rate of disposal; NB = net balance [adapted from ref. 2].

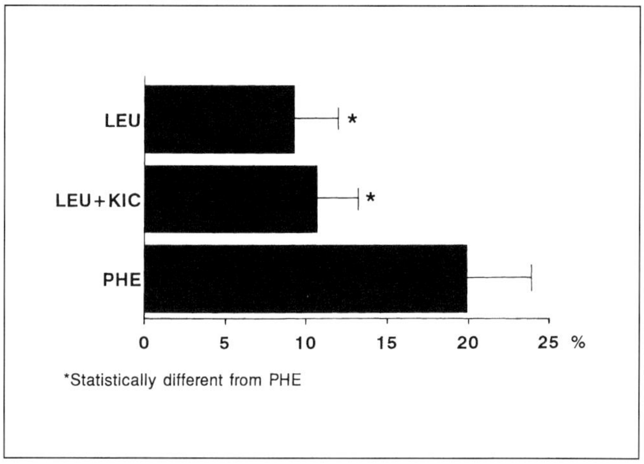

Fig. 2. Fractional uptake of leucine (LEU), ketoisocaproate (KIC), and phenylalanine (PHE) by the kidney [adapted from ref. 2].

oxidized in the medullary thick ascending limb [3]. Leucine oxidation may be coupled with sodium reabsorption in this tubule segment [3].

Fractional amino acid uptake by the kidney has been calculated by dividing the rate of uptake of the amino acid(s) over the respective amino acid delivery to the kidney (the product of arterial concentration times blood flow; fig. 2). The fractional uptake of phenylalanine is higher than that for leucine and ketoisocaproate (ca. 20% vs. ca. 10%). The amount of phenylalanine taken up by the kidney is much higher than the amount filtered, whereas filtered leucine and ketoisocaproate account for their uptake by the kidney. This indicates that unlike leucine, phenylalanine enters the kidney is substantial amounts by peritubular uptake. Unlike the kidney, peripheral tissues show similar extractions of leucine and phenylalanine [2]. The greater extraction of phenylalanine by the kidney might be related to its high phenylalanine hydroxylase activity [4].

Leucine Metabolism and Protein Turnover in the Kidney, Splanchnic Organs, and Leg Muscle

Figure 3 shows leucine kinetics across the leg, splanchnic organs, and the kidney according to a multicompartment model [2]. The rate of appearance (an index of protein breakdown) in the legs is greater than the rate of disposal (an index of protein synthesis) and is consistent with a negative protein balance. Leucine oxidation (not shown) represents about 27% of intracellular leucine incorporation into protein [2]. In the splanchnic organs, there was no difference between the rate of leucine release from protein breakdown and the rate of incorporation into protein. Therefore, in the postabsorptive state, the splanchnic bed is in a neutral protein balance. Leucine oxidation in splanchnic organs is as much as 15% of leucine incorporation into proteins [2]. In the kidney, the release of leucine from proteolysis is about 30% greater than leucine entry into protein synthesis. The net balance of leucine, and thus of protein, is negative. Leucine oxidation by the kidney represents about 50% of leucine incorporation into proteins. The high rate of leucine-carbon oxidation is quantitatively accounted for by net leucine release from proteolysis as well as by the net ketoisocaproate uptake. Similar results were obtained employing other models [2].

Figure 4 depicts whole-body leucine rate of appearance, oxidation and incorporation into protein synthesis. The whole muscle mass contributes as much as 48% to whole-body leucine oxidation and 32–36% to whole-body protein synthesis and degradation, according to previous data [5, 6]. The splanchnic organs account for as much as 22% of whole-body proteolysis, for 27% of whole-body protein synthesis and for 18% of leucine oxidation. The kidney accounts for 11% of whole-body proteolysis, for 10% of whole-body protein synthesis and for 25% of whole-body leucine oxidation. When summed, the individual contributions of these organs accounted for as much as 70% of whole-body proteolysis and for about 90% of whole-body leucine oxidation [2].

Regulation of Kidney Protein Turnover

Studies on animals have shown that adaptive changes in kidney protein content and size occur in response both to physiological and to pathological stimuli such as nutritional status, diabetes mellitus and kidney hypertrophy or decrease in the number of functioning nephrons [1]. There is also evidence that kidney size and protein content in animals are modulated by hormones. Insulin inhibits protein breakdown in isolated proximal tubules of rats [7], while glucocorticoids stimulate protein breakdown [8]. Androgen administration to female mice increases the protein content of the kidney [9]. Furthermore, both chronic metabolic acidosis and hypokalemia stimulate the growth of the kidney [10, 11]. Ammonium-chloride induced metabolic acidosis in rats induces renal growth,

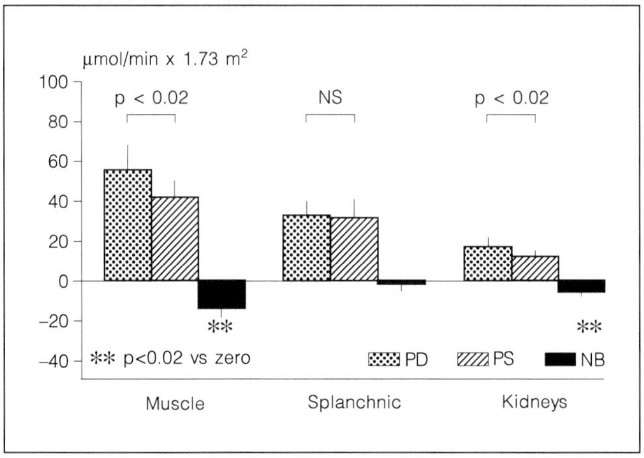

Fig. 3. Protein degradation (PD), synthesis (PS), and net protein balance (NB) across the legs, splanchnic organs, and the kidneys [adapted from ref. 2].

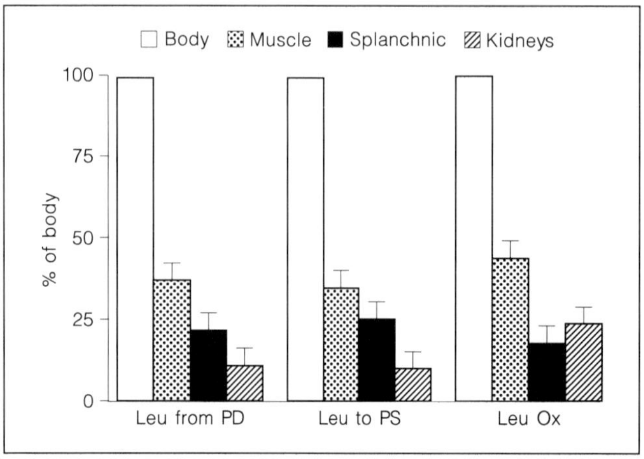

Fig. 4. Percent contribution by kidneys, splanchnic area, and skeletal muscle to whole-body protein degradation (PD), synthesis (PS) and leucine oxidation (Ox) [adapted from ref. 2].

similar to that observed after uninephrectomy [12]. Ammonia may be a mediator for hypertrophy [13]. Moreover, growth hormone, by stimulating elaboration of insulin-like growth factor-I, may play an important role in renal hypertrophy [1]. Depending on the underlying conditions, different mechanisms appear to contribute to the increase in kidney protein content in animals. A reduction in protein degradation, owing to a decrease in the autophagic-lysosomal system, occurs in diabetes mellitus and in ammonium-chloride-induced acidosis [14]. In contrast, in compensatory renal growth after uninephrectomy, the increase in protein content is entirely due to an increase in protein synthesis [14]. However, little is known about protein turnover in the human kidney.

Approximately 15–18% of adult kidney weight is accounted for by proteins [1]. Previous attempts at estimating kidney protein turnover during in vitro studies have indicated that about 30% of rat renal cortical proteins are turned over every day [15]. Our data in humans indicate that the rate of leucine release from intrarenal protein degradation is high, about one-third to one-half of that in skeletal muscle or the splanchnic bed [2]. Assuming that leucine represents 8.5% (on a molar basis) of all amino acid residues in human mixed proteins [unpublished observations from our laboratory] and that proteins account for about 15% of kidney wet weight, one can estimate from leucine kinetics that fractional protein synthesis in the human kidney may be as high as 40% per day. In comparison, fractional protein synthesis is 12% in splanchnic organs and 1.5% in muscle. These latter estimates are in agreement with previous observations [5, 6].

The negative protein balance observed in the human kidney implies that fasting may be followed by significant protein wasting. Animal studies have shown that the kidney protein content depends on dietary factors, such as starvation or the exogenous amino acid supply. In contrast with muscle, in which protein content falls only by 8% during fasting, starvation in rats reduces the protein content in the kidney by 20% [16]. Dietary modulation of protein turnover may include an increase in kidney protein content following a high protein intake [17]. In humans, the kidney increases its use of amino acids, and mainly of leucine, following acute amino acid administration [18]. Moreover, long-term intravenous therapy with amino acids promotes an increase in kidney size [19].

The negative protein balance across the human kidney probably does not result from the catabolism of renal structural proteins but is, rather, derived mainly from intrarenal degradation of peptides and low-molecular-weight proteins. The kidney is an important site for the catabolism of low-molecular-weight (<50,000) plasma proteins and peptides which are hydrolyzed; their amino acid components are returned into the systemic circulation [20].

Studies on kidney protein turnover in the basal, postabsorptive state may serve as a guidepost to comprehension of the factors and mechanisms that regulate kidney protein turnover. Further studies should define the pathways by which hormones and substrates may affect kidney protein balance. Advances in understanding these processes might also lead to a better comprehension of the mechanisms of disease.

Acknowledgements

We are grateful to Mr. Marco Marchelli for his excellent technical assistance and to Francesca Tincani for her assistance in the preparation of the manuscript and figures. The studies were supported by grants from the Ministero dell'Università e della Ricerca Scientifica e Tecnologica (Assegnazione per la ricerca scientifica 40%) and by grants from the National Research Council (CNR)-Targeted project 'Prevention and Control Disease Factors; Subproject Sp1 Alimentazione' and Progetto Finalizzato Invecchiamento.

References

1 Rabkin R, Dahl DC: Factors controlling intracellular protein turnover in the kidney. Semin Nephrol 1990;10:472–480.

2 Tessari P, Garibotto G, Inchiostro S, Robaudo C, Saffioti S, Vettore M, Zanetti M, Russo R, Deferrari G: Kidney, splanchnic and leg protein turnover in humans: Insights from leucine and phenylalanine kinetics. J Clin Invest 1996;98:1481–1492.

3 Tring-Trang-Tan MM, Levillain O, Bankir L: Contribution of leucine to oxidative metabolism of the rat medullary thick ascending limb. Pfluegers Arch 1988;411:676–680.

4 Tourian A, Goddard J, Puck TT: Phenylalanine hydroxylase activity in mammalian cells. J Cell Physiol 1969;73:159–170.

5 Gelfand RA, Glickman MG, Castellino P, Louard RJ, DeFronzo RA: Measurement of *L*-[1-^{14}C]leucine kinetics in splanchnic and leg tissues in man. Diabetes 1988;37:1365–1372.

6 Halliday D, McKeran RO: Measurement of muscle protein synthetic rate from serial muscle biopsies and total body protein turnover in man by continuous intravenous infusion of *L*-[δ-^{15}N]lysine. Clin Sci 1975;49:581–590.

7 Pfeifer U, Scheller H: Inhibition by insulin of cellular autophagy in proximal tubular cells of rat kidney. Am J Physiol 1983;244:E109–E114.

8 Ichii S, Ikeda A: Effect of adrenalectomy and gonadectomy on protein turnover of liver, kidney and spleen of mouse. Endocrinol Jpn 1970;17:273–276.

9 Koths KE, Godchaux W, Doeg LH: The effect of testosterone on the synthesis of mouse kidney mitochondrial and microsomal proteins in vivo and in vitro. Endocrinology 1972;91:125–134.

10 Lotspeich WD: Metabolic aspects of acid-base change. Science 1967;155:1066–1075.

11 Gustafson AB, Shear L, Gabuzda GJ: Protein metabolism in vivo in kidney, liver, muscle and heart of potassium-deficient rats. J Lab Clin Med 1973; 82:287–297.

12 Fine L, Norman J: Cellular events in renal hypertrophy. Annu Rev Physiol 1989;51:19–32.

13 Golchini K, Normal J, Bohman R: Induction of hypertrophy in cultured proximal tubule cells by extracellular NH_4Cl. J Clin Invest 1989;84:1767–1779.

14 Rabkin R, Schechter P, Dong Shi J, Boner J: Protein turnover in the hypertrophying kidney. Miner Electrolyte Metab 1996;22:153–156.

15 Goldspink DF, Kelly FJ: Protein turnover and growth in the whole body, liver and kidney of the rat from the foetus to senility. Biochem J 1984;217: 507–516.

16 Addis T, Poo LJ, Lew W: The quantities of protein lost by the various organs and tissues of the body during a fast. J Biol Chem 1936;115:111–116.

17 Bur JA, Conde RD: Decreased rate of protein breakdown during nutritional recovery of mouse kidney. Am J Physiol 1982;243:E360–E364.

18 Tizianello A, Deferrari G, Garibotto G, Robaudo C, Salvidio G, Saffioti S: Renal ammoniagenesis in the post-prandial period; in Schoolwerth AC, Kurokawa K, Tannen RL, Vinay P (eds): New Advances in Renal Ammonia Metabolism. Contrib Nephrol. Basel, Karger, 1985, vol 47, pp 44–49.

19 Cochran ST, Pagani JJ, Barbaric ZL: Nephromegaly in hyperalimentation. Radiology 1979;130: 603–606.

20 Maak T: Renal handling of low molecular weight proteins. Am J Med 1975;58:57–64.

Mineral and Electrolyte Metabolism

Miner Electrolyte Metab 1997;23:189–193

Martina Franz
Walter H. Hörl

Division of Nephrology,
Department of Internal Medicine,
University of Vienna, Austria

Protein Catabolism in Acute Renal Failure

Key Words
Acute renal failure
Cytokines
Glucocorticoids
Growth factors
Insulin resistance
Metabolic acidosis
Proteinases

Abstract
There is no doubt that acute renal failure (ARF) is associated with enhanced protein breakdown. It has been shown that protein split products can be measured in plasma samples of these patients. On the other hand, ARF frequently occurs in conditions of increased metabolic stress which leads to enhanced protein catabolism. Muscle wasting, loss of lean body weight, and a negative nitrogen balance result in malnutrition which considerably increases morbidity and mortality. Besides the accumulation of uremic toxins, several other factors are involved in the accelerated proteolysis in ARF. Metabolic acidosis appears to be one of the major catabolic factors in chronic renal failure, and probably in ARF as well. Insulin resistance, which is commonly attributed to uremia, also increases protein degradation. However, this derangement of carbohydrate metabolism is not directly accessible to therapy, in contrast to acidosis, which can be easily corrected by bicarbonate administration. There is further evidence that glucocorticoid excess contributes to the enhanced muscle proteolysis in ARF. Moreover, several studies have demonstrated that only in the presence of both glucocorticoids and acidosis could proteolysis occur. Investigation of the cellular mechanism by which muscle proteins are degraded indicates the importance of the cytosolic, soluble ATP- and ubiquitin-dependent proteolytic system. Successful treatment of various catabolic conditions with recombinant human growth hormone and insulin-like growth factor-I seems to be a promising strategy in severely catabolic patients with ARF. Anticytokine therapy appears to be another promising treatment in the course of catabolic illness due to sepsis; however, clinical application is still in its infancy.

Introduction

Patients with renal failure often display a negative nitrogen balance resulting in muscle wasting and loss of lean body weight. Malnutrition is associated with a higher morbidity and mortality. Catabolism in uremia is caused by increased protein degradation, decreased protein synthesis, and altered plasma levels of essential amino acids. The metabolic and nutritional disturbances in renal failure could be explained as the result of both the uremic condition and its nonspecific toxicity, and by several specific mechanisms. End-stage renal disease is frequently complicated by malnutrition, particularly in hemodialysis patients [1]. However, malnourishment may be present at an earlier stage of renal insufficiency, and this condition is usually attributed to inadequate nutrition or superimposed illness. In chronic renal failure (CRF), protein wasting develops due to restriction of protein intake (therapeutic or due to impaired appetite), the inability to maintain protein balance due to specific adaptive responses and, finally, due to the presence of catabolic factors such as metabolic acidosis [2]. Increased degradation and decreased synthesis of protein is also found in acute uremia [3, 4]. Since acute renal failure (ARF) is frequently found in a setting of increased metabolic stress such as sepsis, trauma, or circulatory shock, accelerated protein catabolism and enhanced protein breakdown occur. The value of nutritional therapy is disputed, since the extent of protein catabolism has not been significantly improved [5]. A better understanding of the underlying mechanisms that cause protein catabolism in uremia can help to optimize treatment options and identify new therapeutic strategies.

Mechanisms of Protein Catabolism in ARF

In ARF, protein catabolism and gluconeogenesis are increased. Released amino acids, from protein breakdown in the skeletal muscle, are ineffectively used for protein synthesis, leading to accelerated protein catabolism with an altered plasma amino acid pattern. In contrast, hepatic and cardiac protein synthesis may be increased. Carbohydrate metabolism is also deranged in ARF and is characterized by impaired insulin-mediated actions.

Insulin Resistance

Uremia is a state of insulin resistance. The primary site of insulin resistance resides in the peripherial tissue (muscle and adipose), whereas basal hepatic glucose production and its suppression by insulin are not altered in uremia [6, 7]. There is a large body of evidence that insulin resistance accounts for one of the hormonal

KARGER
E-Mail karger@karger.ch
Fax +41 61 306 12 34
http://www.karger.ch

0378–0392/97/0236–0189 $12.00/0

Prof. W.H. Hörl, MD, PhD
Med. Univ. Klinik III
Klinische Abteilung für Nephrologie und Dialyse
Währinger Gürtel 18-20, A–1090 Wien (Austria)
Tel. +43 1 40400 4390, fax +43 1 40400 4392

abnormalities involved in the deranged muscle protein metabolism in ARF. Experimental studies have demonstrated decreased insulin-mediated glucose uptake, glycogen synthesis, and glucose oxidation in the muscle of acutely uremic rats [8]. Since insulin binding to its receptor was intact [9], the presence of a postreceptor defect was considered. Clark and Mitch [4] suggested that the abnormal net protein degradation in ARF may be linked to the altered carbohydrate metabolism and may be a consequence of defective glucose utilization. After 24 h of ARF both increased protein degradation and altered glucose utilization could be detected, so that after 48 h of ARF, net muscle protein degradation increased further because protein synthesis was depressed.

The hypothesis of Tischler [10] and of Heddon and Buse [11], that the redox state of the muscle changes with a decrease in the cytoplasmic NADH/NAD ratio, would be another potential explanation for the increased net rate of muscle protein degradation. However, all these abnormalities may be linked to a single metabolic defect.

Growth Hormone (GH) and Insulin-Like Growth Factor-I (IGF-I)

The GH-IGF axis is altered in uremia, resulting in peripheral resistance. However, several growth factors have been found to play an important role in recovery from ARF [12–14]. The kidney is an important source of IGF-I and its production is influenced by GH [15, 16]. IGF-I stimulates protein synthesis and other anabolic processes [17]. In ischemic ARF kidney, mRNA levels for IGF-I and several IGF-binding proteins are reduced, whereas kidney IGF-I receptor density is increased. This phenomenon might serve to offset the lowered kidney IGF-I production and to enhance the reparative actions of circulating endogenous or exogenous administered IGF-I by increasing IGF-I bioavailability [13]. GH, as the prototype anabolic hormone, also has great potential in improving catabolic conditions of various origin, including those due to renal failure. Recombinant human GH (rhGH) can improve the nitrogen balance and nutritional parameters with an increase of muscle mass and a decrease of fat mass [18, 19]. Thus rhGH could be utilized to improve the hypercatabolic condition in ARF.

Since many of the effects of rhGH are mediated through IGF-I, and rhGH treatment may fail to increase IGF-I levels in severe sepsis, malnutrition, or uremia [20, 21], direct administration of rhIGF-I seems to be an even more promising strategy in ARF with severe protein catabolism. Experimental studies on rats demonstrated that administration of rhIGF-I or its analogue, des-(1-3)-IGF-I, attenuated muscle protein breakdown and increased protein synthesis in chronically uremic rats [22] and in rats with ischemic ARF [12]. Thus, treatment with rhIGF-I may be particularly useful in the management of ARF because it reduces net catabolism and enhances recovery of renal function, even when ARF is already established, at least in animal experiments.

Glucocorticoids

Uremia causes high corticosterone levels in ARF and CRF, thus this hormone could contribute significantly to increased proteolysis, branched-chain amino acid break- down and, possibly, the inhibition of protein synthesis. The hypothesis that excess glucocorticoids are involved in protein catabolism associated with ARF has been disputed. Although Bondy et al. [23] and later Schaefer et al. [24] had demonstrated that adrenalectomy, as a model of glucocorticoid deficiency, reduced the process of muscle protein breakdown and improved the nitrogen balance, the concomitant removal of a considerable source of catecholamines in this model has been criticized. However, Wernze [25] found a significant increase in plasma corticosterone levels in ARF following bilateral nephrectomy and, furthermore, there is evidence for reduced catabolism following treatment with the glucocorticoid receptor antagonist RU 486 in acutely uremic rats [26].

Cytokines

There is also evidence that cytokines contribute to protein catabolism. Immunologically activated mediators are important in the pathogenesis of ARF in sepsis. Since it was shown that plasma from patients with severe trauma and sepsis contains a peptide which accelerates muscle proteolysis, further investigations by Baracos et al. [27] demonstrated that incubation of rat muscle with interleukin (IL)-1 stimulated proteolysis by increasing prostaglandin E_2 synthesis. Incubation with the cyclo-oxygenase inhibitor indomethacin blocked the prostaglandin E_2 production and abolished increased muscle protein breakdown, as did the lysosomal proteinase inhibitor Ep-475, indicating that lysosomal proteinases were involved. Other inflammatory cytokines, such as tumor necrosis factor (TNF), IL-2 and interferon-γ may also contribute to protein catabolism during severe illness [18].

Metabolic Acidosis

Metabolic acidosis is a common finding in both ARF and CRF. Numerous studies indicate that metabolic acidosis stimulates protein degradation and further highlight the negative impact that metabolic acidosis could have on patients with kidney disease [28]. Increased muscle protein degradation which provides the nitrogen needed to increase glutamine production, so that the kidney can excrete acid, seems to block adaptive responses to preserve the nitrogen balance during catabolic conditions. Chronic metabolic acidosis also seems to decrease albumin synthesis [29]. The mechanisms by which metabolic acidosis stimulates muscle proteolysis have been intensively investigated. Several studies provide evidence that glucocorticoids are involved. May et al. [30] found increased urinary corticosterone excretion in rats with CRF in proportion to the degree of metabolic acidosis. However, although nonacidotic rats also displayed increased levels of urinary corticosterone, only the acidotic rats had accelerated muscle protein degradation. Garibotto et al. [31] showed that protein degradation in the forearm muscle of CRF patients was indirectly related to the serum bicarbonate but directly related to the serum cortisone level. Cell culture work by England et al. [32] demonstrated that acidification of the medium stimulated protein degradation in BC3H-1 myocytes and, recently, Isozaki et al. [33] reported that addition of the glucocorticoid receptor antagonist RU 486 to the medium abolished protein catabolism despite the presence of acidification. Another recently published work [34] confirmed these results by showing that in adrenalectomized acidotic rats, proteolysis increased only when physiologic concentrations of glucocorticoids were supplemented. All these studies have demonstrated that only in the presence of both glucocorticoids and acidosis could proteolysis occur. In the absence of one factor, muscle catabolism failed to increase. Thus acidosis appears to evoke a glucocorticoid-dependent catabolic response in the muscle that can account for the protein wasting associated with uremia.

Skeletal Muscle Proteinases

In skeletal muscle, various proteolytic pathways for muscle protein breakdown are known. Explorations of the cellular mechanisms resulting in protein degradation have suggested that the originally assumed activation of the lysosomal or calcium-dependent pathway seems rather unlikely in acute uremia [28, 35] because suppression of the respective proteinases with specific proteinase inhibitors did not normalize muscle catabolism [36]. In contrast, the cytosolic soluble ATP- and ubiquitin-dependent proteolytic system appears to be the prominent pathway by which muscle proteins are degraded [28, 37]. This system (proteasome) digests proteins that are marked by ubiquitin. The activity of this pathway correlates with the degree of metabolic acidosis and ATP depletion blocks the increased muscle protein degradation [37]. Moreover, the activation of the ATP-dependent pathway was associated with increased ubiquitin subunit mRNAs in rat muscle, indicating stimulation of gene transcription. However, accelerated protein degradation and mRNA expression required conditions of metabolic acidosis and the contribution of glucocorticoids [34] which, on the other hand, further enhanced gene activity [38]. To study the influence of insulin on protein metabolism, Reaich et al. [39] have found that insulin had no effect on the acidosis-induced proteolysis in patients with CRF. These results further support the considerable role of the ATP-dependent ubiquitin pathway in muscle catabolism because this system is not suppressed by insulin.

Finally, it was observed that despite a severely low extracellular pH value, the intracellular pH was well maintained [40]. Thus, the cellular changes associated with muscle proteolysis probably do not occur as a direct result of lowered intracellular pH, and other signal transduction pathways must be involved.

Uncomplicated ARF: Is It a Catabolic Event?

In contrast to ARF with the presence of stress states, it is not clear if uncomplicated ARF per se is a catabolic event. Animal studies have shown that protein metabolism is in fact altered in uncomplicated ARF. Salusky et al. [41] demonstrated a significantly increased catabolic activity in bilateral nephrectomized rats compared to sham-nephrectomized rats suggesting that protein wasting was due to acute uremia. In vitro and in vivo studies of Mitch [42] indicated altered protein metabolism in uncomplicated ARF, with an early increase in protein degradation and later depression of protein synthesis. Maroni et al. [43] studied amino acid transports in rat muscle and found abnormal amino acid concentrations in muscle cells of acutely uremic rats as a result of abnormal metabolism and responsiveness to insulin-mediated transport.

However, it has not yet been demonstrated in humans that isolated acute uremia is necessarily associated with a negative nitrogen balance and therefore an inevitable catabolic process. In patients with uncomplicated ARF, digestion by the protein phosphorylase kinase in their plasma ultrafiltrate was not increased compared to controls [44]. Furthermore, the laboratory evidence that ARF is associated with decreased oxygen consumption could not be confirmed in humans. Energy requirements in patients with postoperative and posttraumatic ARF were similar to those of comparable patients without ARF [45].

Therapeutic Strategies

During catabolic illness, optimal nutrition, usually by the intravenous route, is assumed to prevent body protein loss. However, several studies indicate that aggressive nutritional support does not maintain or replenish body protein during severe catabolic events [5, 18]. Therefore, new techniques to modify or reverse the catabolic changes have been evaluated. Explorations by Wilmore [18] emphasize three general approaches: alteration of the stress response, provision of specific fuels, and administration of growth factors.

The stress response is characterized by neurohormonal events which contribute to the net breakdown of skeletal muscle protein. Attenuation of the stress response may therefore reduce protein catabolism. In particular, inflammation sets off a complex network of responses associated with the release of a number of mediators. Several pathways can be blocked or inhibited simultaneously to the advantage of the host. For instance, investigations with human monoclonal antibodies specific for lipid A of endotoxin in patients with gram-negative sepsis [46] demonstrated a significantly reduced mortality by limiting the complex systemic responses associated with gram-negative bacteremia. Strategies to reduce the level or block the binding of specific cytokines may also hold promise for the treatment of ARF in critically ill patients. In animal models, neutralizing antibodies or soluble receptors to TNF reduced mortality and disease severity in gram-negative sepsis [47]. A similar beneficial effect was reported by blocking IL-1 using soluble IL-1 receptors or IL-1 receptor antagonists. Preliminary studies in humans suggest that anticytokine strategies may become useful in severely ill patients suffering from sepsis [48].

Glucocorticoids can theoretically interrupt cytokine gene transcription. Beneficial effects have been shown in animals when corticosteroids have been administered before or as early as possible after the insult with endotoxin or live bacteria. In humans with septic shock, however, the use of high-dose corticosteroids is controversial because of related secondary infections. Bone et al. [49] did not find a benefit of glucocorticoids in patients with sepsis, suggesting steroid ineffectivity once cytokines have been released. On the other hand, there is evidence that glucocorticoid excess may be a contributing mechanism for enhanced nitrogen appearance and protein catabolism in ARF. The use of the glucocorticoid receptor antagonist RU 486 could normalize enhanced protein catabolism in acutely uremic rats, but concomitant hypoglycemia and hyperkalemia limit its application to experimental studies only.

The suggestion that cytokines could be removed by extracorporeal blood detoxification remains controversial. Bellomo et al. [50] reported removal of some TNF-α and IL-1β from the circulation of septic patients with ARF, as well as extraction of IL-6 and IL-8 in a subsequent study [51]. In contrast, Nagaki et al. [52] and Hoffmann et al. [53] have found no evidence for significant cytokine elimination. However, the latter demonstrated that continuous venovenous hemofiltration was associated with a significant decline in the complement component C3adesArg, emphasizing that hemofiltration may represent a modality for removal of immunomodulatory mediators.

The requirement for specific nutrients during catabolic events may differ from the requirement during health. Although glutamine is considered a nonessential amino acid, the increased demand during catabolic states may outstrip supply from the intracellular amino acid pool of the skeletal muscle and may result in low blood concentrations of glutamine [54]. Glutamine provides several functions in visceral organs and, as mentioned above, serves as an ammonia donor for the excretion of hydrogen ions in the acid-base homeostasis of the kidney [55]. Glutamine deficiency would therefore aggravate metabolic acidosis, which would further enhance protein breakdown, and vice versa. One study has shown that administration of glutamine in the postoperative phase could enhance synthesis of skeletal muscle protein [56]. Furthermore, *L*-arginine deficiency occurs in certain settings in which supplementation with this amino acid may be beneficial [57]. Particularly in severe illness, dietary intervention with *L*-arginine supplementation appears advantageous; this has been related to subsequent increases in GH levels, provided that the hypothalamic-pituitary axis is intact [18].

Since metabolic acidosis is considered to be an important catabolic factor for protein breakdown, the emphasis should be on restoring and maintaining an acid-base homeostasis. Beside bicarbonate dialysis, treatment with sodium bicarbonate is a simple, inexpensive therapy which improves nitrogen balance with considerable beneficial consequences in the course of ARF.

References

1 Bergström J: Why are dialysis patients malnourished? Am J Kidney Dis 1995;26:229–241.
2 Price SR, Mitch WE: Metabolic acidosis and uremic toxicity: Protein and amino acid metabolism. Semin Nephrol 1994;14:232–237.
3 Flügel-Link RM, Salusky IB, Jones MR, Kopple JD: Protein and amino acid metabolism in posterior hemicorpus of acutely uremic rats. Am J Physiol 1983;244:E615–E623.
4 Clark AS, Mitch EW: Muscle protein turnover and glucose uptake in acutely uremic rats: Effects of insulin and the duration of renal insufficiency. J Clin Invest 1983;72:836–845.
5 Sponsel H, Conger JD: Is parenteral nutrition therapy of value in acute renal failure patients? Am J Kidney Dis 1995;25:96–102.
6 Mondon CE, Dolkas CB, Reaven GM: The site of insulin resistance in acute uremia. Diabetes 1978; 27:571–576.
7 DeFronzo RA, Alvestrand A, Smith D, Hendler R: Insulin resistance in uremia. J Clin Invest 1981;67: 563–568.
8 May RC, Clark AS, Goheer A, Mitch WE: Specific defects in insulin-mediated muscle metabolism in acute uremia. Kidney Int 1985;28:490–497.
9 Kahn CR: Insulin resistance, insulin sensitivity and insulin unresponsiveness: A necessary distinction. Metabolism 1978;27(suppl 2):1893–1902.
10 Tischler ME: Is regulation of proteolysis associated with redox-state changes in rat skeletal muscle? Biochem J 1980;192:963–966.
11 Hedden MP, Buse MG: Effects of glucose, pyruvate, lactate, and amino acids on muscle protein synthesis. Am J Physiol 1982;242:E184– E192.
12 Ding H, Kopple JD, Cohen A, Hirschberg R: Recombinant human insulin-like growth factor-1 accelerates recovery and reduces catabolism in rats with ischemic acute renal failure. J Clin Invest 1993; 91:2281–2287.
13 Tsao W, Wang J, Fervenza FC, Vu TH, Jin IH, Hoffman AR, Rabkin R: Renal growth hormone-insulin-like growth factor-I system in acute renal failure. Kidney Int 1995;47:1658–1668.
14 Friedlaender M, Popovtzer MM, Weiss O, Nefesh I, Kopolovic J, Raz I: Insulin-like growth factor-1 (IGF-1) enhances recovery from $HgCl_2$-induced acute renal failure: The effects on renal IGF-1, IGF-1 receptor, and IGF-binding protein-1 mRNA. J Am Soc Nephrol 1995;5:1782–1791.
15 Rogers SA, Miller SB, Hammerman MR: Growth hormone stimulates IGF 1 gene expression in isolated rat renal collecting duct. Am J Physiol 1990; 259:F474–F479.
16 Zumkeller W, Schofield PN: The role of insulin-like growth factors and IGF-binding proteins in the physiological and pathological processes of the kidney. Virchows Arch B Cell Pathol 1992;62:207–220.
17 Tomas FM, Knowles SE, Owens PC, Read LC, Chandler CS, Gargosky SC, Ballard FJ: Effects of full-length and truncated insulin-like growth factor-I on nitrogen balance and muscle protein metabolism in nitrogen-restricted rats. J Endocrinol 1991; 128:97–105.
18 Wilmore DW: Catabolic illness: Strategies for enhancing recovery. N Engl J Med 1991;325:695–702.
19 Schaefer RM, Schaefer L, Hörl WH: Mechanisms for protein catabolism in acute renal failure. Nephrol Dial Transplant 1994;9(suppl 3):44–47.
20 Kopple JD: The rationale for the use of growth hormone or insulin-like growth factor-1 in adult patients with renal failure. Miner Electrolyte Metab 1992;18:269–275.
21 Mehls O, Tonshoff B, Blum WF, Heinrich U, Seidel C: Growth hormone and insulin-like growth factor 1 in chronic renal failure: Pathophysiology and rationale for growth hormone treatment. Acta Paediatr Scand 1990;370(suppl):28–34.
22 Martin AA, Tomas FM, Ballard FJ, Read LC: Insulin-like growth factor-1 and its variant, des-(1-3)-IGF-1, improve nitrogen balance and food utilization in rats with renal failure. Miner Electrolyte Metab 1992;18:264–268.
23 Bondy P, Engel F, Farrar B: The metabolism of amino acids and protein in the adrenalectomized nephrectomized rat. Endocrinology 1949;44:476–483.
24 Schaefer RM, Weipert J, Moser M, Peter G, Heidbreder E, Hörl WH, Heidland A: Reduction of urea generation and muscle protein degradation by adrenalectomy in acutely uremic rats. Nephron 1988;48:149–153.
25 Wenze H: Changes of plasma renin substrate in physiological and pathophysiological states; in Krause DK, Hummerich W, Poulsen K (eds): Radioimmunoassay Renin-Angiotensin. Stuttgart, Thieme, 1978, pp 91–103.
26 Schaefer RM, Teschner M, Kulzer P, Leibold J, Peter G, Heidland A: Evidence of reduced catabolism by the antiglucocorticoid RU 38486 in acutely uremic rats. Am J Nephrol 1987;7:127–131.
27 Baracos V, Rodemann HP, Dinarello CA, Goldberg AL: Stimulation of muscle protein degradation and prostaglandin E2 release by leukocytic pyrogen (interleukin-1): A mechanism for the increased degradation of muscle protein during fever. N Engl J Med 1983;308:553–558.
28 Bailey JL, Mitch WE: Metabolic acidosis as a uremic toxin. Semin Nephrol 1996;16:160–166.
29 Ballmer PE, McNurlan MA, Hulter HN, Anderson SE, Garlick PJ, Krapf R: Chronic metabolic acidosis decreases albumin synthesis and induces negative nitrogen balance in humans. J Clin Invest 1995; 95:39–45.

30 May RC, Hara Y, Kelly RA, Block KP, Buse MG, Mitch WE: Branched-chain amino acid metabolism in rat muscle: Abnormal regulation in acidosis. Am J Physiol 1987;252:E712–E718.

31 Garibotto G, Russo R, Sofia A, Sala MR, Robaudo C, Moscatelli P, Deferrari G, Tizianello A: Skeletal muscle protein synthesis and degradation in patients with chronic renal failure. Kidney Int 1994; 45:1432–1439.

32 England BK, Chastain JL, Mitch WE: Abnormalities in protein synthesis and degradation induced by extracellular pH in BC3H1 myocytes. Am J Physiol 1991;260:C277–C282.

33 Isozaki U, Mitch WE, England BK, Price FR: Protein degradation and increased mRNAs encoding proteins of the ubiquitin-proteasome proteolytic pathway in BC3H-1 myocytes require an interaction between glucocorticoids and acidification. Proc Natl Acad Sci USA 1996;93:1967–1971.

34 May RC, Bailey JL, Mitch WE, Masud T, England BK: Glucocorticoids and acidosis stimulate protein and amino acid catabolism in vivo. Kidney Int 1996; 49:679–683.

35 Lowell BB, Ruderman NB, Goodman MN: Evidence that lysosomes are not involved in the degradation of myofibrillar proteins in rat skeletal muscle. Biochem J 1986;234:237–240.

36 May RC, Kelly RA, Mitch WE: Metabolic acidosis stimulates protein degradation in rat muscle by a glucocorticoid-dependent mechanism. J Clin Invest 1986;77:614–621.

37 Mitch WE, Medina R, Greiber S, May RC, England BK, Price SR, Baily JL, Goldberg AL: Metabolic acidosis stimulates protein degradation by activating the adenosine triphosphate-dependent pathway involving ubiquitin and proteasomes. J Clin Invest 1994;93:2127–2133.

38 Price SR, Endland BK, Bailey JL, Van Vreede K, Mitch WE: Acidosis and glucocorticoids concomitantly increase ubiquitin and proteasome subunit mRNAs in rat muscle. Am J Physiol 1994;267: C955–C960.

39 Reaich D, Graham KA, Channon SM, Hetherington C, Scrimgeour CM, Wilkinson R, Goodship TH: Insulin-mediated changes in protein degradation and glucose uptake after correction of acidosis in humans with CRF. Am J Physiol 1995;268:E121–E126.

40 Bailey JL, England BK, Long RC Jr, Weissman J, Mitch WE: Experimental acidemia and muscle cell pH in chronic acidosis and renal failure. Am J Physiol 1995;269:C706–C712.

41 Salusky I, Flugel-Link R, Jones M, Kopple J: Effect of acute uremia on protein degradation and amino acid release in the rat hemicorpus. Kidney Int 1983; 24(suppl 16):S43–S47.

42 Mitch WE: Amino acid release from the hindquarter and urea appearance in acute uremia. Am J Physiol 1981;241:E415–E419.

43 Maroni BJ, Haesemeyer RW, Kutner MH, Mitch WE: Kinetics of system A amino acid uptake by muscle: Effects of insulin and acute uremia. Am J Physiol 1990;258:F1304–F1310.

44 Hörl W, Stepinski J, Schaefer RM, Wanner C, Heidland A: Role of proteases in hypercatabolic patients with renal failure. Kidney Int 1983;24(suppl 16):S-37–S-42.

45 Wesson DE, Mitch WE, Wilmore D: Nutritional considerations in the treatment of acute renal failure; in Brenner B, Lazarus JM (eds): Acute Renal Failure. Philadelphia, Saunders, 1983, pp 618–641.

46 Ziegler EJ, Fisher CJ Jr, Sprung CL, Straube RC, Sadoff JC, Foulke GE, Wortel CH, Fink MP, Dellinger RP, Teng NNH, Allen IE, Berger HJ, Knatterud GL, LoBuglio AF, Smith GR, and the HA-1A Sepsis Study Group: Treatment of gram-negative bacteremia and septic shock with HA-1A human monoclonal antibody against endotoxin: A randomized, double-blind, placebo-controlled trial. N Engl J Med 1991;324:429–436.

47 Tracey KJ, Fong Y, Hesse DG, Manogue KR, Lee AT, Kuo GC, Lowry SF, Cerami A: Anti-cachectin/TNF monoclonal antibodies prevent septic shock during lethal bacteraemia. Nature 1987;330:662–664.

48 Dinarello CA, Gelfand JA, Wolff SM: Anticytokine strategies in the treatment of the systemic inflammatory response syndrome. J Am Med Assoc 1993; 269:1829–1835.

49 Bone RC, Fisher CJ Jr, Clemmer TP, Slotman GJ, Metz CA, Balk RA: A controlled clinical trial of high-dose methylprednisolone in the treatment of severe sepsis and septic shock. N Engl J Med 1987; 317:653–658.

50 Bellomo R, Tipping P, Boyce N: Continuous veno-venous hemofiltration with dialysis removes cytokines from the circulation of septic patients. Crit Care Med 1993;21:522–526.

51 Bellomo R, Tipping P, Boyce N: Interleukin-6 and interleukin-8 extraction during continuous veno-venous hemodiafiltration in septic acute renal failure. Renal Failure 1995;17:457–466.

52 Nagaki M, Hughes RD, Keane HM, Lau JY, Williams R: In vitro plasma perfusion through absorbents and plasma ultrafiltration to remove endotoxin and cytokines. Circ Shock 1992;38:182–188.

53 Hoffmann JN, Hartl WH, Deppisch R, Faist E, Jochum M, Inthorn D: Hemofiltration in human sepsis: Evidence for elimination of immunomodulatory substances. Kidney Int 1995;48:1563–1570.

54 Parry-Billings M, Evans J, Calder PC, Newsholme EA: Does glutamine contribute to immunosuppression after major burns? Lancet 1990;336:523–525.

55 Welbourne TC: Interorgan glutamine flow in metabolic acidosis. Am J Physiol 1987;253:F1069–F1076.

56 Hammarqvist F, Wernerman J, Ali R, Decken A von der, Vinnars E: Addition of glutamine to total parenteral nutrition after elective abdominal surgery spares free glutamine in muscle, counteracts the fall in muscle protein synthesis, and improves nitrogen balance. Ann Surg 1989;209:455–461.

57 Reyes AA, Karl IE, Klahr S: Role of arginine in health and in renal disease. Am J Physiol 1994;267: F331–F346.

Mineral and Electrolyte Metabolism

Miner Electrolyte Metab 1997;23:194–197

Xiaoyu Ding
S. Russ Price
J.L. Bailey
William E. Mitch

Renal Division,
Emory University School of Medicine,
Atlanta, Ga., USA

Cellular Mechanisms Controlling Protein Degradation in Catabolic States

Key Words
Protein degradation
Ubiquitin
Proteasome
Protein catabolism
Uremia

Abstract
The daily turnover of protein amounts to 280 g in an adult weighing 70 kg but the metabolic processes responsible for protein turnover are only just beginning to be understood. In cells,the major pathway of protein degradation is the ubiquitin-proteasome pathway and protein flux through this pathway is precisely regulated. In catabolic conditions such as uremia, activity of the ubiquitin-proteasome pathway increases, resulting in degradation of muscle protein. In addition to increased protein degradation, gene transcription is activated, resulting in higher levels of the mRNAs encoding ubiquitin and proteasome subunits. The signals activating this pathway include metabolic acidosis and glucocorticoids but must be more diverse since the pathway is also activated in response to starvation, sepsis, cancer, muscle denervation, thermal injury, and acute diabetes. Understanding how the pathway is controlled could lead to the prevention of muscle loss in uremia and other conditions.

Introduction

Individual proteins in the cytosol, endoplasmic reticulum, mitochondria and nucleus of mammalian cells are degraded at widely differing rates. The average rates of protein turnover in cells of different tissues also vary widely, indicating that the proteolytic process must be highly selective and tightly regulated. If this were not true, unregulated destruction of an essential protein or failure to degrade a short-lived regulatory protein would drastically alter cell function.

The amount of cellular protein turned over each day is quite large, about tenfold more than the turnover of plasma proteins (in a normal 70-kg adult, approximately 290 g protein are synthesized and degraded each day [1]). The overall rates of protein synthesis and degradation in each cell are precisely balanced, since even a small decrease in protein synthesis or a small stimulation of degradation, if sustained (e.g., in skeletal muscle), could cause a marked loss of body mass as occurs in AIDS or cancer cachexia.

Why Cells Degrade Proteins

Continued destruction of cell proteins serves important homeostatic functions. (1) It rapidly removes rate-limiting enzymes and regulatory proteins which control the growth and metabolism of cells. (2) Protein degradation also serves as a 'quality-control' system to eliminate proteins with abnormal conformations that arise by mutations, synthetic errors, or damage by oxygen radicals. For example, the mutant chloride channel transmembrane regulator (CFTR) protein in cystic fibrosis patients is degraded in the endoplasmic reticulum and never reaches the cell surface [2]. (3) The degradation of cell proteins is critical for the immune system in its survey for cancer cells or virus-infected cells [3]. (4) The breakdown of cell proteins can provide amino acids as substrates in the production of energy, for gluconeogenesis, or protein synthesis. This also allows the cell to adapt to physiological conditions. For example, short-lived gluconeogenic enzymes accumulate in the liver during fasting, but quickly disappear with refeeding when there is an increase in enzymes that store glucose as glycogen. In mammals, the breakdown of muscle protein is the source of substrates in fasting or disease states because skeletal muscle is the major protein reserve.

Proteolytic Pathways

Mammalian cells contain multiple proteolytic systems, and each appears to serve distinct functions. The lysosomal proteases (e.g., the cathepsins B, D, H, and L) function maximally at the low pH of lysosomes to degrade endocytosed proteins, including phagocytosed bacteria or pinocytosed plasma proteins (e.g., lipoproteins) and membrane receptors. Although a low supply of insulin or amino acids will stimulate the engulfment of intracellular proteins into autophagic vacuoles that fuse with lysosomes where cell proteins are degraded, lysosomal proteolysis is not of major importance in

KARGER
E-Mail karger@karger.ch
Fax +41 61 306 12 34
http://www.karger.ch

0378–0392/97/0236–0194 $12.00/0

William E. Mitch, MD
Renal Division, Emory University
1364 Clifton Road, Atlanta, GA 30322 (USA)
Tel. +1 404 727 2525

the normal turnover of protein or in the accelerated breakdown of muscle proteins in catabolic states [4]. Besides lysosomes, cells contain potent calcium-dependent proteases (calpains) that are activated when cytosolic calcium rises, as with cell injury. The function of calpains in the turnover of protein under normal conditions is unclear. In addition, there is a nonlysosomal system that degrades some proteins (e.g., oxidant-damaged hemoglobins) by a process that does not require energy; its biochemical mechanism and physiological importance are unknown.

The bulk of proteins in mammalian cells is degraded by a pathway that requires ATP [1]. Extensive work, mainly by Hershko and Ciechanover [5, 6], demonstrated that most protein substrates degraded in this pathway are first marked by covalent conjugation to the small protein, ubiquitin. This process is repeated to create a chain of at least five ubiquitin molecules attached to the substrate. The conjugated protein can then be recognized and degraded by a large complex, termed the 26S proteasome, whose function also requires ATP hydrolysis. Studies using pharmacological inhibitors of proteasome function have established that it is the primary site for degradation of most cell proteins, many membrane proteins (e.g., CFTR), and the source of most antigenic peptides presented to the immune system on MHC class 1 molecules [1, 3].

How does the ubiquitin-proteasome pathway achieve remarkable specificity when it is capable of degrading most cell proteins? First, selectivity resides in the process of conjugating ubiquitin to the substrate protein and it is likely that this is the rate-limiting step in the pathway [5, 6]. Secondly, a protein must be recognized, unfolded and then transported into the core of the proteasome before degradation occurs. These processes depend on 19S complexes located on either end of the proteasome so the 19S particles convey selectivity. They contain ATPases and a subunit that binds ubiquitin chains plus a 'ubiquitin isopeptidase' which can disassemble the ubiquitin chain so that ubiquitin can be reused in the degradation of other proteins.

Recent reports concerning the structure of the proteasome particle [1] indicate that its core is formed by two α and two β rings. Each ring is composed of seven subunits and each of the seven subunits of the β rings has an active site that attacks peptide bonds. These active sites are novel because they consist of a threonine at the amino terminus, and this hydroxyl initiates a nucleophilic attack to break the peptide bond.

Activation and Suppression of Muscle Protein Degradation

The mechanisms controlling protein degradation in muscle have been identified primarily from studies of isolated muscles from normal rats or rats responding to pathophysiologic conditions. For example, the loss of muscle in catabolic states largely reflects accelerated breakdown of the long-lived myofibrillar proteins (e.g., actin and myosin), the bulk of protein in muscle. In response to fasting, acidosis, infection, or certain tumors, protein is lost preferentially from skeletal muscle, but not all muscles are affected to the same degree. Muscles composed of pale fibers lose more protein than the continually active dark muscles (the antigravity muscles or the heart) [7]. Interestingly, the accelerated breakdown of proteins and amino acids in muscle in catabolic states can change to a different series of adaptations that conserve cell protein and essential amino acids, as occurs when there is an inadequate intake of amino acids but adequate calories [7]. Our knowledge about the mechanisms signaling conservation of body protein is limited, and it remains unclear why muscle protein degradation and amino acid oxidation are activated in diseases (e.g., in acidotic patients), even if the diet contains an inadequate supply of amino acids which should suppress proteolysis. If we understood the mechanism(s) that suppress muscle protein turnover, it might be possible to activate them and suppress the excessive proteolysis occurring in catabolic states.

The first step in understanding excessive catabolism in disease states is to identify the pathway that is activated to degrade protein. The increase in muscle proteolysis found in rats responding to acidosis, diabetes, fasting, sepsis, or cancer was unaffected when the atrophying muscles were incubated in vitro with agents that blocked the activity of lysosomes and calcium-activated proteases. However, muscle protein degradation decreased to the level measured in control muscles when ATP production was blocked, suggesting that the ubiquitin-proteasome pathway was activated [1, 7, 8]. More definitive evidence for a critical role of the ubiquitin-proteasome pathway was the finding that levels of ubiquitin-protein conjugates were increased in muscles (primarily in the myofibrillar fraction) and the observation that the levels of mRNAs encoding ubiquitin and several subunits of the proteasome were increased [9]. Finally, it was shown that addition of inhibitors of the proteasome will block the excessive muscle proteolysis associated with uremia and diabetes [10, 11]. Together, these findings strongly suggest a coordinated series of biochemical adaptations that enhance the capacity of the ubiquitin-proteasome pathway and lead to muscle wasting.

In certain conditions associated with activation of the ubiquitin-proteasome pathway, there is increased expression of enzymes that are linked metabolically to muscle protein breakdown. For example, increased proteolysis in fasting or acidosis is associated with accelerated oxidation of the branched-chain amino acids in muscle. At least in acidosis and acute uremia, this response is accompanied by an increase in mRNAs encoding subunits of the rate-limiting enzyme in this process, branched-chain ketoacid dehydrogenase [12, 13]. Hopefully, future studies will identify the 'atrophy-related genes' that are induced in muscle and the biochemical mechanisms leading to their coordinated expression in catabolic states.

Signals Regulating Muscle Proteolysis

What factors signal the enhanced muscle protein degradation in these diverse physiological and pathological conditions (table 1)? One important factor is glucocorticoids. For example, the rise in muscle protein degradation in rats fed NH_4Cl to induce metabolic acidosis requires glucocorticoids; if rats are adrenalectomized and then fed NH_4Cl, neither muscle proteolysis nor the mRNAs of

Table 1. Activation of the ubiquitin-proteasome pathway

Signal	Model
Glucocorticoids	acidosis, starvation (ADX)
Cytokines	cancer
TNF	(TNF antibody)
IL-6	IL-6 transgenic mice
Insulinopenia	streptozotocin
Reduced β-agonist	cancer (clenbuterol)
Angiotensin II	angiotensin II infusion

ADX = Adrenalectomy; TNF = tumor necrosis factor; IL = interleukin.

the ubiquitin-proteasome pathway increase unless they are given replacement doses of dexamethasone [14–16]. Likewise, in response to fasting, muscles of adrenalectomized rats do not exhibit increased protein breakdown nor is there an increase in the levels of ubiquitin mRNA or of ubiquitin-protein conjugates; all of these responses are restored with glucocorticoids [7].

Since adrenalectomy may influence muscle only indirectly (e.g., through changes in other hormonal mediators), it is noteworthy that glucocorticoids can directly alter protein turnover in cultured BC3H-1 myocytes. Protein degradation was shown to increase when the pH of the culture medium was reduced, but only if dexamethasone was also added; and this proteolytic response was accompanied by increased levels of mRNAs encoding ubiquitin and subunits of the proteasome [17]. Additional evidence for a key role of glucocorticoids was that RU 486 (a receptor antagonist which prevents activation of gene transcription by glucocorticoids and progesterone) blocked both the increase in proteolysis and mRNA levels in BC3H-1 myocytes exposed to acidification plus glucocorticoids. In short, the cultured cells mimicked the responses to acidosis seen in muscles of rats, but in this case the influence of other hormones was eliminated.

Besides glucocorticoids, other signals are involved in controlling muscle protein breakdown. For example, insulin suppresses protein breakdown in the fed state, indicating that activation of muscle proteolysis in fasting (and probably in diabetes) seems to require two signals, glucocorticoids and a fall in insulin levels. We found that there is striking activation of the ubiquitin-proteasome pathway in muscle of insulinopenic rats [11]; there is also increased glucocorticoid production.

In sepsis and certain cancers, activation of the ubiquitin-proteasome pathway in muscle appears to be signaled by cytokines, and a prime candidate is tumor necrosis factor (TNF). Certainly, sepsis, certain types of cancer, burn injury and, apparently, uremia are all associated with release of large amounts of TNF and other monokines (and glucocorticoids) [1]. Experimentally, injection of large amounts of TNF can activate muscle proteolysis [18]. Pentoxifylline (an inhibitor of TNF release from macrophages) can suppress the muscle proteolytic response to sepsis in rats [19] and injections of a TNF antibody can suppress the excessive proteolysis occurring in muscles of tumor-bearing or septic rats [20, 21].

Thyroid hormone also seems to be important in regulating muscle protein breakdown, as administration of triiodothyronine (T_3) or thyroxin stimulates protein catabolism in muscle, probably by stimulating the ATP-dependent proteolytic pathway. It is likely that the decrease in muscle protein degradation occurring when rats are fed a very low protein diet (but adequate calories) is due to suppression of thyroid function, since such diets reduce T_3 levels.

Implications for Therapy of Protein-Wasting Conditions

Advances in our understanding of accelerated protein breakdown in catabolic states suggest therapeutic strategies. First, catabolic states are largely due to stimulation of protein degradation rather than suppression of protein synthesis. Consequently, diets or infusions with excessive protein or amino acids are unlikely to reverse catabolism. A potential approach that merits study would be to inhibit partially the activity of the ubiquitin-proteasome pathway; complete inhibition would probably by dangerous. Because the proteasome functions by an unusual proteolytic mechanism, selective inhibitors have been developed which can block the increase in protein degradation that occurs in the muscle of rats with insulinopenia or acute or chronic renal failure [10, 11, 13]. In addition, new inhibitors of the proteasome have been given for extended periods to animals with inflammatory diseases and, in preliminary studies, exhibit potent anti-inflammatory effects in several disease models. Another attractive therapeutic strategy would be to inhibit the production or actions of the various catabolic factors that signal the enhanced proteolysis. It may also be possible to identify agents that induce the cellular program that conserves body proteins. Clearly, continued progress in understanding the biochemistry of protein breakdown and its regulation in vivo should provide further insights into these diseases and also new therapeutic approaches.

References

1 Mitch WE, Goldberg AL: Mechanisms of muscle wasting: The role of the ubiquitin-proteasome system. N Engl J Med 1996;335:1897–1905.

2 Ward CL, Omura S, Kopito RR: Degradation of CFTR by the ubiquitin-proteasome pathway. Cell 1995;83:121–127.

3 Rock KL, Gramm C, Rothstein L, Clark K, Stein R, Dick L, Hwang D, Goldberg AL: Inhibitors of the proteasome block the degradation of most cell proteins and the generation of peptides presented on MHC class 1 molecules. Cell 1994;78:761–771.

4 Lowell BB, Ruderman NB, Goodman MN: Evidence that lysosomes are not involved in the degradation of myofibrillar proteins in rat skeletal muscle. Biochem J 1986;234:237–240.

5 Hershko A, Ciechanover A: The ubiquitin system for protein degradation. Annu Rev Biochem 1992; 61:761–807.

6 Ciechanover A: The ubiquitin-proteasome proteolytic pathway. Cell 1994;79:13–21.

7 Wing SS, Goldberg AL: Glucocorticoids activate the ATP-ubiquitin-dependent proteolytic system in skeletal muscle during fasting. Am J Physiol 1993; 264:E668–E676.

8 Mitch WE, Medina R, Greiber S, May RC, England BK, Price SR, Bailey JL, Goldberg AL: Metabolic acidosis stimulates muscle protein degradation by activating the ATP-dependent pathway involving ubiquitin and protea- somes. J Clin Invest 1994;93: 2127–2133.

9 Medina R, Wing SS, Goldberg AL: Increase in levels of polyubiquitin and proteasome mRNA in skeletal muscle during starvation and denervation atrophy. Biochem J 1995;307:631–637.

10 Bailey JL, Wang X, England BK, Price SR, Ding X, Mitch WE: The acidosis of chronic renal failure activates muscle proteolysis in rats by augmenting transcription of genes encoding proteins of the ATP-dependent, ubiquitin-proteasome pathway. J Clin Invest 1996;97:1447–1453.

11 Price SR, Bailey JL, Wang X, Jurkovitz C, England BK, Ding X, Phillips LS, Mitch WE: Muscle wasting in insulinopenic rats results from activation of the ATP-dependent, ubiquitin-proteasome pathway by a mechanism including gene transcription. J Clin Invest 1996;98:1703–1708.

12 England BK, Greiber S, Mitch WE, Bowers BA, Herring WJ, McKean MJ, Ebb RG, Price SR, Danner DJ: Rat muscle branched-chain ketoacid dehydrogenase activity and mRNAs increase with extracellular acidemia. Am J Physiol 1995;268: C1395–C1400.

13 Reaich D, Grewal M, Price SR, England BK, Bailey JL, Mitch WE, Maroni BJ: Catabolism in acute renal failure (ARF): Evidence for stimulation of the ubiquitin-proteasome system and activation of branched-chain a-ketoacid dehydrogenase (BCKAD). J Am Soc Nephrol 1994;5:953.

14 May RC, Kelly RA, Mitch WE: Metabolic acidosis stimulates protein degradation in rat muscle by a glucocorticoid-dependent mechanism. J Clin Invest 1986;77:614–621.

15 Price SR, England BK, Bailey JL, Van Vreede K, Mitch WE: Acidosis and glucocorticoids concomitantly increase ubiquitin and proteasome subunit mRNAs in rat muscle. Am J Physiol 1994;267: C955–C960.

16 May RC, Bailey JL, Mitch WE, Masud T, England BK: Glucocorticoids and acidosis stimulate protein and amino acid catabolism in vivo. Kidney Int 1996; 49:679–683.

17 Isozaki Y, Mitch WE, England BK, Price SR: Interaction between glucocorticoids and acidification results in stimulation of proteolysis and mRNAs of proteins encoding the ubiquitin-proteasome pathway in BC3H-1 myocytes. Proc Natl Acad Sci USA 1996;93:1967–1971.

18 Goodman MN: Tumor necrosis factor induces skeletal muscle protein breakdown in rats. Am J Physiol 1991;260:E727–E730.

19 Breuille D, Farge MC, Rose F, Arnal M, Attaix D, Obled C: Pentoxifylline decreases body weight loss and muscle protein wasting characteristics of sepsis. Am J Physiol 1993;265:E660–E666.

20 Costelli P, Carbo N, Tessitore L, Bagby GJ, Lopez-Soriano JL, Argiles JM, Baccino FM: Tumor necrosis factor-α mediates changes in tissue protein turnover in a rat cancer cachexia model. J Clin Invest 1993;92:2783–2789.

21 Zamir O, Hasselgren P, Kunkel SL, Frederick J, Higashiguchi T, Fischer JE: Evidence that tumor necrosis factor participates in the regulation of muscle proteolysis during sepsis. Arch Surg 1992; 127:170–174.

Mineral and Electrolyte Metabolism

Miner Electrolyte Metab 1997;23:198–200

J.L. Bailey
S.R. Price
B.K. England
C. Jurkovitz
X. Wang
X. Ding
W.E. Mitch

Renal Division,
Emory University School of Medicine,
Atlanta, Ga., USA

Signals Regulating Accelerated Muscle Protein Catabolism in Uremia

Key Words
Chronic renal failure
Ubiquitin
Diabetes mellitus
Insulin
Protein degradation

Abstract
In chronic renal failure (CRF), the ATP-dependent, ubiquitin-proteasome proteolytic pathway is activated with concurrent increases in the transcription of genes encoding proteins of this pathway in muscle. We have shown that the stimuli for these responses include acidosis and glucocorticoids, but other endocrine abnormalities in CRF (e.g., insulin resistance) could contribute to these responses. In fact, a major effect of insulin in muscle is to suppress protein degradation. To examine whether insulin influences the ubiquitin-proteasome pathway, we measured protein degradation in incubated epitrochlearis muscles of diabetic and pair-fed control rats. Muscle proteolysis was increased in pathways that do not involve lysosomes or Ca^{2+}-dependent proteases; but MG132, a protease inhibitor that blocks ATP synthesis, eliminated the accelerated rate of protein degradation in diabetic rat muscles. Diabetes mellitus also increased levels of mRNAs encoding ubiquitin (334%), E2 ubiquitin-conjugating enzyme (247%), and the C3 (320%), C5 (349%), and C9 (216%) proteasome subunits in muscle. Finally, transcription of the ubiquitin gene in diabetic rat muscles was increased. Diabetic rats were acidotic, but eliminating acidemia by giving $NaHCO_3$ did not block the increase in muscle proteolysis. Giving diabetic rats insulin prevented the excessive muscle proteolysis, suggesting that insulin acts as a suppressor of the ubiquitin-proteasome pathway. Thus, the insulin resistance of uremia could contribute to muscle protein wasting in CRF.

Introduction

Anorexia is common in renal failure. Various factors contribute to increased catabolism. Among them, metabolic acidosis [1] plays a quintessential role mediated in part through a glucocorticoid mechanism. It is still not clear if other hormonal factors also alter catabolism in chronic uremia. For instance, insulin resistance is reported in uremia [2]. Work by various authors [3, 4] suggests that insulin would have a favorable effect by promoting anabolism and a positive nitrogen balance. Because there are so many other confounding variables associated with uremia, it is difficult, if not impossible, to isolate one factor associated with the uremic syndrome and to determine its contribution to excessive muscle protein wasting. Other animal models can be of some utility in this regard.

When rats are made insulinopenic, they lose weight despite the fact that they are eating the same if not more than their pair-fed controls. Significant differences in weight begin to appear in as little as 3 days after the rats become insulinopenic. By 7 days, there is profound muscle wasting in all muscle groups. Cellular pathways mediating these catabolic changes remain unidentified.

There are four major cellular pathways by which proteins can be degraded. These include the lysosomal, calcium-dependent, ATP-dependent, and the non-ATP-dependent pathways. We have been interested for some time in the ATP-dependent pathway [5]. Here, proteins are destined for catabolism after they have been bound to ubiquitin in a process requiring ATP. Several enzymes are necessary for this process. The first, designated E1, binds the ubiquitin to ATP. A second, the E2 ubiquitin-conjugating enzyme, fuses the activated ubiquitin to the protein destined for catabolism. Once bound to ubiquitin, the proteins are transported to the proteasome and undergo catabolism. The proteolytic activity of the proteasome degrades the protein to small peptides which are further catabolized to amino acids by cytosolic peptidases.

Using a model of diabetes in which rats are made insulinopenic by injection with streptozotocin, we studied the role that insulin plays in mediating protein catabolism in muscle.

KARGER
E-Mail karger@karger.ch
Fax +41 61 306 12 34
http://www.karger.ch

0378–0392/97/0236–0198 $12.00/0

James L. Bailey, MD
Renal Division, Emory University School of Medicine
Box 19, 1364 Clifton Road, NE
Atlanta, GA 30322 (USA)
Tel. +1 404 727 2525, fax +1 404 727 3425

Methods

Rat Model

Male Sprague Dawley rats (ca. 175 g) were made insulinopenic by injecting their tail veins with streptozotocin. The rats were subsequently pair-fed with vehicle-injected controls for periods of 3 or 7 days. Urine was sampled for glucose and ketones. All animals were fasted from the evening before the experiment.

Muscle Protein Degradation Measurements

Epitrochlearis muscles obtained from the forelimbs of insulinopenic and the pair-fed control rats were incubated in KRB medium containing glucose and cycloheximide to determine rates of protein degradation. Tyrosine released into the incubation medium was measured to calculate the rate of protein degradation. Following measurement of total protein degradation, the contribution of individual proteolytic pathways was determined by incubating one epitrochlearis muscle with one or more inhibitors of the different proteolytic pathways while the contralateral muscle was incubated without these inhibitors. Methylamine and E-64 were used to block lysosomal function while E-64 and a calcium-free medium were used to inhibit calcium-dependent proteases. The ATP-dependent pathway was blocked by adding dinitrophenol and substituting 2-deoxyglucose for glucose in the medium. The proteasome was inhibited by adding MG132.

RNA Isolation

RNA was isolated from the mixed-fiber, gastrocnemius muscle and hybridized with cDNA probes for ubiquitin, the $E2_{14K}$ ubiquitin-conjugating enzyme and proteasome subunits C3, C5, and C9. Autoradiographic bands were quantified using densitometry.

Nuclear Run-On Assays

Nuclei were isolated from rat hind limb muscles and then incubated with radiolabelled UTP as well as ATP, CTP, and GTP. The newly transcribed RNA was hybridized with cDNA probes for ubiquitin and γ-actin.

Results

The basal rate of protein degradation was higher in muscles of insulinopenic rats than in the pair-fed control rats and blocking lysosomal and calcium-dependent proteases did not eliminate this difference. Only when the ATP-dependent proteolytic pathway was blocked was this difference in protein degradation eliminated. To confirm the specificity of this pathway, we inhibited the proteasome with a specific peptide aldehyde inhibitor, MG132, and found no difference in rates of protein degradation in control and insulinopenic rat muscles. In other experiments, we found that maintaining the serum bicarbonate at about 24 m*M* with sodium bicarbonate supplements did not alter rates of protein degradation. When the insulinopenic rats were supplemented with insulin, the rates of protein degradation were the same in muscles of insulinopenic and control rats. Interestingly, the increase in protein degradation was also blocked if the animals were adrenalectomized and allowed to recover prior to injection with streptozotocin.

To provide further evidence that the ATP-dependent ubiquitin pathway is activated in muscles of insulinopenic animals, mRNA levels for ubiquitin, C3, C5, and C9 of the proteasome were measured by Northern blot hybridization. There were significant increases in the abundance of these mRNAs in muscles of insulinopenic rats compared to pair-fed control rats. Specifically, insulinopenia increased the levels of mRNAs encoding ubiquitin (334%), the E2 ubiquitin-conjugating enzyme (247%), and the C3 (320%), C5 (349%), and C9 (216%) proteasome subunits in muscle. Differences in mRNA levels were noted in as little as 3 days and continued to be high 7 days after streptozotocin.

Nuclear run-on assays confirmed that insulinopenia stimulates transcription of the ubiquitin gene, while other genes such as γ-actin did not show the same up-regulation.

Discussion

Our results affirm that insulin plays a significant role in regulating ATP-dependent proteolysis. In contrast to pair-fed control rats which gained weight, insulinopenic rats experienced profound muscle wasting despite an adequate intake of nutrients and calories. This response was not due to the direct effect of streptozotocin on muscle proteolysis since this difference is readily corrected by the addition of insulin. This highlights the role that insulin plays in promoting growth and a positive nitrogen balance. Moreover, it extends the work by Gulve and Dice [6] and England et al. [7] showing that the primary role of insulin in protein balance is to decrease proteolysis.

It is somewhat surprising that inhibition of the lysosomal pathway did not account for differences in proteolysis as it is well known that insulin suppresses lysosomal proteolysis in hepatocytes [8]. Despite the addition of inhibitors of the lysosomal pathway to the muscle incubation milieu, the increase in protein degradation persisted in incubated epitrochlearis muscles taken from insulinopenic rats. Likewise, calcium-dependent proteolysis could not account for these differences. We found that ATP-dependent muscle proteolysis was activated by 3 days and persisted for at least 7 days. Moreover, the specificity of this response involving ubiquitin and components of the proteasome was shown by specifically blocking this pathway with MG132. Thus, the ATP-dependent ubiquitin-proteasome pathway is activated in insulinopenia. Since this pathway is activated in other conditions associated with muscle wasting (e.g., uremia [9], sepsis [10], cancer [11], and burn injury [12]), its importance in disease is highlighted and indicates that this pathway is the major system for degradation of muscle protein.

Because our previous work had shown that metabolic acidosis was a potent activator of the ubiquitin-proteasome catabolic pathway [13], we expected metabolic acidosis to play a similar role. This was not the case. We found that maintaining a neutral acid-base balance by giving insulinopenic rats sodium bicarbonate had little effect on the excessive rate of protein catabolism. Indeed, sodium bicarbonate did not alter muscle weights or the negative growth rates seen in untreated insulinopenic animals. This correlates with previous clinical studies done by Morris [14], who advocated not using sodium bicarbonate for the routine treatment of diabetic ketoacidosis. Moreover, that the increased rates of protein degradation and levels of mRNA and components of the proteasome persisted 7 days after streptozotocin treatment argues against metabolic acidosis being the stimulus for muscle wasting in insulinopenia.

Our previous work with acidotic and uremic rats demonstrated an important role for glucocorticoids in the regulation of protein degradation in muscle. We measured glucocorticoid levels in both control and insulinopenic rats and found them to be high in the latter. Like acidosis, it appears that we can eliminate differences in rates of protein degradation in insulinopenic rats by adrenalectomy (data not shown). However, other endocrine changes could also contribute to protein degradation. It is intriguing that insulin resistance is also associated with uremia and acidosis. How this resistance facilitates the proteolysis present in these conditions is unknown and is ground for further investigation.

References

1 May RC, Kelly RA, Mitch WE: Metabolic acidosis stimulates protein degradation in rat muscle by a glucocorticoid-dependent mechanism. J Clin Invest 1986;77:614–621.

2 May RC, Kelly RA, Mitch WE: Mechanisms for defects in muscle protein catabolism in rats with chronic uremia: The influence of metabolic acidosis. J Clin Invest 1987;79:1099–1103.

3 Nakhooda AF, Wei CN, Marliss EB: Muscle protein catabolism in diabetes: 3-Methyl histidine excretion in the spontaneously diabetic 'BB' rat. Metabolism 1980;29:1272–1277.

4 Marchesini G, Forlani G, Zoli M, Pisi E: Muscle protein breakdown in uncontrolled diabetes as assessed by urinary 3-methyl histidine excretion. Diabetologia 1982;23:456–458.

5 Goldberg AL: Functions of the proteasome: The lysis at the end of the tunnel. Science 1995;268: 522–523.

6 Gulve EA, Dice JF: Regulation of protein synthesis and degradation in L8 myotubes: Effects of serum, insulin, and insulin-like growth factors. Biochem J 1989;260:377–387.

7 England BK, Chastain JL, Mitch WE: Abnormalities in protein synthesis and degradation induced by extracellular pH in BC3H1 myocytes. Am J Physiol 1991;260:C227–C282.

8 Furuno K, Goodman MN, Goldberg AL: Role of different proteolytic systems in the degradation of muscle proteins during denervation atrophy. J Biol Chem 1990;265:8550–8557.

9 Bailey JL, Wang X, England BK, Price SR, Ding X, Mitch WE: The acidosis of chronic renal failure activates muscle proteolysis in rats by augmenting transcription of genes encoding the ATP-dependent, ubiquitin-proteasome pathway. J Clin Invest 1996;97:1447–1453.

10 Tiao G, Fagan JM, Samuels N, James JH, Hudson K, Lieberman M, Fischer JE, Hasselgren PO: Sepsis stimulates nonlysosomal, energy-dependent proteolysis and increases ubiquitin mRNA levels in rat skeletal muscle. J Clin Invest 1994;94:2255–2264.

11 Baracos VE, DeVivo C, Hoyle DHR, Goldberg AL: Activation of the ATP-ubiquitin-proteasome pathway in skeletal muscle of cachectic rats bearing a hepatoma. Am J Physiol 1995;268:E996–E1006.

12 Fanf CH, James HJ, Ogle C, Fischer JE, Hasselgren PO: Burn injury stimulates multiple proteolytic pathways in skeletal muscle, including the ubiquitin-energy-dependent pathway. J Am Coll Surg 1995;180:161–170.

13 Mitch WE, Medina R, Greiber S, May RC, England BK, Price SR, Bailey JL, Goldberg AL: Metabolic acidosis stimulates muscle protein degradation by activating the ATP-dependent, ubiquitin-proteasome pathway. J Clin Invest 1994;93:2127–2133.

14 Morris LR: Bicarbonate therapy in severe diabetic ketoacidosis. Ann Intern Med 1986;105:6.

Mineral and Electrolyte Metabolism

Miner Electrolyte Metab 1997;23:201–205

Effect of Cellular Hydration on Protein Metabolism

Siegfried Waldegger[a]
Gillian L. Busch[a]
Nubia K. Kaba[a]
Günther Zempel[a]
Hong Ling[a]
August Heidland[a]
Dieter Häussinger[b]
Florian Lang[a]

[a] Department of Physiology, University of Tübingen, and
[b] Department of Internal Medicine, University of Düsseldorf, Germany

Key Words
Cell volume
Insulin
Glucagon
TGF-β1
Proteolysis
Protein synthesis

Abstract
In the past few years, the paramount importance of cell volume for the regulation of cell function, including protein metabolism, has been recognized. Among many other effects, cell swelling inhibits proteolysis and stimulates protein synthesis, whereas cell shrinkage stimulates proteolysis and inhibits protein synthesis. Moreover, cell swelling and cell shrinkage influence the expression of a number of genes, including carriers, enzymes, and signaling molecules. Hormones exploit the influence of cell volume on metabolism to exert their effects. Insulin swells hepatocytes by activation of Na^+/H^+ exchange and $Na^+,K^+,2Cl^-$ cotransport, while glucagon shrinks hepatocytes by activation of ion channels. The effects of these hormones on hepatic proteolysis completely depend on their influence on cell volume. The effects of cell volume are mediated in part by alterations of lysosomal pH, which modifies the activity of acidic lysosomal proteases. Transforming growth factor-β1, as other growth factors, activates the Na^+/H^+ exchanger, swells cells, leads to lysosomal alkalinization, inhibits proteolysis and may thus contribute to renal hypertrophy in chronic renal disease. Moreover, a decrease in cell volume correlates with catabolic states in a variety of diseases.

Introduction

The intracellular milieu is not just a dilute solution of enzymes and substrates but is densely packed with macromolecules [1–4]. Interaction of the macromolecules in a crowded medium strongly influences the rate and equilibria of their reactions [5]. The cellular hydration state is expected to modify macromolecular crowding and thus cellular metabolism. Accordingly, cellular homeostasis requires maintenance of the hydration state [6]. Beyond cellular hydration state, cell volume constancy is a prerequisite for the maintenance of cellular structure, such as the cytoskeleton. Excessive cell swelling will eventually disrupt the cell membranes and thus lead to cell death.

A number of mechanisms are tuned to maintain cell volume and thus the cellular hydration state, such as ion transport across the cell membrane and cellular accumulation and disposal of osmolytes, molecules designed to create intracellular osmolarity (see below). Growing evidence indicates that the cellular hydration state resulting from the activity of cell volume regulatory mechanisms is modified in a variety of physiological and pathophysiological conditions.

Determinants of Cellular Hydration

Since most animal cell membranes are highly permeable to water and too fragile to resist hydrostatic pressure gradients, the movement of water depends on osmotic gradients across the cell membrane [7]. Thus, any alteration of intra- and/or extracellular osmolarity leads to alterations in cellular hydration state. Obviously, the osmotic gradient across the cell membrane can be altered by an increase or decrease in extracellular osmolarity. Excessive alterations of extracellular osmolarity are encountered in the kidney medulla, where osmolarity may rapidly change between 300 and 1,200 mOsmol during transition between antidiuretic and diuretic states [8, 9]. Less dramatic alterations of extracellular osmolarity occur during intestinal absorption, which exposes intestinal cells to anisosmotic luminal fluid and may modify portal blood osmolarity and liver cell volume [10]. Other tissues are exposed to altered extracellular osmolarity during a variety of disorders, such as hyponatremic states [11–13], hypernatremic states [13, 14], uncontrolled diabetes mellitus [15] and ethanol poisoning [16].

KARGER
E-Mail karger@karger.ch
Fax +41 61 306 12 34
http://www.karger.ch

0378–0392/97/0236–0201 $12.00/0

Prof. Dr. F. Lang
Physiologisches Institut der Universität Tübingen,
Gmelinstrasse 5, D–72076 Tübingen (Germany)
Tel. +49 7071 292194, fax +49 7071 293073
E-Mail florian.lang@uni-tuebingen.de

Rapid correction of chronically enhanced osmolarity may lead to cell swelling, namely to cerebral edema [15, 17]. Chronic increases in extracellular osmolarity are compensated by cells through accumulation of osmolytes (see below), which may not be rapidly and adequately readjusted [18, 19]. Extracellular osmolarity increases due to accumulation of urea in uremia [20]. Even though urea easily passes cell membranes and thus does not usually cause osmotic gradients across the cell membrane, high extracellular urea concentrations may trigger cell shrinkage by modifying the set point for volume regulatory mechanisms [21–23] .

Cells employ transport systems and osmolytes to adjust cellular osmolarity and thus defend cell volume constancy against alterations in extracellular osmolarity. Most importantly, cell swelling leads to activation of K^+ and anion channels. Among the cloned K^+ channels invoked to serve cell volume regulation are the Kv1.3 (n-type K^+ channel) [24], the Kv1.5 channel [25], and the minK channel [26]. Cloned Cl^- channels invoked in cell volume regulation include the ClC-2 channel [27], BRI-VDAC [28], I_{Cln} [29–31], and the P-glycoprotein (or MDR protein) [32]. However, the role of P-glycoprotein and I_{Cln} in cell volume regulation has been a matter of debate. Besides ion channels, electroneutral KCl cotransport is the most frequently utilized mechanism of volume regulatory ion release [for review see ref. 33].

The major ion transport systems accomplishing electrolyte accumulation in shrunken cells are the Na^+, K^+, $2Cl^-$ cotransport [34–37] and the Na^+/H^+ exchangers NHE-1 [38], NHE-2 [38, 39], and NHE-4 [40]. The latter carriers alkalinize the cell leading to parallel activation of the Cl^-/HCO_3^- exchanger. The H^+ and HCO_3^- exchanged for NaCl by the Na^+/H^+ exchanger and the Cl^-/HCO_3^- exchanger form CO_2 within the cell, which diffuses out of the cell and is thus osmotically not relevant. In some cells, electrolyte accumulation during RVI is accomplished by activation of Na^+ channels and/or unselective cation channels [for review see ref. 33].

The efficacy of volume regulatory mechanisms may be compromised by altered ion gradients across the cell membrane, as during hyperkalemia, energy depletion, and acidosis [for review see ref. 7]. Moreover, as pointed out below, hormones may shift the cell volume regulatory set point of the transporters and thus modify cell volume.

Besides stimulation of ion transport, shrunken cells produce so-called osmolytes, molecules specifically designed to create osmolarity without compromising other cell functions [41–46]. Unlike ions, organic osmolytes are, even at high concentrations, compatible with normal macromolecular function. Three groups of osmolytes are used in mammalian cells: polyalcohols, such as sorbitol and inositol; methylamines, such as glycerophosphorylcholine and betaine; amino acids and amino acid derivatives such as glycine, glutamine, glutamate, aspartate, and taurine [41, 43, 44, 47, 48].

Cellular Hydration and Hepatic Protein Metabolism

As first shown in hepatocytes, the cellular hydration state has profound effects on protein and amino acid metabolism. Hepatocyte swelling stimulates protein synthesis [49] and inhibits protein degradation [50–61]. Furthermore, cell swelling stimulates amino acid uptake [56, 62], glutamine breakdown [56] as well as glycine and alanine oxidation [59, 63]. Cell swelling stimulates ornithine decarboxylase activity and expression [64, 65] and favors urea synthesis from amino acids [66] but inhibits urea synthesis from NH_4^+ [66]. Cell swelling stimulates glutathione efflux [67] and inhibits glutathione-S-conjugate release into bile [68].

Both cell swelling and cell shrinkage modify the expression of a number of genes, as shown in various cell types. Genes expressed preferably by swollen cells include c-*jun* [69], arginine succinate lyase [70], ornithine decarboxylase [64, 65, 71, 72], tissue plasminogen activator [73], β-actin [74], and tubulin [75]. Genes expressed in shrunken cells include those for aldose reductase [76, 77], the Na^+,inositol cotransporter, SMIT [41, 78], the Na^+,betaine cotransporter BGLT1 [79], the Na^+,taurine cotransporter [80], the organic solute transporter ROSIT [81], amino acid transport system A [82–85], Egr-1 [86, 87], c-*fos* [86–88, heat shock proteins [86], cyclooxygenase-2 [89], PEPCK [90], tyrosine amino transferase [70], the serine/threonine kinase hsgk [91], B crystallin [92], matrix proteins [93], while laminin B_2 expression is inhibited [94].

Cellular Hydration as a Second Message in Hormonal Regulation of Hepatic Protein Metabolism

The volume sensitivity of cellular metabolism is exploited by hormones. Insulin swells hepatocytes by activation of both Na^+/H^+ exchange and Na^+, K^+, $2Cl^-$ cotransport [52, 95, 96], and glucagon shrinks hepatocytes, presumably by activation of ion channels [52–54]. The effect of these hormones on cell volume accounts for several of the effects on hepatocyte metabolism, including their effect on proteolysis [53, 54, 59]. Similar to insulin, several growth factors increase cell volume in a variety of cells by stimulation of Na^+/H^+ exchange and in some cases of Na^+, K^+, $2Cl^-$ cotransport [for a review see ref. 97]. As discussed below, an increase in cell volume appears to be required for cell proliferation.

Similar to insulin, several amino acids may cause cell swelling. Amino acids swell hepatocytes by cellular accumulation [50, 98–102]. The antiproteolytic effect of glutamine and glycine in liver has been shown to be completely accounted for by their influence on cell volume [50]. The antiproteolytic and swelling effect of glycine is potentiated following starvation [50], which up-regulates the glycine transporting system A [103]. However, the antiproteolytic action of other amino acids such as phenylalanine, serine, alanine, and proline cannot be fully explained by their effects on cell volume. Thus, mechanisms other than cell swelling contribute to the antiproteolytic action of amino acids. While the inhibitory effect of cell swelling, cAMP, and alcohol dehydrogenase on protein synthesis are well correlated, this is not true for the effects of insulin and phenylephrine [49]. Thus, cell volume may play a less prominent role in hormonal regulation of protein synthesis than in proteolysis.

Lysosomal pH – A Third Message?

As shown for proximal renal tubules [104], liver cells [105–110], vascular smooth muscle cells [111], glial cells [112], neurons [113], MDCK cells, cardiac myocytes, macrophages, fat cells, dentritic cells of the skin, and fibroblasts [114], cell swelling leads to a pH increase in acidic cellular compartments. This effect is expected to decrease the activity of pH-sensitive proteases in a swollen cell. In fact, lysosomal alkalinization has previously been shown to impede hepatic proteolysis [115]. However, it is not known to what extent lysosomal alkalinization contributes to the antiproteolytic action of cell swelling.

Transforming Growth Factor (TGF)-β1, Cell Volume, and Progression of Renal Failure

Enhanced activity of the growth factor TGF-β1 has been observed in experimental glomerulonephritis and streptozotocin-induced diabetic nephropathy [116, 117] and it has been suggested that the antiproteolytic action of TGF-β1 is involved in the progression of renal failure [118, 119]. The antiproteolytic action of TGF-β1 is paralleled by cytosolic alkalinization, an increase in cell volume, and an increase in lysosomal pH [118]. Apparently, TGF-β1 swells cells by stimulating Na^+/H^+ exchange, leading not only to cytosolic alkalinization, but also to cell swelling and lysosomal alkalinization [118]. Accordingly, the influence of TGF-β1 on cytosolic pH, lysosomal pH, and proteolysis is reversed by inhibition of the Na^+/H^+ exchanger [G.L. Busch et al., submitted].

Cell Volume in Catabolic States

In a variety of catabolic states, such as liver tumors, polytrauma, acute pancreatitis, and burns, a decrease in muscle cell volume was found to correlate with hypercatabolism, as demonstrated by the nitrogen balance [120]. Given the above discussion, cell shrinkage could well have accounted for the enhanced proteolysis in these patients. Moreover, the anticatabolic effect of glutamine [121] may well result from its ability to swell cells. At this point, we do not know whether cell shrinkage is restricted to muscle cells or involves other cells as well, nor do we know the mechanisms accounting for cell shrinkage in these patients.

Conclusions

Alterations in cell volume influence a wide variety of cellular functions including protein metabolism. The influence of cell volume accounts for regulation of proteolysis and contributes to regulation of protein synthesis by hormones, such as insulin, glucagon, and growth factors, as well as amino acids, such as glutamine. Additional experimental efforts are needed to elucidate the pathophysiological role of cell volume in disordered protein metabolism.

Acknowledgements

The authors acknowlege the support by grants from the Deutsche Forschungsgemeinschaft (La 315/4–3) and from Baxter Foundation.

References

1 Brown GC: Total cell protein concentration as an evolutionary constraint on the metabolic control distribution in cells. J Theor Biol 1991;153:195–203.
2 Clegg JS: Properties and metabolism of the aqueous cytoplasm and its boundaries. Am J Physiol 1984; 246:R133–R151.
3 Fulton AB: How crowded is the cytoplasm? Cell 1982;30:345–347.
4 Minton AP, Colclasure GC, Parker JC: Model for the role of macromolecular crowding in regulation of cellular volume. Proc Natl Acad Sci USA 1992; 89:10504–10506.
5 Zimmerman SB, Minton AP: Macromolecular crowding: Biochemical, biophysical, and physiological consequences. Annu Rev Biophys Biomol Struct 1993;22:27–65.
6 Parker JC: In defense of cell volume? Am J Physiol 1993;265:C1191–C1200.
7 Lang F, Busch GL, Ritter M, Völkl H, Waldegger S, Gulbins E, Häussinger D: The functional significance of cell volume regulatory mechanisms. Physiol Rev, in press.
8 Beck FX, Dörge A, Thurau K: Cellular osmoregulation in renal medulla. Renal Physiol Biochem 1988;11:174–186.
9 Roy DR, Layton HE, Jamison RL: Countercurrent mechanism and its regulation, in Seldin DW, Giebisch G (eds): The Kidney: Physiology and Pathophysiology, ed 2. New York, Raven, 1992, pp 1649–1692.
10 Haberich FJ, Aziz O, Nowacki PE: Über einen osmoreceptorisch tätigen Mechanismus in der Leber. Pfluegers Arch 1965;285:73–89.
11 McManus ML, Churchwell KB: Clinical significance of cellular osmoregulation; in Strange K (ed): Cellular and Molecular Physiology of Cell Volume Regulation. Boca Raton, CRC, 1994, pp 63–77.
12 Anderson RJ: Hospital-associated hyponatremia. Kidney Int 1986;29:1237–1247.
13 Bichet DG, Kluge R, Howard RL, Schrier RW: Hyponatremic states; in Seldin DW, Giebisch G (eds): The Kidney: Physiology and Pathophysiology, ed 2. New York, Raven, 1992, pp 1727–1751.
14 Howard RL, Bichet DG, Schrier RW: Hypernatremic and polyuric states; in Seldin DW, Giebisch G (eds): The Kidney: Physiology and Pathophysiology, ed 2. New York, Raven, 1992, pp 1753–1778.
15 Arieff AI, Kleeman CR: Cerebral edema in diabetic comas. II. Effects of hyperosmolarity, hyperglycemia and insulin in diabetic rabbits. J Clin Endocrinol Metab 1974;38:1057–1067.
16 Robinson AG, Loeb JN: Ethanol ingestion – Commonest cause of elevated plasma osmolarity? N Engl J Med 1971;284:1253–1255.
17 Sterns RH, Thomas DJ, Herndon RM: Brain dehydration and neurologic deterioration after rapid correction of hypernatremia. Kidney Int 1989;35: 69–75.
18 Lien YH, Shapiro JI, Chan L: Study of brain electrolytes and organic osmolytes during correction of chronic hyponatremia: Implications for the pathogenesis of central pontine myelinolysis. J Clin Invest 1991;88:303–309.
19 Thurston JH, Sherman WR, Hauhart RE, Kloepper RF: Myoinositol: A newly identified nonnitrogenous osmoregulatory molecule in mammalian brain. Pediatr Res 1989;26:482–485.
20 Martinez-Maldonado M, Benabe JE, Cordova HR: Chronic clinical intrinsic renal failure; in Seldin DW, Giebisch G (eds): The Kidney: Physiology and Pathophysiology, ed 2. New York, Raven, 1992, pp 3227–3288.
21 Dunham PB: Effects of urea on volume sensitive K-Cl cotransport in LK sheep red blood cells – Evidence for two different signals of swelling. Am J Physiol 1995;268:C1026–C1032.

22 Parker JC: Urea alters set point volume for K-Cl cotransport, Na-H exchange, and Ca-Na exchange in dog red blood cells. Am J Physiol 1993;265:C447–C452.
23 Hallbrucker C, Vom Dahl S, Ritter M, Lang F, Häussinger D: Effects of urea on K^+ fluxes and cell volume in perfused rat liver. Pfluegers Arch 1994; 428:552–560.
24 Deutsch C, Chen LQ: Heterologous expression of specific K^+ channels in T lymphocytes: Functional consequences for volume regulation. Proc Natl Acad Sci USA 1993;90:10036–10040.
25 Felipe A, Snyders DJ, Deal KK, Tamkun MM: Influence of cloned voltage-gated K^+ channel expression on alanine transport, Rb^+ uptake, and cell volume. Am J Physiol 1993;265:C1230–C1238.
26 Busch AE, Varnum M, Adelman JP, North RA: Hypotonic solution increases the slowly activating potassium current I_{SK} expressed in *Xenopus* oocytes. Biochem Biophys Res Commun 1992;184:804–810.
27 Gründer S, Thiemann A, Pusch M, Jentsch TJ: Regions involved in the opening of the ClC-2 chloride channel by voltage and cell volume. Nature 1992;360:759–762.
28 Dermietzel R, Hwang TK, Buettner R, Hofer A, Dotzler E, Krmer M, Deutzmann R, Thinnes FP, Fishman GI, Spray DC, Siemen D: Cloning and in situ localization of a brain derived porin that constitutes a large conductance anion channel in astrocytic plasma membranes. Proc Natl Acad Sci USA 1994;91:499–503.
29 Gschwentner M, Nagl UO, Wöll E, Schmarda A, Ritter M, Paulmichl M: Antisense oligonucleotides suppress cell-volume-induced activation of chloride channels. Pfluegers Arch 1995;430:464–470.
30 Paulmichl M, Li Y, Wickman K, Ackerman M, Peralta E, Clapham D: New mammalian chloride-channel identified by expression cloning. Nature 1992;356:238–241.
31 Paulmichl M, Gschwentner M, Wöll E, Schmarda A, Ritter M, Kanin G, Ellemunter H, Waitz W, Deetjen P: Insight into the structure-function relation of chloride channels. Cell Physiol Biochem 1993;3:374–387.
32 Valverde MA, Diaz M, Sepúlveda FV, Gill DR, Hyde SC, Higgins CF: Volume-regulated chloride channel associated with the human multidrug resistant P-glycoprotein. Nature 1992;355:830–833.
33 Lang F, Busch GL, Völkl H: The diversity of cell volume regulatory mechanisms. Cell Physiol Biochem, in press.
34 Delpire E, Rauchmann M, Beier DR, Hebert SC, Gullans SR: Molecular cloning and chromosome localization of a putative basolateral Na^+-K^+-$2Cl^-$ cotransporter from mouse inner medullary collecting duct (mIMCD-3) cells. J Biol Chem 1994; 269:25677–25683.
35 Gamba G, Miyanoshita A, Lombardi M, Lytton J, Lee WS, Hedinger MA, Hebert SC: Molecular cloning, primary structure and characterization of two members of the mammalian electroneutral sodium-(potassium)-chloride cotransporter family expressed in kidney. J Biol Chem 1994;269:17713–17722.
36 Payne JA, Forbush B III: Alternatively spliced isoforms of the putative renal Na-K-Cl cotransporter are differentially distributed within the rabbit kidney. Proc Natl Acad Sci USA 1994;91:4544–4548.
37 Xu JC, Lytle C, Zhu TT, Payne JA, Benz E Jr, Forbush B: Molecular cloning and functional expression of the bumetanide-sensitive Na-K-Cl cotransporter. Proc Natl Acad Sci USA 1994;91: 2201–2205.
38 Demaurex N, Grinstein S: Na^+/H^+ antiport: Modulation by ATP and role in cell volume regulation. J Exp Biol 1994;196:389–404.
39 Kapus A, Grinstein S, Wasan S, Kandasamy R, Orlowski J: Functional characterization of three isoforms of the Na^+/H^+ exchanger stably expressed in chinese hamster ovary cells. J Biol Chem 1994; 269:23544–23552.
40 Bookstein C, Musch MW, Depaoli A, Xie Y, Villereal M, Rao MC, Chang EB: A unique sodium-hydrogen exchange isoform (NHE-4) of the inner medulla of the rat kidney is induced by hyperosmolarity. J Biol Chem 1994;269:29704–29709.
41 Burg MB: Molecular basis for osmoregulation of organic osmolytes in renal medullary cells. J Exp Zool 1994;268:171–175.
42 Handler JS, Kwon HM: Regulation of renal cell organic osmolyte transport by tonicity. Am J Physiol 1993;265:C1449–C1455.
43 Kinne RK, Czekay RP, Grunewald JM, Mooren FC, Kinne-Saffran E: Hypotonicity-evoked release of organic osmolytes from distal renal cells: Systems, signals, and sidedness. Renal Physiol Biochem 1993;16:66–78.
44 Law RO: Amino acids as volume-regulatory osmolytes in mammalian cells. Comp Biochem Physiol 1991;99:263–277.
45 Nakanishi T, Balaban RS, Burg MB: Survey of osmolytes in renal cell lines. Am J Physiol 1988; 255:C181–C191.
46 Somero GN: Protons, osmolytes, and fitness of internal milieu for protein function. Am J Physiol 1986;251:R197–R213.
47 Garcia-Perez A, Burg MB: Renal medullary organic osmolytes. Physiol Rev 1991;71:1081–1115.
48 Yancey PH, Clark ME, Hand SC, Bowlus RD, Somero GN: Living with water stress: Evolution of osmolyte systems. Science 1982;217:1214–1222.
49 Stoll B, Gerok W, Lang F, Häussinger D: Liver cell volume and protein synthesis. Biochem J 1992;287: 217–222.
50 Hallbrucker C, Vom Dahl S, Lang F, Häussinger D: Control of hepatic proteolysis by amino acids: The role of cell volume. Eur J Biochem 1991;197: 717–724.
51 Hallbrucker C, Vom Dahl S, Lang F, Gerok W, Häussinger D: Inhibition of hepatic proteolysis by insulin. Role of hormone-induced alterations of the cellular K^+ balance. Eur J Biochem 1991;199:467–474.
52 Hallbrucker C, Vom Dahl S, Lang F, Gerok W, Häussinger D: Modification of liver cell volume by insulin and glucagon. Pfluegers Arch 1991;418: 519–521.
53 Vom Dahl S, Hallbrucker C, Lang F, Häussinger D: Regulation of cell volume in the perfused rat liver by hormones. Biochem J 1991;280:105–109.
54 Vom Dahl S, Hallbrucker C, Lang F, Gerok W, Häussinger D: Regulation of liver cell volume and proteolysis by glucagon and insulin. Biochem J 1991;278:771–777.
55 Vom Dahl S, Stoll B, Gerok W, Häussinger D: Inhibition of proteolysis by cell swelling in liver requires intact microtubular structures. Biochem J 1995;308:529–536.
56 Häussinger D, Hallbrucker C, Vom Dahl S, Lang F, Gerok W: Cell swelling inhibits proteolysis in perfused rat liver. Biochem J 1990;272:239–242.
57 Häussinger D, Hallbrucker C, Vom Dahl S, Decker S, Schweizer U, Lang F, Gerok W: Cell volume is a major determinant of proteolysis control in liver. FEBS Lett 1991;283:70–72.
58 Häussinger D, Gerok W, Lang F: Cell volume and hepatic metabolism. Adv Comp Environ Physiol 1993;14:33–65.
59 Häussinger D, Lang F, Gerok W: Regulation of cell function by the cellular hydration state. Am J Physiol 1994;267:E343–E355.
60 Häussinger D, Newsome W, Vom Dahl S, Stoll B, Noe B, Schreiber R, Wettstein M, Lang F: Control of liver cell function by the hydration state. Biochem Soc Trans 1994;22:497–502.
61 Luiken JJFP, Blommaart EFC, Boon L, Van Woerkom GM, Meijer AJ: Cell swelling and the control of autophagic proteolysis in hepatocytes: Involvement of phosphorylation of ribosomal protein S6? Biochem Soc Trans 1994;22:508–511.
62 Bode BP, Kilberg MS: Amino acid-dependent increase in hepatic system N activity is linked to cell swelling. J Biol Chem 1991;266:7376–7381.
63 Häussinger D, Stoll B, Morimoto Y, Lang F, Gerok W: Anisoosmotic liver perfusion: Redox shifts and modulation of alpha-ketoisocaproate and glycine metabolism. Biol Chem Hoppe Seyler 1992;373: 723–734.
64 Lundgren DW: Effect of hypotonic stress on ornithine decarboxylase mRNA expression in cultured cells. J Biol Chem 1992;267:6841–6847.
65 Tohyama Y, Kameji T, Hayashi S: Mechanisms of dramatic fluctuations of ornithine decarboxylase activity upon tonicity changes in primary cultured rat hepatocytes. Eur J Biochem 1991;202:1327–1331.
66 Häussinger D, Lang F: Exposure of perfused liver to hypotonic conditions modifies cellular nitrogen metabolism. J Cell Biochem 1990;43:355–361.
67 Häussinger D, Lang F, Bauers K, Gerok W: Control of hepatic nitrogen metabolism and glutathione release by cell volume regulatory mechanisms. Eur J Biochem 1990;193:891–898.
68 Saha N, Stoll B, Lang F, Häussinger D: Effect of anisotonic cell volume modulation on glutathione-S-conjugate release, t-butylhydroperoxide metabolism and the pentose-phosphate shunt in perfused rat liver. Eur J Biochem 1992;209:437–444.
69 Finkenzeller G, Newsome W, Lang F, Häussinger D: Increase of c-*jun* mRNA upon hypoosmotic cell swelling of rat hepatoma cells. FEBS Lett 1994;340: 163–166.
70 Häussinger D: The role of cellular hydration for the regulation of cell function. Biochem J, in press.
71 Benis RC, Lundgren DW: Sodium-dependent cotransported analogues of glucose stimulate ornithine decarboxylase mRNA expression in LLC-PK_1 cells. Biochem J 1993;289:751–756.
72 Poulin R, Wechter RS, Pegg AE: An early enlargement of the putrescine pool is required for growth in L1210 mouse leukemia cells under hypoosmotic stress. J Biol Chem 1991;266:6142–6151.
73 Levin EG, Santell L, Saljooque F: Hyperosmotic stress stimulates tissue plasminogen activator expression by a PKC-independent path- way. Am J Physiol 1993;265:C387–C396.
74 Theodoropoulos PA, Stournaras C, Stoll B, Markogiannakis E, Lang F, Gravanis A, Häussinger D: Hepatocyte swelling leads to rapid decrease of the G-/total actin ratio and increases actin mRNA levels. FEBS Lett 1992;311:241–245.
75 Häussinger D, Stoll B, Vom Dahl S, Theodoropoulos PA, Markogiannakis E, Gravanis A, Lang F, Stournaras C: Effect of hepatocyte swelling on microtubule stability and tubulin mRNA levels. Biochem Cell Biol 1994;72:12–19.
76 Garcia-Perez A, Martin B, Murphy HR, Uchida S, Murer H, Cowley BD, Handler JS, Burg MB: Molecular cloning of cDNA coding for kidney aldose reductase: Regulation of specific mRNA accumulation by NaCl-mediated osmotic stress. J Biol Chem 1989;264:16815–16821.
77 Smardo FL, Burg MB, Garcia-Pérez A: Kidney aldose reductase gene transcription is osmotically regulated. Am J Physiol 1992;262:C776–C782.

78 Kwon HM, Yamauchi A, Uchida S, Preston AS, Garcia-Perez A, Burg MB, Handler JS: Cloning of the cDNA for a Na^+/myo-inositol cotransporter, a hypertonicity stress protein. J Biol Chem 1992;267: 6297–6301.
79 Yamauchi A, Kwon HM, Uchida S, Preston AS, Handler JS: Myo-inositol and betaine transporters regulated by tonicity are basolateral in MDCK cells. Am J Physiol 1991;261:F197–F202.
80 Uchida S, Kwon HM, Preston AS, Handler JS: Expression of Madin-Darby kidney cell Na^+- and Cl^--dependent taurine transporter in *Xenopus laevis* oocytes. J Biol Chem 1991;266:9605–9609.
81 Wasserman JC, Delpire E, Tonidandel W, Kojima R, Gullans SR: Molecular characterization of ROSIT, a renal osmotic stress-induced Na(+)-Cl(−)-organic solute cotransporter. Am J Physiol 1994;267:F688–F694.
82 Chen JG, Klus LR, Steenbergen DK, Kempson SA: Hypertonic upregulation of amino acid transport system A in vascular smooth muscle cells. Am J Physiol 1994;267:C529–C536.
83 Gazzola GC, Dall'Asta V, Nucci FA, Rossi PA, Bussolati O, Hoffman EK, Guidotti GG: Role of amino acid transport system A in the control of cell volume in cultured human fibroblasts. Cell Physiol Biochem 1991;1:131–142.
84 Soler C, Felipe A, Casado FJ, McGivan JD, Pastor-Anglada M: Hyperosmolarity leads to an increase in derepressed system A activity in the renal epithelial cell line NBL-1. Biochem J 1993;289:653–658.
85 Yamauchi A, Miyai A, Yokoyama K, Itoh T, Kamada T, Ueda N, Fujiwara Y: Response to osmotic stimuli in mesangial cells: Role of system A transporter. Am J Physiol 1994;267:C1493–C1500.
86 Cohen DM, Wasserman JC, Gullans SR: Immediate early gene and HSP70 expression in hyperosmotic stress in MDCK cells. Am J Physiol 1991; 261:C594–C601.
87 Wollnik B, Kubisch C, Maass A, Vetter H, Neyses L: Hyperosmotic stress induces immediate-early gene expression in ventricular adult cardiomyocytes. Biochem Biophys Res Commun 1993;194: 642–646.
88 Giovannelli L, Shiromani PJ, Jirikowski GF, Bloom FE: Expression of c-Fos protein by immunohistochemically identified oxytocin neurons in the rat hypothalamus upon osmotic stimulation. Brain Res 1992;588:41–48.
89 Zhang F, Wettstein M, Warskulat U, Henninger P, Schreiber R, Decker K, Häussinger D: Hyperosmolarity stimulates prostaglandin synthesis and cyclooxygenase-2 expression in activated rat Kupffer cells. Biochem J 1995;312:135–143.
90 Newsome WP, Warskulat U, Noe B, Wettstein M, Stoll B, Gerok W, Häussinger D: Modulation of phosphoenolpyruvate carboxykinase mRNA levels by the hepatocellular hydration state. Biochem J 1994;304:555–560.
91 Waldegger S, Barth P, Raber G, Lang F: Cloning and characterization of a putative human serine/threonine protein kinase transcriptionally modified during anisotonic and isotonic alterations of cell volume. Proc Natl Acad Sci USA 1997;94: 4440–4445.
92 Dasgupta S, Hohman TC, Carper D: Hypertonic stress induces alpha B-crystallin expression. Exp Eye Res 1992;54:461–470.
93 Urban JP, Hall AC, Gehl KA: Regulation of matrix synthesis rates by the ionic and osmotic environment of articular chondrocytes. J Cell Physiol 1993;154:262–270.
94 Katayama S, Abe M, Tanaka K, Omoto A, Negishi K, Itabashi A, Ishii J: High glucose concentration suppresses mesangial laminin B2 gene expression. J Diabetes Complic 1991;5:118–120.
95 Agius L, Peak M, Beresford G, Al-Habori M, Thomas TH: The role of ion content and cell volume in insulin action. Biochem Soc Trans 1994; 22:516–522.
96 Peak M, Al-Habori M, Agius L: Regulation of glycogen synthesis and glycolysis by insulin, pH and cell volume: Interactions between swelling and alkalinization in mediating the effects of insulin. Biochem J 1992;282:797–805.
97 Ritter M, Wöll E: Modification of cellular ion transport by the Ha-Ras oncoprotein: Steps toward malignant transformation. Cell Physiol Biochem, in press.
98 Bakker-Grunwald T: Potassium permeability and volume control in isolated rat hepatocytes. Biochim Biophys Acta 1983;731:239–242.
99 Bear CE, Peterson OH: *L*-Alanine evokes opening of single Ca^{2+}-activated K^+ channels in rat liver cells. Pfluegers Arch 1987;410:342–344.
100 Häussinger D, Lang F, Bauers K, Gerok W: Interactions between glutamine metabolism and cell-volume regulation in perfused rat liver. Eur J Biochem 1990;188:689–695.
101 Kristensen LO: Associations between transports of alanine and cations across cell membrane in rat hepatocytes. Am J Physiol 1986;251:G575–G584.
102 Wettstein M, Vom Dahl S, Lang F, Gerok W, Häussinger D: Cell volume regulatory responses of isolated perfused rat liver: The effect of amino acids. Biol Chem Hoppe Seyler 1990;371:493–501.
103 Hayes MR, McGivan JD: Differential effects of starvation on alanine and glutamine transport in isolated rat hepatocytes. Biochem J 1982;204:365–368.
104 Völkl H, Rehwald W, Waitz W, Häussinger D, Lang F: Acridine orange fluorescence in renal proximal tubules: Effects of NH_3/NH_4^+ and cell volume. Cell Physiol Biochem 1993;3:28–33.
105 Völkl H, Friedrich F, Häussinger D, Lang F: Effect of cell volume on acridine orange fluorescence in hepatocytes. Biochem J 1993;295:11–14.
106 Völkl H, Busch GL, Häussinger D, Lang F: Alkalinization of acidic cellular compartments following cell swelling. FEBS Lett 1994;338:27–30.
107 Lang F, Busch GL, Völkl H, Häussinger D: Lysosomal pH – A link between cell volume and metabolism. Biochem Soc Trans 1994;22:502–505.
108 Busch GL, Schreiber R, Dartsch PC, Völkl H, Vom Dahl S, Häussinger D, Lang F: Involvement of microtubules in the link between cell volume and pH of acidic cellular compartments in rat and human hepatocytes. Proc Natl Acad Sci USA 1994;91:9165–9169.
109 Schreiber R, Stoll B, Lang F, Häussinger D: Effects of aniso-osmolarity and hydroperoxides on intracellular pH in isolated rat hepatocytes as assessed by (2′,7′)-bis(carboxyethyl)-5(6)-carboxyfluroescein and fluorescein isothiocyanate-dextran fluorescence. Biochem J 1994;303:113–120.
110 Schreiber R, Häussinger D: Characterization of the swelling-induced alkalinization of endocytotic vesicles in fluorescein isothiocyanate-dextran-loaded rat hepatocytes. Biochem J 1995;309:19–24.
111 Busch GL, Lang HJ, Lang F: Studies on the mechanism of swelling-induced lysosomal alkalinization in vascular smooth muscle cells. Pfluegers Arch 1996;431:690–696.
112 Busch GL, Gulbins E, Wiesinger H, Hamprecht B, Lang F: Effect of astroglial cell swelling on pH of acidic intracellular compartments. Biochim Biophys Acta, in press.
113 Busch GL, Günther E, Hewig B, Zrenner E, Lang F: Effect of cell swelling, NH_4CL and glutamate on acridine orange fluorescence in retinal ganglion cells. Cell Physiol Biochem, in press.
114 Busch GL, Völkl H, Haller T, Ritter M, Siemen D, Moest J, Koch F, Lang F: Vesicular pH is sensitive to changes in cell volume. Cell Physiol Biochem 1997;7:25–34.
115 Mortimore GE, Pösö AR: Intracellular protein catabolism and its control during nutrient deprivation and supply. Annu Rev Nutr 1987;7:539–564.
116 Okuda S, Languino LR, Ruoslahti E, Border WA: Elevated expression of transforming growth factor-β and proteoglycan production in experimental glomerulonephritis – Possible role in expansion of the mesangial extracellular matrix. J Clin Invest 1990;86:453–462.
117 Shankland S, Scholey JW: Expression of transforming growth factor-β_1 during diabetic renal hypertrophy. Kidney Int 1994;46:430–442.
118 Ling H, Vamvakas S, Busch G, Dämmrich J, Schramm L, Lang F, Heidland A: Suppressing role of transforming growth factor-β1 on cathepsin activity in cultured kidney tubule cells. Am J Physiol 1995;269:F911–F917.
119 Sharma K, Ziyadeh FN: The transforming growth factor-β system and the kidney. Semin Nephrol 1993;13:116–128.
120 Häussinger D, Roth E, Lang F, Gerok W: Cellular hydration state: An important determinant of protein catabolism in health and disease. Lancet 1993; 341:1330–1332.
121 Roth E, Karner J, Ollenschläger G: Glutamine: An anabolic effector? J Parent Ent Nutr 1990;14: 130S–136S.

Mineral and Electrolyte Metabolism

Miner Electrolyte Metab 1997;23:206–209

Xiaonan Wang
Claudine Jurkovitz
S. Russ Price

Renal Division, Department of Medicine, Emory University, Atlanta, GA., USA

Branched-Chain Amino Acid Catabolism in Uremia: Dual Regulation of Branched-Chain α-Ketoacid Dehydrogenase by Extracellular pH and Glucocorticoids

Key Words
Acidosis
Branched-chain amino acids
Gene expression
Muscle

Abstract
In muscles of rats with metabolic acidosis, branched-chain α-ketoacid dehydrogenase (BCKAD) activity is increased but the signal initiating this response is unknown. Potential signals include intracellular responses to acidemia and/or glucocorticoids. It is not known whether the signal activates BCKAD by changing the proportion of inactive (phosphorylated) or active (dephosphorylated) enzyme or increases the amount of enzyme. To separate the effects of extracellular pH and glucocorticoids on BCKAD, enzyme activity was measured in two cell types: (1) LLC-PK_1 cells that do not express glucocorticoid receptors and (2) LLC-PK_1-GR101 cells that express rat glucocorticoid receptors. Extracellular acidification (pH 6.95) increased the maximal BCKAD activity, the percentage of active enzyme, and the amounts of BCKAD protein. Dexamethasone also increased the percentage of active BCKAD but the time required for activation was longer than with acidification. These results demonstrate that BCKAD is dually regulated by extracellular pH and glucocorticoids.

Critical nutritional adaptations which act to maintain protein balance when dietary protein is restricted include a reduction in the degradation of branced-chain amino acids (BCKAAs) [1, 2]. This adaptation could be beneficial because the availability of BCAA for protein synthesis increases, and because leucine and its ketoacid, α-ketoisocaproate, regulate protein degradation in skeletal and heart muscle [3, 4]. If pathologic conditions (e.g., uremia) stimulate BCAA catabolism and interfere with these adaptations, then protein wasting could occur [5].

Defective BCAA Metabolism in Uremia

In patients with chronic renal failure (CRF), a condition associated with muscle protein wasting, plasma and muscle levels of the BCAAs are low [6]. Bergström et al. [6] reported a significant relationship ($r = 0.82$) between plasma bicarbonate and the intracellular concentration of valine in muscles of hemodialysis patients, suggesting that the metabolic acidosis of uremia increased muscle BCAA oxidation. Support for this conclusion was provided by Reaich et al. [7], who showed that giving CRF patients sodium bicarbonate to correct their metabolic acidosis normalized whole-body leucine oxidation and the levels of BCAAs in muscle. Similarly, leucine oxidation was higher in incubated muscle from CRF rats than from pair-fed control rats and administration of sodium bicarbonate to CRF rats blocked the increased BCAA catabolism [8]. Thus, uremia accelerates BCAA degradation in muscle.

Acidosis Stimulates Branched-Chain α-Ketoacid Dehydrogenase (BCKAD)

In humans and rats with normal kidneys, $NHCl_4$-induced acidosis stimulates whole-body leucine oxidation [9, 10] indicating the activity of the rate-limiting enzyme for BCAA oxidation, BCKAD, is increased. BCKAD is a multienzyme complex located on the inner mitochondrial membrane. There are three enzyme activities associated with the BCKAD complex: (1) branched-chain α-ketoacid decarboxylase (E1) composed of E1α and E1β subunits; (2) dihydrolipoyl transacylase (E2), and (3) dihydrolipoyl dehydro-

KARGER
E-Mail karger@karger.ch
Fax +41 61 306 12 34
http://www.karger.ch

S. Russ Price, PhD
Renal Division, Emory University, WMRB Room 338, 1639 Pierce Drive
Atlanta, GA 30322 (USA)
Tel. +1 404 727 2525, fax +1 404 727 3425
E-Mail medrp@emory.edu

Table 1. Metabolic acidosis increases BCKAD activity in rat muscle

	Control	Acidosis
Basal activity, nmol·min⁻¹·g protein⁻¹	9±2	100±30*
Total activity, nmol·min⁻¹·g protein⁻¹	1,330±185	2,081±179*
Percent active	0.7±0.2	4.6±1.2*

Basal and total BCKAD activities were measured in gastrocnemius muscle of six acidotic rats given NH_4Cl, and in pair-fed control rats. Results are reported as the mean ± SE. The percent BCKAD in the active state is calculated by dividing the basal BCKAD activity by the total activity × 100.

* $p<0.05$ vs. pair-fed controls using paired Student's t test.

genase (E3). BCKAD activity in muscle can be regulated in two ways: (1) BCKAD is inactivated by phosphorylation and activated by dephosphorylation of the E1α subunit, and (2) the amount of BCKAD enzyme can be altered to influence the total enzyme activity as well as the percentage of BCKAD in the active state (percent active). In rats fed a low-protein or no-protein diet, BCKAD activity is reduced in liver, muscle, and heart [11, 12] and levels of immunoprecipitable BCKAD subunits and their mRNAs are also decreased in the liver of these rats [11, 13]. These responses are evidence that regulation of BCKAD activity is part of the adaptional response to a reduction in protein.

When BCKAD activity was measured directly in muscle, the basal level of activity and the total BCKAD activity (measured after BCKAD is completely dephosphorylated) were greater in acidotic rats than in pair-fed control rats (table 1); the percentage of active enzyme (the ratio of basal activity to total activity) was also increased (table 1) [8, 14, 15]). Consistent with the increase in total BCKAD activity (i.e., the amount of enzyme), there was a concomitant increase in the levels of mRNAs encoding the BCKAD subunit in acidotic rat muscle.

What Is the Signal for Increased BCKAD Activity?

Is a change in intracellular pH responsible for the increase in BCKAD activity? This is unlikely since the intracellular pH in muscle changes minimally in metabolic acidosis or uremia [16]. Another potential stimulus is glucocorticoids since their production is increased by uremia [17] or acidosis [18] and rats given pharmacologic doses of dexamethasone have higher muscle BCKAD activity than untreated rats [19]. Moreover, when acidotic adrenalectomized rats were studied, whole-body leucine oxidation was increased only when 'stress levels' of glucocorticoids were given simultaneously [20]; glucocorticoids alone did not increase BCAA oxidation [20].

To determine whether extracellular acidification and glucocorticoids independently regulate BCKAD, we measured BCKAD activity in two types of $LLC\text{-}PK_1$ kidney cells: control cells that do not express glucocorticoid receptors, $LLC\text{-}PK_1$-GR101 cells which are stably transfected with a rat glucocorticoid receptor gene and express glucocorticoid receptors. It is important to emphasize that our goal in studying these cells was not to determine how BCKAD

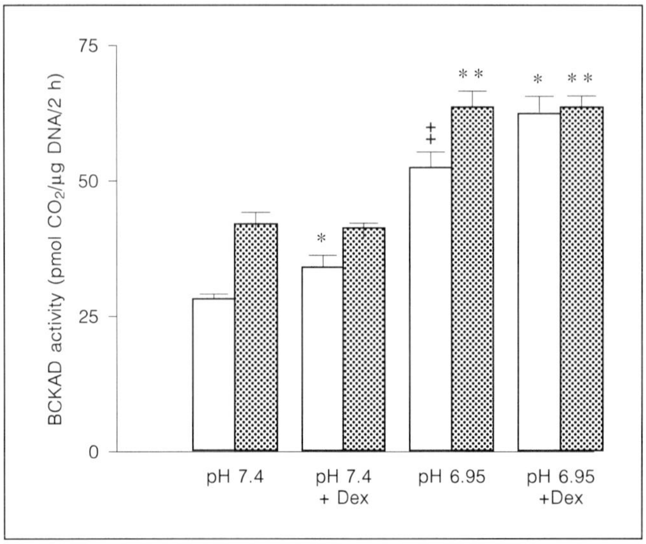

Fig. 1. Extracellular acidification and dexamethasone increase BCKAD activity. Basal (open bars) and total (shaded bars) BCKAD activity were measured in $LLC\text{-}PK_1$-GR101 cells incubated in pH 6.95 or 7.4 media for 24 h; 50 n*M* dexamethasone (Dex) was included as indicated. By Student's t test ($n=6$ for each group): * $p<0.05$ vs. same pH without dexamethasone; ‡ $p<0.05$ vs. pH 7.4, no dexamethasone; ** $p<0.05$ vs. pH 7.4 with or without dexamethasone.

responds in the kidney. The kidney is heterogeneous in cell composition and the intact organ response would reflect the 'net response' of the different cell types. Instead, our goal was to design a cellular system that permits us to distinguish between the responses of BCKAD to glucocorticoids and extracellular acidification in intact cells.

To ensure stable expression of the glucocorticoid receptor, $LLC\text{-}PK_1$-GR101 cells were transiently transfected with a glucocorticoid-responsive luciferase reporter gene and luciferase activity was measured in cells incubated with or without 50 n*M* dexamethasone for 24 h. Luciferase activity was increased 25-fold ($p<0.0001$) in cells treated with dexamethasone and this increase was blocked by the steroid receptor antagonist RU 486.

To separate the influences of extracellular acidification and glucocorticoids on BCAA degradation, we measured BCKAD activity in control $LLC\text{-}PK_1$ cells (lacking glucocorticoid receptors) incubated at pH 7.4 or 6.95 with or without 50 n*M* dexamethasone for 24 h. Acidification increased both the basal activity (53%, $p<0.05$) and the total activity (47%, $p<0.05$; measured under V_{max} conditions). Predictably, dexamethasone did not influence BCKAD activity at either pH.

To compare the influences of glucocorticoids and extracellular acidification, we measured BCKAD activity in the $LLC\text{-}PK_1$-GR101 cells that stably express rat glucocorticoid receptors. As in control cells, acidification increased basal and total BCKAD activity (fig. 1). Basal BCKAD activity was also increased in these cells after addition of 50 n*M* dexamethasone (24 h) regardless of whether the cells were incubated at pH 6.95 or 7.4 (fig. 1). In

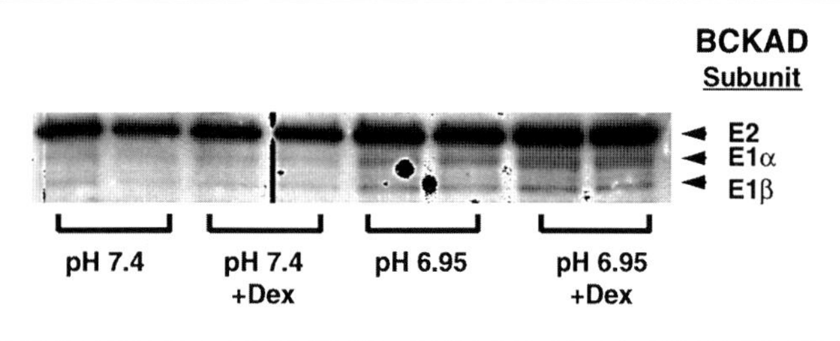

Fig. 2. Extracellular acidification increases BCKAD subunit proteins. A mitochondrial fraction was purified from LLC-PK_1-GR101 cells after incubation (24 h) at pH 6.95 or 7.4 with or without 50 n*M* dexamethasone (Dex). A polyclonal BCKAD complex antibody was used to detect the E1α, E1β, and E2 subunits in equal amounts of mitochondrial protein. Results are shown for two separate mitochondrial preparations from the indicated treatment groups.

contrast, dexamethasone did not influence total BCKAD activity at pH 7.4 nor was there an additional increase in total activity at pH 6.95 (fig. 1). To ensure the stimulation of BCKAD activity by glucocorticoids was receptor mediated, basal BCKAD activity was measured in cells incubated with equimolar concentrations of dexamethasone and RU 486 at pH 7.4 and 6.95 (we did not measure total activity since it is not influenced by glucocorticoids). RU 486 prevented the glucocorticoid-induced increase in basal BCKAD activity at pH 7.4 and 6.95. Interestingly, RU 486 did not prevent the increase in BCKAD activity at pH 6.95, indicating that the response to extracellular acidification does not involve the glucocorticoid receptor.

When cells were incubated at pH 6.95, total BCKAD activity was increased (70%, $p < 0.05$) within 2 h. This increase was sustained for 24 h and did not rise further with glucocorticoids at pH 6.95 or 7.4. In contrast, the percent active BCKAD did not increase until 6 h after addition of dexamethasone. The increase in the percent active BCKAD by dexamethasone was maintained for 24 h.

An increase in total BCKAD activity in response to lowering extracellular pH in LLC-PK_1-GR101 cells suggests that the amount of BCKAD enzyme is increased. To evaluate this possibility, we measured BCKAD protein in equal amounts of mitochondrial protein from cells incubated at pH 6.95 or 7.4 (24 h) using a polyclonal antibody that reacts with BCKAD E1α, E1β, and E2 subunits. Extracellular acidification increased all three BCKAD subunits but dexamethasone (50 n*M*) did not appear to elicit a significant increase (fig. 2). Thus, the increase in BCKAD protein is consistent with the increase in the levels of mRNAs encoding BCKAD subunits in muscles of rats with metabolic acidosis [15].

How do these findings in cultured cells relate to animals? Similar to our findings in LLC-PK_1 cells, BCAA degradation increases in adrenalectomized rats only when they are rendered acidotic and given glucocorticoids [20]. If the amount of BCKAD enzyme is increased by metabolic acidosis and glucocorticoids increase the percentage of the enzyme in the active form, then the combined effects could significantly increase 'net' BCKAD activity. The mechanism by which metabolic acidosis activates BCKAD and initiates other catabolic responses (e.g., protein degradation) in rat muscle is unknown but it does not depend on a reduction in intracellular pH. We find this signal may be independent of glucocorticoids. Since other catabolic states (e.g., insulinopenia, sepsis, starvation, cancer) are also associated with increased glucocorticoid production and accelerated BCAA degradation [21–25], our findings may also be applicable to conditions unrelated to uremia or metabolic acidosis.

Acknowledgments

We are grateful to Belinda Sullivan for her assistance in the preparation of this manuscript. These studies were supported by NIH grant DK-50740.

References

1 Choi E, Bailey JL, May RC, Masud T, Maroni BJ: Metabolic responses to nephrosis: Effect of a low protein diet. Am J Physiol 1994,266:F432–F438.
2 Choi EJ, May RC, Bailey JL, Masud T, Dixon A, Maroni BJ: Mechanisms of adaptation to proteinuria in adriamycin nephrosis. Am J Physiol 1993;265:F257–F263.
3 Mitch WE, Clark AS: Specificity of the effect of leucine and its metabolites on protein degradation in skeletal muscle. Biochem J 1984;222:579–586.
4 Mitch WE, Walser M, Sapir DG: Nitrogen-sparing induced by leucine compared with that induced by its keto-analogue, alpha-ketoisocaproate, in fasting obese man. J Clin Invest 1981;67:553–562.
5 Price SR, Mitch WE: Metabolic acidosis and uremic toxicity: Protein and amino acid metabolism. Semin Nephrol 1994;14:232–237.
6 Bergström J, Alvestrand A, Furst P: Plasma and muscle free amino acids in maintenance hemodialysis patients without protein malnutrition. Kidney Int 1990;38:108–114.
7 Reaich D, Channon SM, Scrimgeour CM, Daley SE, Wilkinson R, Goodship THJ: Correction of acidosis in humans with CRF decreases protein degradation and amino acid oxidation. Am J Physiol 1993;265:E230–E235.
8 Hara Y, May RC, Kelly RA, Mitch WE: Acidosis, not azotemia, stimulates branched-chain amino acid catabolism in uremic rats. Kidney Int 1987;32: 808–814.
9 May RC, Masud T, Logue B, Bailey J, England B: Chronic metabolic acidosis accelerates whole body proteolysis and oxidation in awake rats. Kidney Int 1992;41:1535–1542.
10 Reaich D, Channon SM, Scrimgeour CM, Goodship THJ: Ammonium chloride-induced acidosis increases protein breakdown and amino acid oxidation in humans. Am J Physiol 1992;263:E735–E739.
11 Patston PA, Espinal J, Randle PJ: Effects of diet and of alloxan-diabetes on the activity of branched-chain 2-oxo acid dehydrogenase complex and of activator protein in rat tissues. Biochem J 1984;222: 711–719.
12 Miller RH, Eisenstein RS, Harper AE: Effects of dietary protein intake on branched-chain keto acid dehydrogenase actvity in the rat: Immunochemical analysis of the enzyme complex. J Biol Chem 1998; 263(7):3454–3461.
13 Chinsky JM, Bohlen LM, Costeas PA: Noncoordinated responses of branched-chain α-ketoacid dehydrogenase subunit genes to dietary protein. FASEB J 1994;8:114–120.

14 May RC, Hara Y, Kelly RA, Brock KP, Buse MG, Mitch WE: Branched-chain amino acid metabolism in rat muscle: Abnormal regulation in acidosis. Am J Physiol 1987;252:E712–E718.
15 England BK, Greiber S, Mitch WE, Bowers BA, Herring WJ, McKean M, Ebb RG, Price SR, Danner DJ: Rat muscle branched-chain ketoacid dehydrogenase activity and mRNAs increase with extracellular acidemia. Am J Physiol 1995;268: C1395–C1400.
16 Bailey JL, England BK, Long RC Jr, Weissman J, Mitch WE: Experimental acidemia and muscle cell pH in chronic acidosis and renal failure. Am J Physiol 1995;269:C706–C712.
17 May RC, Kelly RA, Mitch WE: Mechanisms for defects in muscle protein metabolism in rats with chronic uremia: The influence of metabolic acidosis. J Clin Invest 1987;79:1099–1103.
18 May RC, Kelly RA, Mitch WE: Metabolic acidosis stimulates protein degradation in rat muscle by a glucocorticoid-dependent mechanism. J Clin Invest 1986;77:614–621.
19 Block KP, Richmond WB, Mehard WB, Buse MG: Glucocorticoid-mediated activation of muscle branched-chain a-keto acid dehydrogenase in vivo. Am J Physiol 1987;252:E396–E407.
20 May RC, Bailey JL, Mitch WE, England BK: Glucocorticoids and acidosis stimulate protein and amino acid catabolism in vivo. Kidney Int 1996;49: 679–683.
21 Aftring RP, Miller WJ, Buse MG: Effects of diabetes and starvation on skeletal muscle branched-chain α-keto acid dehydrogenase activity. Am J Physiol 1988;254:E292–E300.
22 Argilés JM, Lopez-Soriano FJ: The enzymatic activities of branched-chain amino acid catabolism in rumor-bearing rats. Cancer Lett 1992;1991:239–242.
23 García-Martínez C, Llovera M, López-Soriano FJ, Santo B del, Argilés JM: Lipopolysaccharide (LPS) increases the in vivo oxidation of branched-chain amino acids in rat: A cytokine-mediated effect. Mol Biochem 1995;148:9–15.
24 Nawabi MD, Block KP, Chakrabarti MC, Buse MG: Administration of endotoxin, tumor necrosis factor, or interleukin 1 to rats activates skeletal muscle branched-chain α-keto acid dehydrogenase. J Clin Invest 1990;85:256–263.
25 Patston PA, Espinal J, Randle PJ: Effects of diet and of alloxan-diabetes on the activity of branched-chain 2-oxo acid dehydrogenase complex and of activator protein in rat tissues. Biochem J 1984;222: 711–719.

Mineral and Electrolyte Metabolism

Miner Electrolyte Metab 1997;23:210–213

Gerald Cohen
Michael Rudnicki
Walter H. Hörl

Division of Nephrology,
Department of Medicine,
University of Vienna, Austria

A Peptide Isolated from a Patient on Continuous Ambulatory Peritoneal Dialysis Has Homology to Ubiquitin and Inhibits the Chemotactic Response of Polymorphonuclear Leukocytes

Key Words
Chemotaxis
Continuous ambulatory peritoneal dialysis
Infection
Polymorphonuclear leukocytes
Ubiquitin
Uremic toxins

Abstract
Susceptibility to bacterial infections in the peritoneal cavity is a serious problem for peritoneal dialysis patients. Uremic toxins are among the causes of the diminished function of polymorphonuclear leuckocytes (PMNLs), cells of the local immune defense. We isolated from the peritoneal dialysis effluent of a uremic patient, a peptide that inhibits PMNL chemotaxis in a concentration-dependent manner. The amino terminus of this factor is identical to that of ubiquitin. The peptide shows the same size as ubiquitin on SDS-polyacrylamide gels, but has a more acidic isoelectric point than intact ubiquitin, which has no effect on PMNL chemotaxis. We conclude that we have isolated a modified form of ubiquitin and that this modification is responsible for its inhibiting effect on PMNL chemotaxis.

Introduction

One of the main problems encountered by uremic patients undergoing peritoneal dialysis (PD) is peritonitis. These bacterial infections are partly responsible for the increased risk of morbidity and mortality in this group of patients. Cellular defense mechanisms in the peritoneal cavity play a crucial role in the prevention of infection [1]. One main component of the local immune defense system during infection are polymorphonuclear leukocytes (PMNLs). In a stable situation, most cells within the peritoneal cavity of a continuous ambulatory peritoneal dialysis (CAPD) patient are macrophages. However, the most obvious change at the onset of peritonitis is the massive and rapid influx of cells into the peritoneal cavity and the fact that over 85% of these cells are PMNLs [2].

Chemotaxis is the first step in nonspecific immune defense [3]. The movement of PMNLs along a chemotactic gradient and their accumulation at sites of infection are crucial leukocyte functions and involve, like other functions, a variety of coordinated mechanisms. After the PMNLs have arrived at the site of infection, they ingest bacteria by phagocytosis and kill them with proteolytic enzymes and with toxic oxygen radicals produced during the oxidative burst. It is obvious that impairment of any of these essential functions leads to an increased risk of bacterial infections.

It has been shown that many functions and metabolic parameters, such as carbohydrate metabolism, chemotactic activity, intracellular ATP levels, oxidative metabolism, intracellular killing, and the response to phagocytotic stimuli of PMNLs from uremic patients are disturbed [4–8]. The fact that most of these changes can be corrected by PD treatment leads to the conclusion that uremic toxins are excreted into the peritoneal cavity. Another, more direct indication for the presence of uremic toxins in the peritoneal cavity comes from experiments showing that PMNL phagocytosis is markedly depressed upon addition of CAPD effluent to PMNLs from healthy donors [9].

One focus of our research is the isolation and characterization of PMNL-inhibiting substances from the peritoneal cavity of uremic patients. The purpose of these experiments is to learn more about uremic toxins and to use this knowledge for future diagnosis and therapy. In the present study, we characterized a peptide isolated from a CAPD patient and we describe its inhibitory effect on the chemotactic response of PMNLs.

KARGER
E-Mail karger@karger.ch
Fax +41 61 306 12 34
http://www.karger.ch

0378–0392/97/0236–0210 $12.00/0

Dr. Gerald Cohen
Med. Univ. Klinik III,
Währinger Gürtel 18–20, A–1090 Wien (Austria)

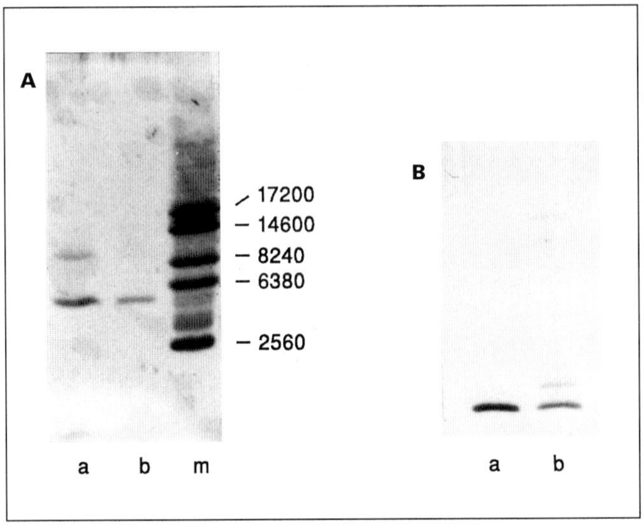

Fig. 1. A SDS-PAGE of CIP. Lane a = Reduced; lane b = nonreduced; m = molecular weight marker. **B** SDS-PAGE of ubiquitin (lane a) and CIP (lane b).

CIP: MQIFVKTLTGKTITLE
Ubiquitin: MQIFVKTLTGKTITLE

Fig. 2. Amino-terminal amino acid sequence (one-letter code) of the isolated factor CIP and of ubiquitin.

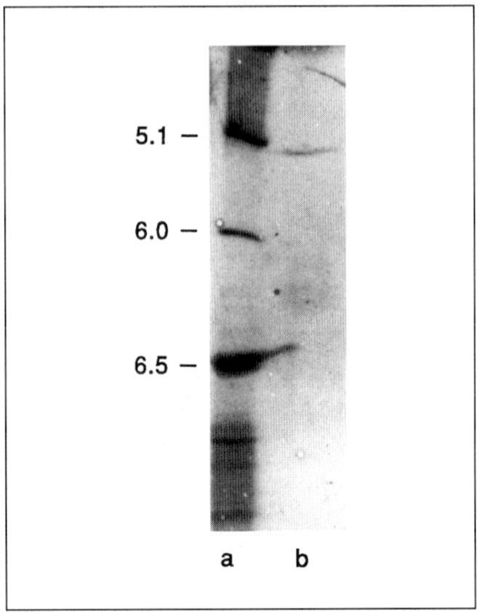

Fig. 3. IFF of CIP (lane b). Lane a = pI marker.

Materials and Methods

Informed consent was obtained from the CAPD patient who did not have any infection interfering with our study. The CAPD effluent was concentrated as previously described [10]. Isolation of the chemotaxis-inhibiting peptide (CIP) was performed as described elsewhere [11].

PMNLs were isolated from the venous blood of healthy volunteers mainly as described by Nauseef et al. [12] and Metcalf et al. [13]. Ficoll-Paque (Pharmacia) (5 ml) was overlaid by 10 ml blood anticoagulated with 10 μl heparin sodium (Liquemin Roche iv, Hoffmann-La Roche, Basel, Switzerland). Most of the erythrocytes sedimented into the Ficoll-Paque layer after about 45 min. A Percoll (Pharmacia) density gradient was used to separate PMNLs, lymphocytes, monocytes, and thrombocytes that remained in the plasma supernatant. The PMNL band was isolated and the PMNLs were washed twice in phosphate-buffered saline (PBS, pH 7.4). The number of isolated cells was determined in a cell counter (Cell Dyn 610, Abbott, Wiesbaden, Germany). The viability of the PMNLs isolated by this procedure was greater than 95% as determined by the trypan blue (Sigma) exclusion assay.

PMNL chemotaxis was determined by the under-agarose method [14]. N-formyl-methionyl-leucyl-phenylalanine (fMLP, Sigma) dissolved in Hank's buffer with Ca^{2+} and Mg^{2+} (Seromed, Biochrom, Berlin, Germany) was used as a chemoattractant at a concentration of 2.5×10^{-7} *M*. PMNLs were resuspended at a concentration of approximately 1×10^{6} cells/10 μl either in PBS or in PBS containing the isolated peptide. During the PMNL migration step, the agarose plates were incubated for 2 h at 37 °C. After fixation with methanol and paraformaldehyde, the cells were stained with Giemsa (Merck, Darmstadt, Germany). The distance migrated under the agarose was determined under the microscope.

Sodium dodecyl sulphate polyacrylamide gel electrophoresis (SDS-PAGE, 8–25% gradient gels) and isoelectric focusing (IEF, IEF 3–9 gels) were performed using the Phast System (Pharmacia LKB Biotechnology, Uppsala, Sweden). For the reducing SDS-PAGE 10% (v/v) β-mercaptoethanol was added to the sample buffer and the samples were boiled for 3 min prior to loading. Protein concentrations were determined by the bicinchroninic acid protein assay (Pierce, Rockford, Ill., USA). As a control, ubiquitin from bovine red blood cells (U-6253, Sigma) was used.

Results

Characterization of the CIP

After concentrating the peritoneal dialysis effluent of a uremic patient, we isolated a peptide that inhibited the chemotactic movement of PMNLs from healthy donors using three different chromatographic methods [11].

Both reducing (fig. 1A, lane a) and nonreducing (fig. 1A, lane b) SDS-PAGE of CIP showed a single band indicating that our isolated factor is pure and does not consist of subunits connected by disulfide bridges.

The result of sequencing the N-terminal amino acids of CIP is shown in figure 2. Within the first 16 amino acids, the primary structure of CIP is identical to that of ubiquitin. As CIP (fig. 1B, lane b) and commercially available ubiquitin (fig. 1B, lane a) comigrate on a SDS-polyacrylamide gel, one can conclude that both substances have the same size. IEF shows that CIP has an isoelectric point of 5.1 (fig. 3). This is much more acidic than the isoelectric point of ubiquitin (about 7) [15]. This indicates that we have isolated a modified form of ubiquitin from the peritoneal effluent.

Effect of CIP and Ubiquitin on PMNL Chemotaxis

Figure 4 shows that CIP significantly inhibits the chemotactic movement of PMNLs, whereas unmodified ubiquitin has no effect. Figure 5 shows that the inhibitory effect of CIP is concentration dependent.

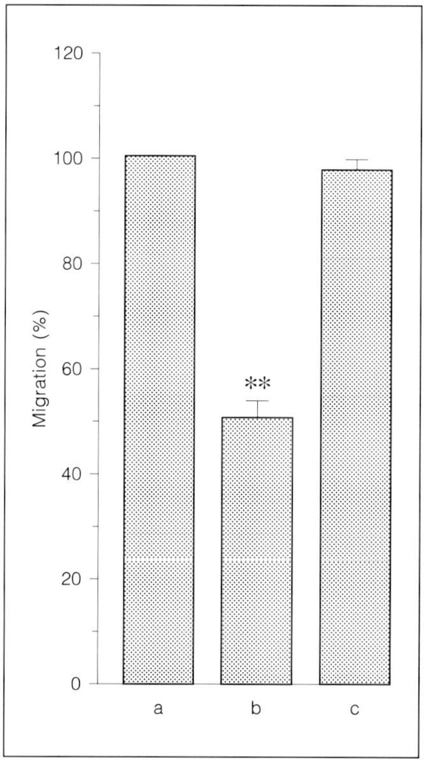

Fig. 4. Effect of buffer (a), CIP (b), and ubiquitin (c) on the chemotactic movement of PMNLs from healthy subjects toward fMLP. Mean ± SE. ** $p < 0.01$.

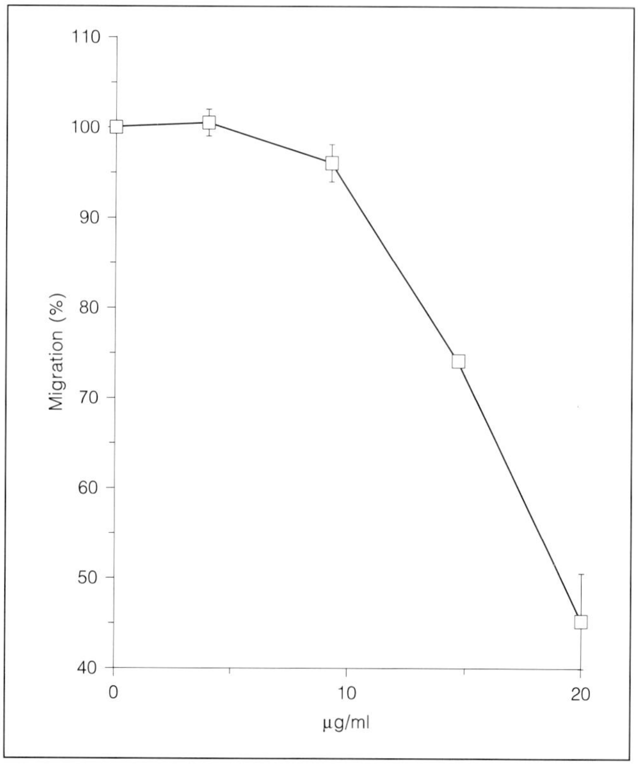

Fig. 5. Effect of different concentrations of CIP on the chemotactic movement of PMNLs from healthy subjects toward fMLP.

Discussion

Uremic toxins in the peritoneal cavity of CAPD patients are at least partially responsible for the diminished functions of PMNLs and consequently for the high risk of peritonitis in CAPD patients. In the present study, we focused on the chemotactic response of PMNLs as the first important step of the nonspecific immune response. We have described a CIP isolated from the peritoneal effluent of a CAPD patient. The inhibition was shown to be concentration dependent. As N-formyl-methionyl peptides are released by microorganisms and represent the main chemoattractants in bacterial infections, we used fMLP as a chemoattractant in our in vitro chemotaxis assay.

Determination of the amino-terminal amino acid sequence of CIP showed complete identity to the primary structure of ubiquitin. Ubiquitin has 76 amino acids and a molecular weight of about 8.5 kD and is highly conserved: bovine ubiquitin, used as a control in our experiments, and human ubiquitin have the same amino acid sequence [16]. Due to an incomplete unfolding ubiquitin shows a faster migration during SDS-PAGE (fig. 1). As indicated by its name, ubiquitin is found in all cells so far, investigated, including bacteria, yeast, and cells in higher plants and animals [17].

The main function of ubiquitin is the degradation of intracellular protein [18]. A variety of short-lived proteins are degraded in a ubiquitin-dependent manner. Ubiquitin is also responsible for the degradation of misfolded proteins: its carboxy terminus is attached to lysine residues of the target protein by an isopeptide linkage and is used as a degradation marker for a protease. Furthermore, ubiquitin is involved in the regulation of gene expression, in the stress response, and in receptor functions [19].

We isolated an obviously modified version of ubiquitin from the peritoneal cavity; however, all the activities of ubiquitin described above take place within the cell. Extracellular ubiquitin was also observed by Pancré et al. [20] and Okada et al. [21]. Pancré et al. [20] found ubiquitin in the culture supernatant of oocytes and they point out that this was the first time that ubiquitin excretion from cells had been observed. The authors also showed that purified ubiquitin can inhibit the cytotoxic properties of platelets. Okada et al. [21] showed that the plasma concentration of ubiquitin in patients with chronic renal failure and in hemodialysis patients is significantly increased. They also demonstrated that ubiquitin has amyloid-enhancing factor activity.

CIP has a more acidic isoelectric point than ubiquitin, as shown by IEF data. We conclude that CIP is a modified form of ubiquitin and that this modification is responsible for the chemotaxis-inhibitory activity. Modifications of ubiquitin that would be compatible with our findings include carbamoylation and glycation. Urea accumulates in the plasma of uremic patients and is spontaneously converted into cyanate (approximately 0.8% of the molar concentration of urea under physiological conditions) [22]. Cyanate can react with the epsilon-amino group of lysins [23]. This reaction causes loss of positive charges in the carbamoylated protein product. Another modification leading to a more acidic protein is glycation [24]: epsilon-amino groups of lysins can react nonenzymatically with sugar aldehydes or ketone groups. It has been shown [25] that protein glycation can occur within the peritoneum during CAPD treatment due to the high glucose concentrations. Furthermore, there are experiments indicating that glycated proteins accumulate preferentially in the CAPD effluent as discussed in Lamb et al. [24]. Since ubiquitin contains seven lysine residues, both carbamoylation and glycation will have a profound effect on its isoelectric point and on its biological activity.

References

1 Holmes CJ: Peritoneal host defense mechanisms in peritoneal dialysis. Kidney Int 1994;46(suppl 48): S58–S70.
2 Ho-Dac-Pannekeet MM, Krediet RT: Inflammatory changes in vivo during CAPD: What can the effluent tell us? Kidney Int 1996;50(suppl 56):S12–S16.
3 Downey GP: Mechanisms of leukocyte motility and chemotaxis. Curr Opin Immunol 1994;6:113–124.
4 Metcoff J, Lindeman R, Baxter D, Pederson J: Cell metabolism in uremia. Am J Clin Nutr 1978;31: 1627–1634.
5 Siriwatratananonta P, Sinsakul V, Stern K, Slavin RG: Defective chemotaxis in uremia. J Lab Clin Med 1978;92:402–407.
6 Mansell MA, Grimes AJ, Jones NF: Leucocyte ATP and renal failure. Clin Science 1981;61:43–46.
7 Ritchey EE, Wallin JD, Shah SV: Chemoluminescence and superoxide anion production by leukocytes from chronic hemodialysis patients. Kidney Int 1981;19:349–358.
8 Cannistra SA, Griffin JD: Regulation of the production and function of granulocytes and monocytes. Semin Hematol 1988;25:173–188.
9 Vanholder R, Lameire N, Waterloos MA, Ringoir S: Depression of PMNL-response to phagocytic challenge in the presence of CAPD-effluent (abstract). Am J Soc Nephrol 1993;4:419A.
10 Cohen G, Haag-Weber M, Mai B, Deicher R, Hörl WH: Effect of immunoglobulin light chains from hemodialysis and continuous ambulatory peritoneal dialysis patients on polymorphonuclear leukocyte functions. J Am Soc Nephrol 1995;6:1592– 1599.
11 Cohen G, Rudnicki M, Hörl WH: Isolation of modified ubiquitin as neutrophil chemotaxis inhibitor from uremic patients. J Am Soc Nephrol, in press.
12 Nauseef WM, Metcalf JA, Root RK: Role of myeloperoxidase in the respiratory burst of human neutrophils. Blood 1983;61:483–492.
13 Metcalf JA, Gallin JI, Nauseef WM, Root RK: Laboratory Manual of Neutrophil Function. New York, Raven, 1986; pp 2–5.
14 Nelson RD, Quie PG, Simmons RL: Chemotaxis under agarese: A new and simple method for measuring chemotaxis and spontaneous migration of human polymorphonuclear leukocytes and monocytes. J Immunol 1975;115:1650–1655.
15 Haas AL, Murphy KE, Bright PM: The inactivation of ubiquitin accounts for the inability to demonstrate ATP, ubiquitin-dependent proteolysis in liver extracts. J Biol Chem 1985;260:4694–4703.
16 Schlesinger DH, Goldstein G: Molecular conservation of 74 amino acid sequence of ubiquitin between cattle and man. Nature 1975;255:423–424.
17 Goldstein G, Scheid M, Hammerling U, Boyse EA, Schlesinger DH, Niall HD: Isolation of a polypeptide that has lymphocyte-differentiating properties and is probably represented universally in living cells. Proc Natl Acad Sci USA 1975;72:11–15.
18 Hershko A: Ubiquitin-mediated protein degradation. J Biol Chem 1988;263:15237–15240.
19 Hochstrasser M: Protein degradation or regulation: Ub the Judge. Cell 1996;84:813–815.
20 Pancré V, Pierce RJ, Foutnier F, Methali M, Delanoye A, Capron A, Auriault C: Effect of ubiquitin on platelet functions: Possible identity with platelet activity suppressive lymphokine (PASL). Eur J Immunol 1991;21:2735–2741.
21 Okada M, Miyazaki S, Hirasawa Y: Increase in plasma concentration of ubiquitin in dialysis patients: Possible involvement in beta2-microglobulin. Clin Chim Acta 1993;220:135–144.
22 Roxborough HE, Millar CA, Mac Eneny J, Young IS: Carbamylation inhibits the ferroxidase activity of caeruloplasmin. Biochem Biophys Res Commun 1995;214:1073–1078.
23 Stark GR, Stein WH, Moore S: Reactions of the cyanate present in aqueous urea with aminoacids and proteins. J Biol Chem 1960;235:3177–3181.
24 Schleicher E, Wieland OH: Protein glycation: Measurement and clinical relevance. J Clin Chem Clin Biochem 1989;27:577–587.
25 Lamb E, Cattel WR, Dawnay A: Glycated albumin in serum and dialysate of patients on continuous ambulatory peritoneal dialysis. Clin Sci 1993;84: 619–626.

Mineral and Electrolyte Metabolism

Miner Electrolyte Metab 1997;23:214–217

Modulation of Protein Kinetics in Chronic Renal Failure

Gianfranco Guarnieri
Gabriele Toigo
Roberta Situlin
Beniamino Ciocchi
Gianni Biolo

Istituto di Clinica Medica,
University of Trieste, Italy

Key Words
Chronic renal failure
Amino acid metabolism
Protein synthesis
Protein degradation
Growth hormone
Cytokines
Tumor necrosis factor-α
Tumor necrosis factor-α soluble receptors
Pentoxifylline

Abstract
Many factors are involved in inducing muscle wasting and derangements of protein metabolism in chronic renal failure. Anorexia, low protein intake, hormonal abnormalities (increased cortisol and parathyroid hormone secretion, and insulin resistance), acidosis, abnormalities of the cytokine system, and other unidentified uremic toxins create a negative nitrogen balance and stimulate protein catabolism. The protein turnover rate (i.e., synthesis and degradation) is generally decreased in non-acidotic uremic patients. However, in uncorrected acidosis, protein degradation is accelerated and a rapid loss of body proteins supervenes. Nutrition can be improved in chronically uremic patients through pharmacological therapy that can control protein catabolism. Administration of growth hormone and insulin-like growth factor-1 has been shown to be effective. These hormones influence both glucose and glutamine metabolism. Another approach is the administration of pentoxifylline, which may have an anticatabolic effect by interfering with the tumor necrosis factor-α system.

Introduction

It is now well established that chronic uremia is a catabolic state associated with significant perturbations of many aspects of protein and amino acid metabolism [1, 2]. Advanced renal failure is characterized by a negative nitrogen balance, impaired growth, and reduced body protein mass [1, 2]. Depletion of lean body mass may contribute to poor rehabilitation, impaired wound healing, and increased susceptibility to infections, and adversely affect morbidity and mortality in chronically uremic patients [3]. Thus, a primary goal of the treatment of patients with renal failure should be to maintain or to improve body protein mass.

Malnutrition

Protein-energy malnutrition is often present in chronically uremic patients, particularly when the glomerular filtration rate falls to about 5–10 ml/min [4]. Symptoms of uremic intoxication include anorexia, nausea, and vomiting. These symptoms often disappear during an optimal dialytic treatment but they are common for patients who are underdialyzed. Additional causes of anorexia in patients with chronic uremia include unpalatable diets, medications, other complicating illnesses, and various psychosocial factors. Furthermore, the problem of malnutrition in the predialysis patients has recently become even more critical, because many of them are treated with long-term low-protein diets in the attempt to slow down the progression of renal disease [5].

It is likely, however, that the causes of protein wasting in patients with advanced renal failure are more complex than simply impaired nutrient intake. The pathogenesis of the abnormalities of protein metabolism in renal failure appears to be multifactorial (table 1). In these patients, the interactions of nitrogenous wastes, uremic toxins, metabolic acidosis, catabolic hormones and cytokines with malnutrition and physical inactivity [1, 4] may lead to protein wasting and abnormalities of amino acid metabolism.

Amino Acid Metabolism in Chronic Uremia

Chronically uremic patients often exhibit selective alterations in free amino acid concentrations, both in plasma and skeletal muscle [1, 2, 6–9]. Concentrations of some essential amino acids (especially leucine, valine, isoleucine, threonine, and tryptophan) are often decreased in uremia. The ratio of tyrosine to phenylalanine is also reduced due to decreased phenylalanine hydroxylation to tyrosine [10]. Postprandial plasma amino acid levels and the clearance of parenterally infused amino acids [6] are also abnormal in patients with chronic renal failure. Some of these abnormalities are similar to those observed in malnutrition, while others are induced by renal failure per se because they occur in well-nourished patients with relatively preserved renal function [9].

KARGER
E-Mail karger@karger.ch
Fax +41 61 306 12 34
http://www.karger.ch

0378–0392/97/0236–0214 $12.00/0

Gianfranco Guarnieri, MD
Istituto di Clinica Medica, University of Trieste
c/o Ospedale di Cattinara, I–34149 Trieste (Italy)
Tel. +39 40 3994533, fax +39 40 3994593
E-Mail segreteria@clmed.univ.trieste.it

Table 1. Pathogenesis of protein metabolism abnormalities in chronically uremic patients

Poor dietary intake (anorexia, nausea, medications, psychosocial factors, unpalatable diet, excessive dietary restrictions)
Reduced muscular activity
Uremic toxins
Acidosis
Anemia
Cytokines
Hormonal abnormalities
Secondary hyperparathyroidism
Insulin resistance
Increased cortisol
Increased prolactin
Decreased conversion of thyroxine to 3,5,3'-triiodothyronine
Decreased growth hormone
Decreased testosterone
Abnormal vitamin D metabolism
Abnormalities of lipid and carbohydrate metabolism
Abnormal pattern of plasma and muscle amino acid concentrations
Intercurrent illnesses (infections)
Amino acid losses during dialysis
'Dialysis catabolism'

Limb and splanchnic arteriovenous catheterization studies suggest that the alterations in the plasma amino acid pattern may stem from an abnormal interorgan flow of amino acids [11, 12]. The imbalance in plasma amino acid profile may be harmful and impair transmembrane amino acid transport, cerebral neurotransmitter synthesis and overall protein synthesis.

Protein Turnover in Chronic Uremia

Body proteins are continuously broken down and resynthesized. The difference between the rates of protein synthesis and breakdown equals the net balance of body proteins which is measured from the difference between nitrogen intake and total nitrogen loss. The absolute rates of protein synthesis and breakdown can be directly measured in vivo using isotopically labelled amino acids. This technique may be used to measure the kinetics of protein in the whole body or in distinct tissues, such as skeletal muscle, by limb arteriovenous catheterization. It is important to emphasize that similar changes in net protein balance may result from different combinations of changes in protein synthesis and degradation. Thus, the decreased body protein mass in uremia [13, 14] may theoretically be the consequence of diminished protein synthesis and/or increased protein degradation. It has been reported that in the skeletal muscle of uremic rats, protein synthesis is decreased and protein degradation is increased or unchanged [15–18]. Studies on muscle and whole-body protein turnover in chronically uremic patients on conservative or hemodialytic treatment were generally in agreement, showing decreased whole-body and muscle protein synthesis and normal or decreased whole-body and muscle protein degradation [19–25]. The low protein synthesis and degradation in patients with chronic renal failure could depend on uremia and/or an adaptive response to undernutrition. For example, the low muscle cathepsin D activity observed in patients with early renal insufficiency (mean serum creatinine 2.6 ± 1.7 mg/dl) on long-term low-protein diets [13] cannot be ascribed to uremic toxins, and is probably part of the normal adaptive response to low protein intake. In vitro studies have shown, however, that extracts of uremic serum can directly inhibit amino acid transport [26] and protein synthesis [27]. On the other hand, accelerated protein degradation has been demonstrated in rats and uremic patients in the presence of metabolic acidosis [28–31].

The role of acidosis as mediator of protein losses has been recently investigated [28–31]. Acidosis stimulates protein catabolism by increasing branched-chain amino acid oxidation and catabolic hormone secretion (e.g., cortisol), and by promoting proteolytic enzyme synthesis (e.g., the ubiquitin system). In chronically uremic patients with metabolic acidosis, the rates of muscle protein synthesis and degradation were both accelerated [31]. Furthermore, the rate of protein degradation directly correlated with plasma bicarbonate concentrations [31]. The involvement of acidosis in increasing protein degradation was directly demonstrated in humans with chronic renal failure [28].

It is well known that insulin-mediated glucose metabolism is impaired in uremia [2]. However, it is still uncertain whether insulin resistance in uremia may also involve the hormone effects on amino acid and protein metabolism and contribute to the protein loss of these patients [2]. Insulin is an anabolic hormone, because it stimulates synthesis and inhibits degradation of selected proteins. It has been reported that protein breakdown is poorly suppressed and protein synthesis is only slightly stimulated by insulin in acutely and chronically uremic rats, possibly as a consequence of a postreceptor defect [32]. In contrast, the suppressive effect of physiological hyperinsulinemia on protein degradation was normal in patients with chronic renal failure [11, 25].

In addition to malnutrition, insulin resistance and acidosis, a number of other direct consequences of uremia can cause protein catabolism, such as excessive glucocorticoid production [2], hyperparathyroidism [2, 23] and other still unidentified uremic toxins. Furthermore, newer evidence suggests that tumor necrosis factor-α (TNF-α) and interleukin-1 may be increased in undialyzed and dialyzed patients with chronic renal failure [34–36]. These cytokines might be able to mediate some of the metabolic dysfunctions of uremia, as the administration of TNF-α and interleukin-1 to experimental animals can induce a protein catabolic response at the whole-body level and in muscle. Secretion of TNF-α is associated with increased circulating levels of soluble TNF-α receptors. Plasma concentrations of soluble TNF-α receptors constantly increase in chronic uremia and correlate with serum creatinine concentration [34, 36]. We have found that, in the plasma of undialyzed chronically uremic patients, the ratio of soluble TNF-α receptors/TNF-α was 50- to 60-fold greater than in normal subjects (unpublished data). Soluble TNF-α receptors have been suggested to function as physiological inhibitors of TNF-α activity. However, it has been recently reported that TNF-α soluble receptors may stabilize TNF-α thereby prolonging its biological actions [37].

The anabolic response to nutrient intake has been recently evaluated in chronically uremic patients. Castellino et al. [25] showed that, in these patients, acute hyperaminoacidemia was less effective than in controls in enhancing protein synthesis [25]. Nonetheless, the normal adaptative response to low-protein diectary intake (decreased protein degradation, lower amino acid oxidation and better utilization of amino acids for protein synthesis) was not impaired in chronic renal failure both in short- and long-term studies [24, 38, 39].

Modulation of Protein Catabolism

Some drugs are potentially useful for improving the efficiency of dietary protein utilization in chronically uremic patients.

Growth Hormone (GH)

Biosynthetic human GH, produced using DNA technology, is now available for clinical practice. GH promotes amino acid transport, stimulates protein synthesis and decreases amino acid oxidation. GH shows both direct and indirect effects via stimulation of insulin-like growth factor-I synthesis. The use of GH in chronically uremic patients has been recently investigated. In these patients, GH may decrease urea production and protein catabolism and improve the anabolic efficacy of nutrition [40].

We report here some results obtained with GH infusion in hypercatabolic patients with severe trauma [41], because the effects of GH on amino acid metabolism in these patients could be relevant for chronically uremic patients. In trauma patients, protein degradation and amino acid oxidation are greatly accelerated. Thus, the efficacy of an anabolic intervention can be tested during acute studies. We infused human recombinant GH into a femoral artery for 24 h and studied leg muscle amino acid exchange and protein metabolism using the leg arteriovenous balance technique; changes of muscle free amino acid concentrations were directly evaluated by biopsy. GH infusion decreased phenylalanine release from protein degradation and the rate of catabolism of the branched-chain amino acids. GH infusion decreased glutamine release from the leg muscle by about 60%. This effect was completely accounted for by a suppression of de novo glutamine synthesis in muscle tissue. It is well known that glutamine is an important precursor for urea synthesis. Thus, the GH-mediated suppression of muscle glutamine synthesis observed in our study may contribute to decreased urea synthesis in the liver. On the other hand, glutamine is necessary for supplying most of the ammonium produced in the kidney and therefore plays a major role in the maintenance of acid-base balance. Furthermore, circulating glutamine is largely utilized as a source of energy in rapidly dividing cells, such as enterocytes, lymphocytes, reticulocytes, fibro- blasts, and keratinocytes. Thus, a GH-mediated decrease of systemic glutamine availability might have negative effects on tissues with rapid cell turnover, as well as on the maintenance of acid-base balance. Another potential negative effect of GH administration is a decrease of insulin sensitivity.

Pentoxifylline

Recent evidence indicates that some drugs, which are clinically used for their cardiovascular effects, may counteract protein catabolism. Pentoxifylline and amrinone, which are phosphodiesterase inhibitors, reduced protein catabolism in septic rats [42, 43], possibly by inhibiting TNF-α secretion from activated macrophages [44]. Furthermore, pentoxifylline administration improved survival after a lethal dose of endotoxin in mice [45]. Since chronic uremia is characterized by alterations in the TNF-α/soluble TNF-α receptor system [34–36], we performed a preliminary study in patients with chronic renal failure to test the hypothesis that therapeutic doses of pentoxifylline decrease proteolysis in humans [46]. Stable isotopes were used to determine leucine kinetics in six patients with stable chronic renal failure (plasma creatinine concentration about $7\ mg \cdot dl^{-1}$) in the basal postabsorptive state, during intravenous infusion of a balanced amino acid mixture ($0.1\ g \cdot kg^{-1} \cdot h^{-1}$) and during combined amino acid and pentoxifylline ($1\ mg \cdot kg^{-1}$ in 3 h) infusions. This study was designed to evaluate the interaction between pentoxifylline and hyperaminoacidemia. Intravenous amino acid infusion decreased the leucine rate of appearance (Ra) from proteolysis with respect to the basal value, whereas combined pentoxifylline and amino acid infusions resulted in a greater suppression of leucine Ra from proteolysis than amino acid infusion alone. In conclusion, pentoxifylline enhanced the amino-acid-mediated suppression of whole-body protein breakdown in uremia. In this study, pentoxifylline infusion during hyperaminoacidemia did not significantly modify TNF-α concentrations.

Acknowledgments

This study was supported by grants from M.U.R.S.T. and C.N.R.

References

1 Guarnieri G, Toigo G, Situlin R, Russo M, Tamaro G, Sergiani G: Protein metabolism in uremia. Issues Nephrosci 1991;1:87–97.
2 May RC: Effects of renal insufficiency on nutrient metabolism and endocrine function; in Mitch WE, Klahr S (eds): Nutrition and the Kidney. New York, Little, Brown 1993, pp 35–60.
3 Harris LE, Luft FC, Rudy DW, Tierney WM: Clinical correlates of functional status in patients with chronic renal insufficiency. Am J Kidney Dis 1993; 21:161–166.
4 Maroni BJ: Requirements for protein, calories, and fat in the predialysis patient; in Mitch WE, Klahr S (eds): Nutrition and the Kidney. New York, Little, Brown, 1993, pp 185–212.
5 Guarnieri G, Toigo G, Situlin R, Carraro M, Tamaro G, Lucchesi A, Oldrizzi L, Rugiu C, Maschio G: Nutritional state in patients on long-term low-protein diet or with nephrotic syndrome. Kidney Int 1989;36:S195–S200.
6 Takala J: Total plasma clearance of intravenous essential amino acids: Evidence of abnormal metabolism of amino acids in chronic renal failure. J Parent Enter Nutr 1983;7:146–150.
7 Deferrari G, Garibotto G, Robaudo C, Canepa A, Bagnaso S, Tizianello A: Leg metabolism of amino acids and ammonia in patients with chronic renal failure. Clin Sci 1985;69:143–151.
8 Toigo G, Situlin R, Tamaro G, Del Bianco A, Giuliani V, Dardi F, Vianello S, Toffoletti P, Faccini L, Guarnieri G: Effect of intravenous supplementation of a new essential amino acid formulation in hemodialysis patients. Kidney Int 1989;36:S278–S281.
9 Bergström J, Alvestrand A, Fürst P: Plasma and muscle free amino acids in maintenance hemodialysis patients without protein malnutrition. Kidney Int 1990;38:108–112.
10 Young GA, Parsons FM: Impairment of phenylalanine hydroxylation in chronic renal insufficiency. Clin Sci Mol Med 1973;45:89–92.
11 Alvestrand A, DeFronzo RA, Smith D, Wahren J: Infuence of hyperinsulinaemia on intracellular amino acid levels and amino acid exchange across splanchnic and leg tissues in uraemia. Clin Sci 1988; 74:155–163.
12 Garibotto G, Russo R, Sofia A, Sala MR, Robaudo C, Moscatelli P, Deferrari G, Tizianello A: Skeletal muscle protein synthesis and degradation in patients with chronic renal failure. Kidney Int 1994; 45:1432–1439.
13 Guarnieri G, Toigo G, Situlin R, Faccini L, Coli U, Landini S, Bazzato G, Dardi F, Campanacci L: Muscle biopsy studies in chronically uremic patients: Evidence for malnutrition. Kidney Int 1983; 24:S187–S193.
14 Ericsson F, Carlmark B: Potassium in whole body, skeletal muscle and erythrocytes in chronic renal failure. Nephron 1983;33:173–178.
15 Mitch WE: Amino acid release from the hindquarter and urea appearance in acute uremia. Am J Physiol 1981;241:E415–E419.
16 Flügel-Link RM, Salusky IB, Jones MR, Kopple JD: Protein and amino acid metabolism in posterior hemicorpus of acutely uremic rats. Am J Physiol 1983;244:E615–E623.
17 Li JB, Wassner SJ: Protein synthesis and degradation in skeletal muscle of chronically uremic rats. Kidney Int 1986;29:1136–1143.
18 May RC, Kelly RA, Mitch WE: Mechanisms for defects in muscle protein metabolism in rats with chronic uremia: Influence of metabolic acidosis. J Clin Invest 1987;79:1099–1103.
19 Conley SB, Rose GM, Robson AM, Bier DM: Effects of dietary intake and hemodialysis on protein turnover in uremic children. Kidney Int 1980; 17:837–846.
20 Guarnieri G, Toigo G, Situlin R, Faccini L, Rustia R, Dardi F: Muscle cathepsin D activity and RNA, DNA and protein content in maintenance hemodialysis patients; in Horl WH, Heidland A (eds): Proteases: Potential role in health and disease. New York, Plenum, 1984, pp 533–543.
21 Hou JC, Zhou J, Zhu H, Zu J, Zang M: Dynamic aspects of whole-body nitrogen metabolism in uremic patients on dietary therapy. Nephron 1986;44: 288–294.
22 Goodship THJ, Lloyd S, Clague MB, Bartlett K, Ward MK, Wilkinson R: Whole body leucine turnover and nutritional status in continuous ambulatory peritoneal dialysis. Clin Sci 1987;73:463–469.
23 Berkelhammer CH, Baker JP, Leiter LA, Uldall PR, Whittal R, Slater A, Wolman SL: Whole-body protein turnover in adult hemodialysis patients as measured by ^{13}C-leucine. Am J Clin Nutr 1987;46: 778–783.
24 Goodship THJ, Mitch WE, Hoerr RA, Wagner DA, Steinmann TI, Young VR: Adaptation to low protein diets in renal failure: Leucine turnover and nitrogen balance. J Am Soc Nephrol 1990;1:66–75.
25 Castellino P, Solini A, Luzi L, Barr JG, Smith DJ, Petrides A, Giordano M, Carroll C, DeFronzo RA: Glucose and amino acid metabolism in chronic renal failure: Effect of insulin and amino acids. Am J Physiol 1992;31:F168–F176.
26 Cernecek P, Spustova V, Dzurik R: Inhibitors of protein synthesis in uremic serum and urine: Partial purification and relationship to amino acid transport. Biochem Med 1982;27:305–316.
27 Delaporte C, Gross F, Anagnostopoulos T: Inhibitory effects of plasma dialysate on protein synthesis in vitro: Influence of dialysis and transplantation. Am J Clin Nutr 1980;33:1407–1410.
28 Reaich D, Channon SM, Scrimgeour CM, Daley SE, Wilkinson R, Goodship THJ: Correction of acidosis in humans with CRF decreases protein degradation and amino acid oxidation. Am J Physiol 1993;265:E230–E235.
29 Price SR, England BK, Bailey JL, Van Vreede K, Mitch WE: Acidosis and glucocorticoids concomitantly increase ubiquitin and proteasome subunit mRNAs in rat muscle. Am J Physiol 1994;267: C955–C960.
30 Mitch WE, Medina R, Grieber S, May RC, England BK, Russ Price S, Bailey JL, Goldberg AL: Metabolic acidosis stimulates muscle protein degradation by activating the adenosine triphosphate-dependent pathway involving ubiquitin and proteasomes. J Clin Invest 1994;93:2127–2133.
31 Garibotto G, Russo R, Sofia A, Sala MR, Robaudo C, Moscatelli P, Deferrari G, Tizianello A: Skeletal muscle protein synthesis and degradation in patients with chronic renal failure. Kidney Int 1994; 45:1432–1439.
32 Mitch WE, May RC, Clark AS, Maroni BJ, Kelly RA: Influence of insulin resistance and amino acid supply on muscle protein turnover in uremia. Kidney Int 1987;32:S104–S108.
33 Garber AJ: Effects of parathyroid hormone on skeletal muscle protein and amino acid metabolism in the rat. J Clin Invest 1983;71:1803–1807.
34 Herbelin A, Nguyen N, Zingraff J, Urena P, Descamps-Latscha B: Influence of uremia and hemodialysis on circulating interleukin-1 and tumor necrosis factor α. Kidney Int 1990;37:116–125.
35 Brockhaus M, Bar-Khayim Y, Gurwicz S, Frensdorff A, Haran N: Plasma tumor necrosis factor soluble receptors in chronic renal failure. Kidney Int 1992;42:663–667.
36 Pereira BJG, Shapiro L, King AJ, Falagas ME, Strom JA, Dinarello CA: Plasma levels of IL-1β, TNFα and their specific inhibitors in undialyzed chronic renal failure, CAPD and hemodialysis patients. Kidney Int 1994;45:890–896.
37 Aderka D, Engleman H, Maor Y: Stabilization of the bioactivity of tumor necrosis factor by its soluble receptors. J Exp Med 1992;175:323–329.
38 Masud T, Young VR, Chapman T, Maroni BJ: Adaptive responses to very low protein diets: The first comparison of ketoacids to essential amino acids. Kidney Int 1994;45:1182–1192.
39 Tom K, Young VR, Chapman T, Masud T, Akpele L, Maroni BJ: Long-term adaptive responses to dietary protein restriction in chronic renal failure. Am J Physiol 1995;268:E668–E677.
40 Zigler TR, Lazarus JM, Young LS, Hakim R, Wilmore DW: Effects of recombinant human growth hormone in adults receiving maintenance hemodialysis. J Am Soc Nephrol 1991;2:1130–1135.
41 Biolo G, Iscra F, Toigo G, Ciocchi B, Situlin R, Gullo A, Guarnieri G: The effects of growth hormone administration on skeletal muscle glutamine metabolism in severely traumatized patients: A preliminary report. Clin Nutr 1997;16:89–91.
42 Beuillé D, Farge MC, Rosé F, Arnal M, Attaix D, Obled C: Pentoxifylline decreases body weight loss and muscle protein wasting characteristics of sepsis. Am J Physiol 1993;265:E660–E666.
43 Jurasinski C, Kilpatrick L, Vary TC: Amrinone prevents muscle protein wasting during chronic sepsis. Am J Physiol 1995;268:E491–E500.
44 Strieter RM, Remick DG, Ward PA: Cellular and molecular regulation of tumor necrosis factor-α production by pentoxifylline. Biochem Biophys Res Commun 1988;155:1230–1236.
45 Schade UF: Pentoxifylline increases survival in murine endotoxin shock and decreases formation of tumor necrosis factor. Circ Shock 1990;31:171–181.
46 Biolo G, Toigo G, Ciocchi B, Situlin R, Vasile A, Carraro M, Faccini L, Guarnieri G: Pentoxifylline acutely decreases proteolysis in uremic patients. Clin Nutr 1996;15:11.

Mineral and Electrolyte Metabolism

Miner Electrolyte Metab 1997;23:218–224

George A. Kaysen
Jane Yeun
Thomas Depner

University of California Davis, Calif., and the Department of Veterans Affairs Northern California Health Care System, Pleasant Hill, Calif., USA

Albumin Synthesis, Catabolism and Distribution in Dialysis Patients

Key Words
Acute-phase reaction
C-reactive protein
Inflammation
Kt/V
Malnutrition
Peritoneal dialysis

Abstract
Hypoalbuminemia predicts mortality in hemodialysis patients with end-stage renal disease (ESRD) and is assumed to result from malnutrition. We report here that plasma albumin levels are reduced significantly in both hemodialysis and peritoneal dialysis (PD) patients compared to normal subjects. Among hemodialysis patients with normal albumin levels (the upper quartile of albumin distribution), albumin synthesis rates, fractional catabolic rates, and distribution between the vascular and extravascular pool are normal, suggesting that ESRD per se does not derange albumin metabolism. Albumin synthesis is reduced in hemodialysis patients who are hypoalbuminemic, but the fractional albumin catabolic rate decreases normally, and albumin distribution between the vascular and extravascular space remains normal, suggesting that hypoalbuminemia results primarily from decreased synthesis. Using multiple linear regression analysis, the strongest correlates to plasma albumin concentration in 115 hemodialysis patients were the plasma level of the acute-phase C-reactive protein (CRP) and the normalized protein catabolic rate (PCRn). These two independent predictors of plasma albumin concentrations are markers of inflammation and of protein intake, respectively. CRP levels correlate more strongly with albumin concentration than does low PCRn. Activity of the acute-phase response is an important predictor of low plasma albumin concentration in hemodialysis patients independently of nutritional factors. External loss of albumin in the dialysate is an additional factor that contributes significantly to hypoalbuminemia in PD patients.

Introduction

Hypoalbuminemia is the strongest independent risk factor for death in hemodialysis patients [1, 2]. Albumin levels prior to initiation of dialysis also predict mortality in patients who initiate peritoneal dialysis (PD) [3]. To understand the relationship between mortality and albumin levels it is important to understand both the effects of hypoalbuminemia as well as the processes that lead to low albumin levels.

Albumin comprises approximately 50% of plasma protein, contributes greatly to plasma colloid osmotic pressure, and acts as a high-capacity binding protein for free fatty acids and divalent cations [4]. Albumin also binds other biologically relevant lipophilic molecules, such as steroid hormones and bilirubin [4]. Despite these seemingly indispensable functions, hereditary analbuminemia is compatible with life. Rats with hereditary analbuminemia reproduce normally and have normal renal function and blood pressure [5, 6]. These observations suggest that the causes of reduced albumin concentration are more likely responsible for the morbidity and mortality associated with hypoalbuminemia than is the reduction in albumin concentration per se.

A change in the concentration of any plasma protein can be a consequence of a change in its rate of synthesis, a change in its rate of catabolism, the development of external losses, and/or a change in the distribution volume of the protein. To establish the cause of hypoalbuminemia it is necessary to measure external losses, synthetic rates, distribution volumes, and catabolic rate.

Albumin is synthesized exclusively in the liver where it is transported into the plasma [4]. The two most important physiologic processes that cause a decrease in albumin synthesis are a nutritional deficiency in amino acid availability [7] and inflammation [8]. Starvation, and especially protein malnutrition causes a decrease in plasma albumin concentration [7]. As a consequence, plasma albumin concentration is frequently used as a clinical marker of nutritional status.

Albumin synthesis is also suppressed by inflammation [8–10] and varies inversely with the rates of synthesis of positive acute-phase reactant proteins [11, 12]. This response to inflammation, the acute-phase response, is mediated by lymphokines, tumor necrosis factor (TNF), interleukin (IL)-1, and IL-6, acting at least in part at the transcriptional level [13]. The synthesis of positive acute-phase proteins is increased by inflammation. These include fibrinogen, C-reactive protein (CRP), α_1 acid glycoprotein, serum amyloid A (the

KARGER
E-Mail karger@karger.ch
Fax +41 61 306 12 34
http://www.karger.ch

George A. Kaysen, MD, PhD
Chief Division of Nephrology,
University of California Davis, TB 136
Davis, CA 95616 (USA)

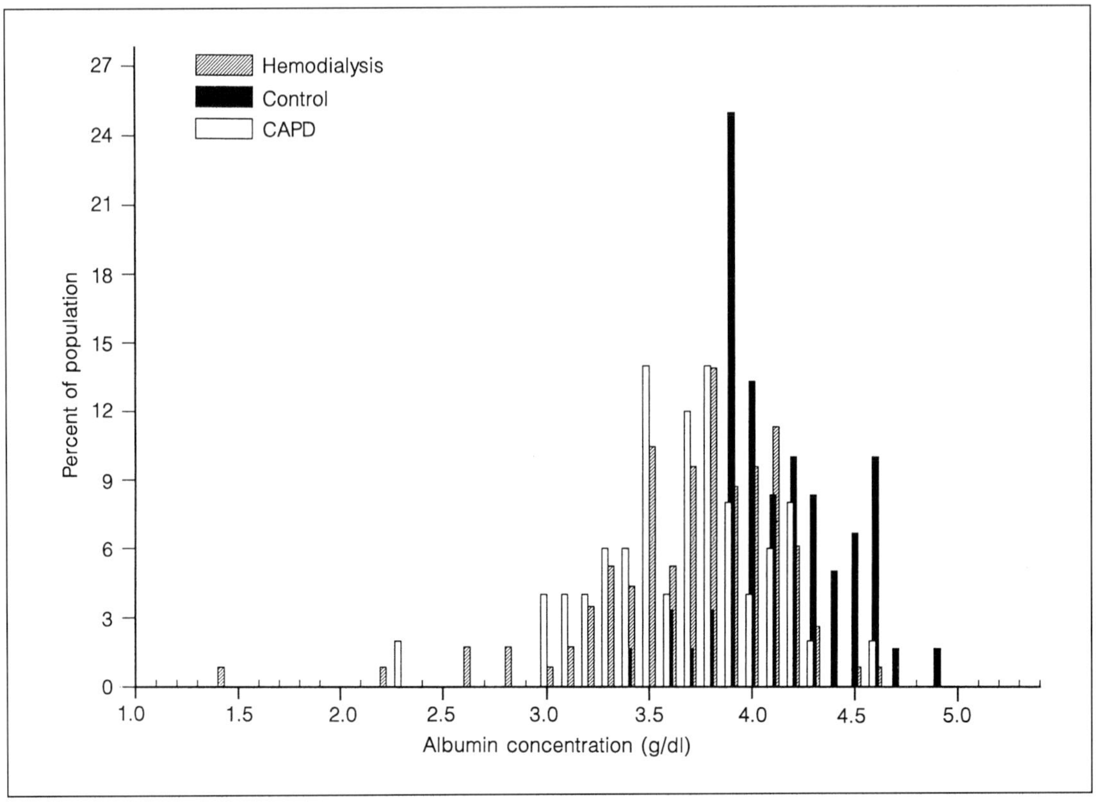

Fig. 1. Distribution of plasma albumin concentrations among 115 hemodialysis patients, 54 PD patients (CAPD), and 60 normal subjects.

protein responsible for secondary amyloidosis in humans), elements of the complement cascade, and α_2-macroglobulin. In addition to albumin, synthesis of several other proteins, including prealbumin (transthyretin) and apolipoprotein A-I is also suppressed by inflammation. These negative acute-phase proteins have, like albumin, been used as markers of nutritional status. Since their synthesis, like that of albumin, is also suppressed during inflammation, measuring their plasma concentrations does not distinguish between an inflammatory basis and a nutritional basis for hypoalbuminemia.

A decrease in the rate of albumin synthesis is followed by a decrease in serum albumin concentration. Since albumin is catabolized by a first-order process, the decreased albumin concentration is followed by a decrease in the absolute rate of albumin degradation. But the decrease in catabolism is caused not only by the decline in albumin concentration but also by a decrease in the rate constant governing albumin catabolism [7]. The decrease in the rate constant causes a decrease in the fraction of the albumin pool that is catabolized per unit of time. This normal physiologic process tends to stabilize albumin levels despite the decrease in synthesis. This process fails in patients with the nephrotic syndrome. The rate of albumin catabolism is not decreased appropriately. Although the absolute rate of albumin catabolism decreases, the fractional rate of albumin catabolism (the fraction of the vascular pool catabolized per unit time) is increased rather than decreased, adding to the burden imposed by urinary losses. The mechanism for this has not been absolutely established, but is hypothesized to be a consequence of renal albumin catabolism [14–16]. While the primary cause of hypoalbuminemia is a consequence of external (urinary) albumin losses, the increased rate of albumin synthesis [14, 17, 18] does not adequately compensate for the combined effect of urinary albumin losses and increased fractional catabolic rate.

Thus hypoalbuminemia in a dialysis patient population could be a consequence of altered distribution, increased fractional catabolism, decreased synthesis, or increased external losses. If albumin levels are reduced in part by a decrease in the rate of albumin synthesis, the fall in synthesis could be a consequence of reduced protein intake, the inflammatory response, or a combination of these processes, each of which could contribute to the high mortality that has been closely associated with hypoalbuminemia in these patients.

Methods

Plasma albumin concentrations in 115 hemodialysis and in 54 PD patients were compared to albumin concentrations in 60 normal subjects, in the same laboratory. Albumin concentration was measured monthly by the bromcresol green method during the 5 months preceding initial measurement of CRP.

Table 1. Variables that predict serum albumin levels in hemodialysis patients

Variable	Coefficient	F to enter	r^2	p
CRP	−0.0421	31.487	0.307	<0.001
PCRn	0.245	12.448	0.404	0.001

Variables subject to analysis in 115 hemodialysis patients were analyzed by multiple forward stepwise regression. Variables excluded by this analysis were Kt/V and fibrinogen.

Kt/V and normalized protein catabolic rate (PCRn) were measured monthly from predialysis and immediate postdialysis blood urea nitrogen (BUN) levels using a formal single-compartment model of hemodialysis urea kinetics [19]. Care was taken when drawing the postdialysis BUN to avoid errors from access recirculation and urea rebound [20]. CRP was measured by end-point nephelometry.

Albumin synthesis was previously measured in 5 normal control subjects [21], 6 hemodialysis patients with albumin values in the upper quartile of albumin concentrations (≥4.0 g/dl), 6 hemodialysis patients with albumin values in the lowest quartile of albumin concentrations (≤3.5 g/dl) [22], and in 9 subjects on continuous ambulatory peritoneal dialysis [21]. The kinetic approach used allowed an estimate of total albumin mass (TAM); plasma albumin mass (PAM) was calculated from the plasma volume, measured by isotope dilution, and the plasma albumin concentration. The ratio of PAM/TAM was obtained.

External albumin loss was measured in the PD patients by measuring albumin in dialysate and urinary albumin losses [21]. External albumin loss was determined in hemodialysis patients by measuring urinary losses and losses of albumin across the dialysis membrane as previously described [22]. The albumin catabolic rate was calculated as the difference between albumin turnover (represented as albumin synthesis) and measured external loss. The fractional rate of albumin catabolism was then calculated as the percent of the plasma albumin pool (PAM) catabolized per unit of time.

Statistical Methods

Data were first analyzed for normality of distribution using the Kolmogorov-Smirnov test for normality [using Sigma Stat, a commercial statistics program (Jandel, San Rafael, Calif., USA)]. Normally distributed data are presented as means ± SD and nonnormally distributed data as median values. If data were nonnormally distributed, they were analyzed by the Kruskal-Wallis one-way analysis of variance on ranks. If differences were detected, multiple groups were then analyzed using Dunn's method for multiple comparisons. The Mann-Whitney sum test was used to make comparisons when only two groups were analyzed. Differences were considered statistically significant at $p<0.05$. Correlative data were analyzed using multiple stepwise linear regression. Linear relationships were plotted using the method of least squares following log transformation.

Results

The median value for albumin concentration was significantly reduced in both dialysis patient groups compared to normal subjects ($p<0.001$, median 4.10, mean 4.133 ± 0.308 g/dl, n = 60), but there was no difference between hemodialysis patients (median 3.800, mean 3.704 ± 0.451 g/dl, n = 115) and PD patients (median 3.711, mean 3.743 ± 0.326 g/dl, n = 54) (fig. 1). Multiple regression analysis showed that the average albumin concentration in hemodialysis patients was a function of both plasma CRP levels and PCRn. Of the two, CRP was a more powerful predictor of albumin concentration

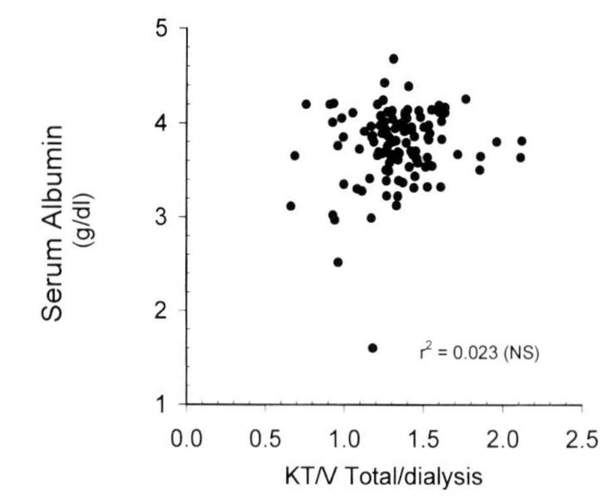

Fig. 2. Relationship between 5-month average plasma albumin concentration and average Kt/V total/dialysis delivered during that period.

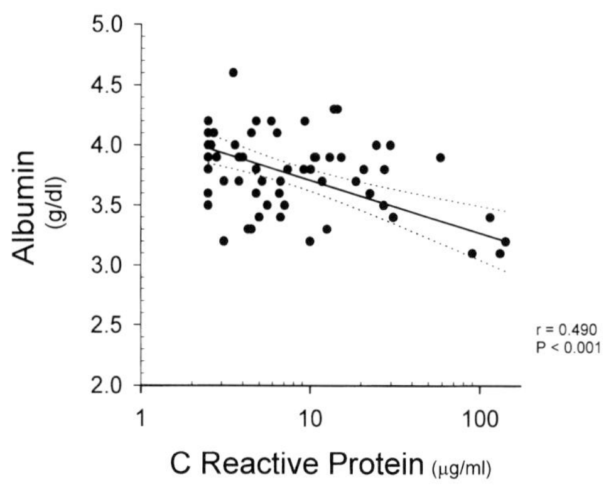

Fig. 3. Relationship between plasma albumin concentration and CRP concentration in hemodialysis patients.

(table 1). Kt/V did not correlate with either albumin concentration (fig. 2; r = 0.153, NS) or with PCRn (r = 0.156, NS).

CRP levels were highly skewed in the hemodialysis patient population and correlated negatively with plasma albumin levels (fig. 3). The distribution of albumin concentrations in the hemodialysis population was also skewed when compared with a normal population (fig. 1). To factor out the effect of CRP and PCRn on albumin, patients were ranked by CRP levels. Those patients with serum CRP levels in the highest quartile were removed from the analysis. The remaining patients were then ranked by PCRn values and those in the lowest quartile were similarly removed. After eliminating the serum albumin values from these patients, the albumin distribution was now normally distributed and fell within the range for normal control subjects (fig. 4). However, serum albumin levels were still significantly lower than in controls (median value 3.900 g/dl, mean 3.926 ± 0.260 g/dl, p = 0.003).

Albumin concentration, synthesis rate, and fractional catabolic rate were the same in hemodialysis patients with plasma albumin

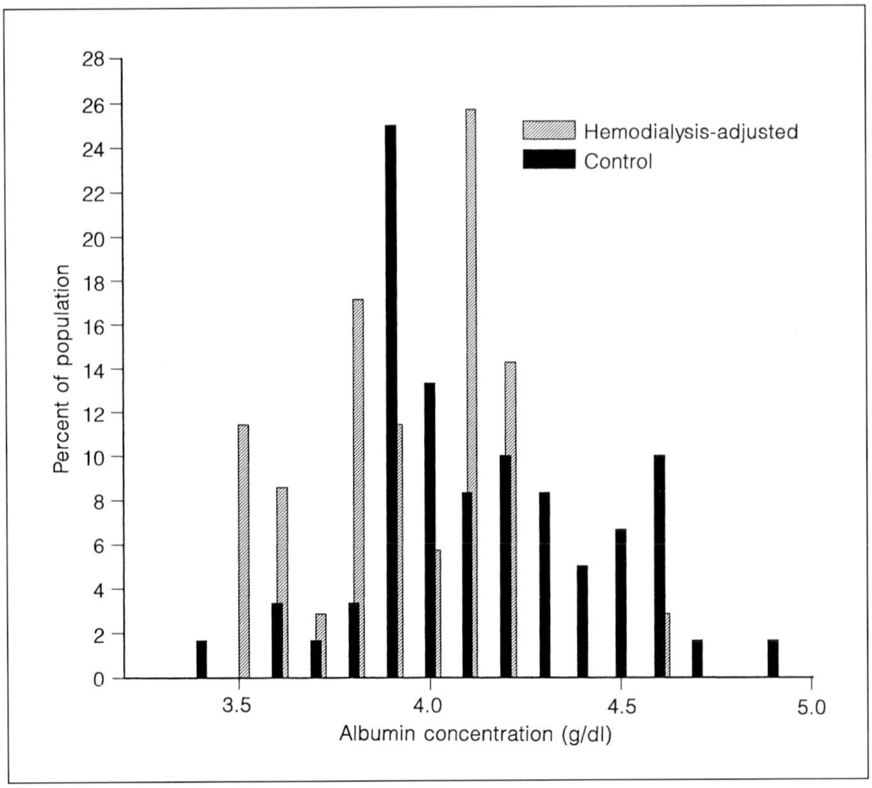

Fig. 4. Distribution of plasma albumin concentrations among hemodialysis patients compared to normal subjects. Patients with CRP concentrations in the upper quartile and patients with PCRn in the lowest quartile were removed.

concentrations in the upper quartile compared to control subjects. Hemodialysis patients with albumin levels in the lowest quartile had a significantly reduced rate of albumin synthesis compared both to normal control subjects and to hemodialysis patients with albumin concentrations in the upper quartile. The distribution of albumin between the plasma and extraplasma space was the same in hemodialysis patients with hypoalbuminemia compared to patients with normal albumin levels and no different from control subjects (table 2). External albumin losses both in urine and across the hemodialysis membrane were significant but there was no difference between the low-albumin and normal-albumin groups (table 2). Losses were significantly less than the transperitoneal losses observed in the companion PD population.

The PD patients presented in table 2 were not chosen by plasma albumin concentration. Despite greater external losses of albumin, their mean concentrations were similar to that in the current PD patient population as a whole and no different from the mean value in the hemodialysis population. Albumin synthesis was not reduced in this group of 9 PD patients. The fractional rate of albumin catabolism was significantly reduced (table 2) compared to control subjects because several patients had large external losses [21]. Redistribution of albumin from the vascular to the exravascular compartment did not contribute to hypoalbuminemia in these patients. Indeed, the fraction of the albumin mass in the plasma compartment tended to be increased rather than reduced, although this difference was not statistically significant.

Discussion

Albumin levels are reduced in some patients treated with either hemodialysis or PD, yet hypoalbuminemia is not inevitable. Studies of albumin turnover have shown that end-stage renal disease (ESRD) in and of itself does not lead to reduced albumin synthesis, to an alteration in albumin fractional catabolic rate, or to an abnormal distribution of the albumin pools. Hypoalbuminemia in dialysis patients is a consequence of well-defined processes, principally reduced albumin synthesis with a substantial contribution by external losses in some PD patients [21].

Among those patients with hypoalbuminemia, there is an appropriate decrease in the fractional rate of albumin catabolism. This contrasts to the nephrotic state where the fractional rate of albumin catabolism increases as the albumin concentration declines [15, 17, 23]. Perhaps the key to understanding the morbidity and mortality associated with hypoalbuminemia in the hemodialysis patient population can be found in the biological basis for decreased albumin synthesis.

There are several possible processes that might suppress the albumin synthetic rate in hemodialysis patients. Much attention has been focused on protein calorie malnutrition as the cause [1]. It is uncontested that protein malnutrition decreases albumin synthesis [7] by both transcriptional [24] and posttranscriptional mechanisms [25, 26] and is a potential cause of reduced albumin synthesis, decreased albumin levels, and increased morbidity and mortality in

Table 2. Relationship between serum albumin concentration and fractional albumin catabolic rate in normal subjects and hemodialysis patients with normal and reduced serum albumin levels

Group	Albumin concentration g/dl	Albumin synthesis rate g/1.73 m² per day	Albumin fractional catabolic rate % plasma pool/day	External albumin loss g/1.73 m² per day	PAM/TAM
Control (n = 5)	4.63 ± 0.06	13.84 ± 0.77	12.80 ± 0.74	0	0.41 ± 0.027
Hemodialysis patients					
Upper quartile (n = 6)	4.19 ± 1.10	14.70 ± 2.34	12.67 ± 3.84	0.49 ± 0.35	0.43 ± 0.078
Lower quartile (n = 6)	3.23 ± 0.46*	11.30 ± 1.10*	9.10 ± 0.65*	1.51 ± 0.91 [a]	0.41 ± 0.057
PD patients (n = 9)	3.74 ± 0.503	13.99 ± 2.52	8.43 ± 1.87*	4.23 ± 1.42 [a]	0.48 ± 0.081

Values are means ± SD. * $p < 0.05$ vs. control.

[a] Urinary losses plus loss across dialysis membrane (transperitoneal loss for PD patients, loss across hemodialysis dialyzer for hemodialysis patients).

any chronically ill population, including those with ESRD. Malnutrition has been clearly documented in this patient population [27, 28], and it has been generally assumed that hypoalbuminemia in this population is purely nutritionally based. The prevalent hypothesis explaining the link between albumin levels and increased mortality suggests that insufficient delivery of dialysis decreases appetite, which then reduces protein intake and ultimately leads to decreased albumin concentration as a consequence of the resulting decrease in albumin synthetic rate. Kopple et al. [29] propose that dietary calorie and protein intake should be 35 kcal/kg per day and between 1.0 and 1.2 g/kg per day, respectively, in patients maintained with hemodialysis. The latter is greater than the 1.0 g/kg per day recommended for the rest of the population and generally more protein than many hemodialysis patients consume, as judged by urea generation rates.

The rate of albumin synthesis normally responds more rapidly to acute changes in dietary protein [4, 30] than do improvements in general nutritional status. When severely malnourished animals or people are refed, the rate of albumin synthesis increases promptly, even while total body protein stores are still severely depleted [31, 32]. The albumin synthetic rate increases within 15 min after feeding rats a complete mixture of amino acids following a 10-day period of protein depletion. In contrast to normal protein malnutrition, oral nutritional supplementation does not reverse hypoalbuminemia in the dialysis patient population [33–35]. Intradialytic parenteral nutrition has also been either ineffective [36] or only partially effective [37] in attempts to repair hypoalbuminemia in hemodialysis patients. The failure of nutritional intervention to correct hypoalbuminemia suggests that other factors may be responsible for reduced albumin concentration in some of these patients despite the statistical relationship between PCRn and albumin levels. Indeed, we found no nutritional differences, either in dietary intake, PCRn, or by anthropometric measurements in hemodialysis patients with either normal or low albumin concentrations in whom albumin synthesis was measured [22].

Since no significant correlation was found in this group of 115 patients between Kt/V and PCRn measured either monthly or averaged over 3 months, the concept that inadequate dialysis caused reduced appetite and reduced intake of protein and calories is not supported either. However, since dialysis was carefully tailored to deliver a minimum Kt/V of 1.2/dialysis in each patient, this conclusion is not firm and interventional studies would be required to answer this question. It is of interest that the previously described mathematical correlation between Kt/V and PCRn was also not seen in these patients [38].

Albumin synthesis is also regulated as part of the acute-phase response [7, 8]. Among the usual influences on plasma albumin concentration, the factor having the greatest influence in our hemodialysis patient population was the plasma concentration of the acute-phase protein, CRP. Furthermore, while no nutritional differences were evident in the two groups of hemodialysis patients in whom albumin synthesis was determined, there were differences in the plasma levels of acute-phase proteins [22], again providing evidence that the inflammatory response may be causally linked to the reduction of albumin synthesis that lowered albumin levels in these patients. In preliminary studies reported in abstract form, Bergström et al. [39] reported that plasma CRP concentration was the most powerful predictor of plasma albumin, and was a better predictor of death at 1 year of dialysis than was albumin concentration. When albumin levels were statistically corrected for CRP, albumin lost its value as a predictor of death.

Supporting the hypothesis that the acute-phase response may in part be responsible for hypoalbuminemia in this patient population are the observations that plasma levels of a number of lymphokines, TNF-α and IL-1 [40], as well as a group of acute-phase reactive proteins, α_1 acid glycoprotein, ceruloplasmin, and C4 complement component [41, 42], are increased in hemodialysis patients compared to other patients with chronic renal failure. Cytokine levels were also recently reported to correlate with albumin levels in hemodialysis patients, and cytokine levels, like albumin, predicted survival [43]. Thus substantial evidence, including that presented here, supports a hypothesis that the acute-phase response is either activated in part by hemodialysis, or that a significant subpopulation of hemodialysis patients are in a chronic inflammatory state which plays a role in reducing plasma albumin levels.

Increased cytokine production causes an increase in metabolic rate and is associated with altered energy metabolism and intake, despite a theoretically adequate diet [44]. Cytokines may also cause patients to decrease food intake, superimposing malnutrition upon a primary inflammatory process. When these two processes occur simultaneously, the decrease in body weight is greater than expected from the decreased food intake alone. Thus the inflammatory response may cause hypoalbuminemia and reduced muscle mass by reducing nutritional intake, adding to the direct effect of cytokines that depletes the muscle and plasma protein pool.

In patients maintained with PD, hypoalbuminemia is in part a consequence of transperitoneal losses of albumin. As in nephrotic patients, albumin synthesis increases in response to external losses in these patients [21]. However, unlike the nephrotic syndrome, the fractional rate of albumin catabolism is not increased, and actually tends to fall as plasma albumin concentration decreases [23], and was significantly reduced in the patients presented here. Preservation of the normal physiologic decrease in albumin fractional catabolic rate in response to hypoalbuminemia in these patients is most probably responsible for the higher albumin concentration in PD patients compared to nephrotic patients with similar extracorporeal albumin losses [23]. Similar to hypoalbuminemic states caused by malnutrition and the nephrotic syndrome and similar to hypoalbuminemic hemodialysis patients, redistribution of the albumin pool does not contribute to hypoalbuminemia in PD patients.

Hypoalbuminemia in the hemodialysis patient population is primarily a consequence of reduced albumin synthesis, which itself is at least in part a consequence of activation of the acute-phase response. The cause of the inflammatory response may be related to dialysis, but this has by no means been established. Unrecognized or undocumented episodes of infection or other processes may cause or contribute to this inflammatory process. When present, inflammation may prevent normal anabolic processes from occurring during attempts to provide adequate nutrition and may contribute to the increased morbidity and mortality in this population.

Acknowledgments

This work was supported in part by the research service of the United States Department of Veterans Affairs, in part by a grant from the National Institutes of Health RO1 DK 42297, and in part by a grant from Dialysis Clinics Incorporated.

References

1 Lowrie EG, Lew NL: Death risk in hemodialysis patients: The predictive value of commonly measured variables and an evaluation of death rate differences between facilities. Am J Kidney Dis 1990;15:458–482.
2 Lowrie EG, Lew NL, Huang WH: Race and diabetes as death risk predictors in hemodialysis patients. Kidney Int 1992;(suppl 38)42:S22–S31.
3 Avram MM, Goldwasser P, Erroa M, Fein PA: Predictors of survival in continuous ambulatory peritoneal dialysis patients: The importance of prealbumin and other nutritional and metabolic markers. Am J Kidney Dis 1994;23:91–98.
4 Rothschild MA, Oratz M, Evans CD, Schreiber SS: Albumin synthesis; in Rosemoer M, Oratz M, Rothschild A (eds): Albumin Structure, Function and Uses. New York, Permagon, 1977, pp 227–255.
5 Fujihara CK, Limongi DM, De Oliveira HC, Zatz R: Absence of focal glomerulosclerosis in aging analbuminemic rats. Am J Physiol 1992;262:R947–R954.
6 Fujihara CK, Limongi DM, Falzone R, Graudenz MS, Zatz R: Pathogenesis of glomerular sclerosis in subtotally nephrectomized analbuminemic rats. Am J Physiol 1991;261:F256–F264.
7 Kirsch R, Frith L, Black E, Hoffenberg R: Regulation of albumin synthesis and catabolism by alteration of dietary protein. Nature 1968;217:578–579.
8 Moshage HJ, Janssen JAM, Franssen JH, Hafkenscheid JCM, Yap SH: Study of the molecular mechanisms of decreased liver synthesis of albumin in inflammation. J Clin Invest 1987;79:1635–1641.
9 Koj A, Gauldie J, Sweeney GD, Regoeczi E, Sauder DN: A simple bioassay for monocyte-derived hepatocyte stimulating factor: Increased synthesis of α_2-macroglobulin and reduced synthesis of albumin by cultured rat hepatocytes. J Immunul Methods 1985;76:317–327.
10 Bauer J, Weber W, Tran-Thi T, Northoff G, Decker K, Gerok W, Heinrich PC: Murine interleukin 1 stimulates α_2-macroglobulin synthesis in rat hepatocyte primary cultures. FEBS Lett 1985;190:271–274.
11 Jamieson JC, Kaplan HA, Woloski BMNJ, Hellman M, Ham K: Glycoprotein biosynthesis during the acute-phase response to inflammation. Can J Biochem Cell Biol 1983;61:1041–1048.
12 Koj A, Gauldie J, Regoecsi E, Sauder DN, Sweeney GD: The acute-phase response of cultured rat hepatocytes. Biochem J 1984;224:505–514.
13 Perlmutter DH, Dinarello CA, Punsal PI, Colten HR: Cachectin/tumor necrosis factor regulates hepatic acute-phase gene expression. J Clin Invest 1986;78:1349–1355.
14 Jensen H, Rossing N, Anderson SB, Jarnum S: Albumin metabolism in the nephrotic syndrome in adults. Clin Sci 1967;33:445–457.
15 Katz J, Bonorris G, Sellers AL: Albumin metabolism in aminonucleoside nephrotic rats. J Lab Clin Med 1963;62:910–934.
16 Katz J, Sellers AL, Bonorris G: Effect of nephrectomy on plasma albumin catabolism in experimental nephrosis. J Lab Clin Med 1964;63:680–686.
17 Kaysen GA, Gambertoglio J, Jiminez I, Jones H, Hutchison FN: Effect of dietary protein intake on albumin homeostasis in nephrotic patients. Kidney Int 1986;29:572–577.
18 Ballmer PE, Weber BK, Roy-Chaudhury P, McNurlan MA, Watson H, Power DA, Garlick PJ: Elevation of albumin synthesis rates in nephrotic patients measured with [1-^{13}C]leucine. Kidney Int 1992;41:132–138.
19 Depner TA: Single-compartment model; in Prescribing Hemodialysis: A Guide to Urea Modeling. Boston, Kluwer 1991, pp 65–89.
20 Depner TA: Pitfalls in quantitating hemodialysis. Semin Dial 1993;6:127–133.
21 Schoenfeld PY, Henry RR, Laird NM, Roxe DM: Assessment of nutritional status of the national cooperative dialysis study population. Kidney Int 1983;23(suppl 13):80–88.
22 Kaysen GA, Rathore V, Shearer GC, Depner TA: Mechanisms of hypoalbuminemia in hemodialysis patients. Kidney Int 1995;48:510–516.
23 Kaysen GA: Hypoalbuminemia in dialysis patients. Semin Dial 1996;9:249–256.
24 Pain VM, Clemens MJ, Garlick PJ: The effect of dietary protein deficiency on albumin synthesis and on the concentration of active albumin messenger ribonucleic acid in rat liver. Biochem J 1978;172:129–135.
25 Yap SH, Strair RK, Shafritz DA: Effect of a short term fast on the distribution of cytoplasmic albumin messenger ribonucleic acid in rat liver. J Biol Chem 1978;253:4944–4950.
26 Yap SH, Strair RK, Shafritz DA: Identification of albumin mRNPs in the cytosol of fasting rat liver and influence of tryptophan or a mixture of amino acids. Biochem Biophys Res Commun 1978;83:427–433.
27 Blumenkrantz MJ, Kopple JD, Gutman RA, Chan YK, Barbour GL, Roberts C, Schen GH, Gandi VC, Tucker CT, Curtis FK, Coburn JW: Methods for assessing nutritional status of patients with renal failure. Am J Clin Nutr 1980;33:1567–1585.
28 Acchiardo SR, Moore LW, Latour AP: Malnutrition as the main factor in morbidity and mortality of hemodialysis patients. Kidney Int 1983;24(suppl 16):S199–S203.
29 Kopple J, Shinaberger J, Coburn J, Sorensen M, Rubini M: Optimal dietary protein treatment during chronic hemodialysis. ASAIO Trans 1989;15:302–308.

30 Morgan EH, Peters T Jr: The biosynthesis of rat serum albumin. J Biol Chem 1971;246:3500–3507.
31 Hoffenberg R, Black E, Brock JF: Albumin and gamma-globulin tracer studies in protein depletion states. J Clin Invest 1966;45:143–152.
32 James WP, Hay AM: Albumin metabolism: Effect of the nutritional state and the dietary protein intake. J Clin Invest 1968;47:1958–1972.
33 Phillips ME, Havard J, Howard JP: Oral essential amino supplementation in patients in maintenance hemodialysis. Clin Nephrol 1978;9:241–248.
34 Hecking E, Kohler H, Zobel R, Lamellel EM, Mader H, Opferkuch W, Prellwitz W, Keim HF, Muller D: Treatment with essential amino acids in patients on chronic hemodialysis: A double blind crossover study. Am J Clin Nutr 1987;31:1821–1826.
35 Compher C, Mullen J, Barker C: Nutritional support in renal failure. Surg Clin North Am 1991;71: 597–608.
36 Wolfson M: Use of nutritional supplements in dialysis patients. Semin Dial 1992;5:285–290.
37 Capelli JP, Kushner H, Camiscioli TC, Chen SM, Tores MA: Effect of intradialytic parenteral nutrition on mortality rates in end-stage renal disease care. Am J Kidney Dis 1994;23:808–816.
38 Uehlinger DE: Another look at the relationship between protein intake and dialysis dose. J Am Soc Nephrol 1996;7:166–168.
39 Bergström J, Heimbürger O, Lindholm B, Qureshi AR: Elevated serum CRP is a strong predictor of increased mortality and low serum albumin in hemodialysis (HD) patients (abstract). J Am Soc Nephrol 1995;6:573.
40 Herbelin A, Nguyen AT, Zingraff J, Urena P, Descamps-Latscha B: Influence of uremia and hemodialysis on circulating interleukin-1 and tumor necrosis factor alpha. Kidney Int 1990;37: 116–125.
41 Docci D, Bilancioni R, Baldrati L, Capponcini C, Turci F, Feletti C: Elevated acute phase reactants in hemodialysis patients. Clin Nephrol 1990;34:88–91.
42 Trznadel K, Luciak M, Paradowski M, Kubasiewicz-Ujma B: Hemodialysis and the acute-phase response in chronic uremic patients. Int J Artif Organs 1989;12:762–765.
43 Bologa RM, Levine DM, Parker TS, Serur D, Stenzel KH, Rubin A, Cheigh JS: High cytokine levels may mediate the high mortality of hemodialysis (HD) patients with low albumin and/or cholesterol (TC) (abstract). J Am Soc Nephrol 1995;6:573.
44 Roubenoff R, Roubenoff RA, Cannon JG, Kehayias JJ, Zhuang H, Dawson-Hughes B, Dinarello CA, Rosenberg IH: Rheumatoid cachexia: Cytokine-driven hypermetabolism accompanying reduced body cell mass in chronic inflammation. J Clin Invest 1994;93:2379–2386.

Mineral and Electrolyte Metabolism

Miner Electrolyte Metab 1997;23:225–228

Bradley J. Maroni

Renal Division,
Emory University School of Medicine,
Atlanta, GA., USA

Protein Restriction and Malnutrition in Renal Disease: Fact or Fiction?

Key Words
Chronic renal failure
Low-protein diet
Malnutrition
Nitrogen balance

Abstract
The protein and energy requirements of chronic renal failure (CRF) patients are similar to normal subjects and evidence indicates that both nephrotic *and* nonnephrotic CRF patients can activate normal homeostatic responses allowing them to achieve a neutral nitrogen balance when dietary protein intake is restricted. The benefits of low-protein (and phosphorus) diets (LPDs) include the amelioration of uremia symptoms and some of its metabolic complications and possibly a slowing of the rate of progression of renal failure. When LPDs are prescribed, patients should be monitored to assess dietary compliance and to ensure nutritional adequacy. Recent evidence that the protein intake of patients with progressive CRF decreases when they consume unrestricted diets should not be interpreted as an argument against the use of LPDs. Rather, it is a persuasive argument to restrict dietary protein intake in order to minimize complications of renal failure while maintaining nutritional status.

Introduction

It is well established that a protein-restricted diet can ameliorate many uremic symptoms and prevent some of the complications of chronic renal failure (CRF) including renal osteodystrophy, insulin resistance, and metabolic acidosis [1, 2]. Evidence also suggests that low-protein diets (LPDs) may slow the rate of progression of renal failure and prolong the time until end-stage renal failure [3–6]. On the other hand, concern has been raised that LPDs cause malnutrition, and it has been suggested that these diets should be used cautiously in CRF patients [7, 8]. These fears arise from two observations: (1) evidence of a spontaneous decrease in protein intake and worsening of some indices of nutritional status when CRF patients consume unrestricted diets, and (2) the association between hypoalbuminemia and mortality in hemodialysis patients [9, 10]. There is evidence that both concerns are misplaced and that LPDs can be used safely in CRF patients.

Normal Compensatory Responses to Dietary Protein Restriction

During fasting, body protein stores (primarily skeletal muscle) are degraded and the amino acids utilized for hepatic gluconeogenesis [11]. In response to feeding, anabolism (i.e., protein synthesis > protein degradation) is achieved by a suppression of protein degradation, with or without an accompanying stimulation of protein synthesis [12, 13]. The net response over 24 h is neutral nitrogen balance (B_N) and preservation of lean body mass. When protein intake is restricted, the principal compensatory response is a dramatic reduction in amino acid oxidation, resulting in more efficient utilization of dietary essential amino acids (EAAs) [14, 15]. Assuming the protein intake is adequate, waste nitrogen excretion decreases and neutral B_N is achieved at the lower intake. Thus, in healthy adults, the adaptive metabolic responses to dietary protein restriction include a marked reduction in amino acid oxidation and postprandial suppression of protein degradation ± stimulation of protein synthesis.

Can LPDs Be Used Safely in CRF Patients?

Two dietary regimens have been used to treat patients with progressive CRF: (1) a conventional LPD providing about 0.6 g protein/kg per day, or (2) a very low protein diet (VLPD) providing about 0.3 g protein/kg per day, supplemented with a mixture of EAAs or their nitrogen-free ketoanalogs (i.e., ketoacids, or KAs). Since energy expenditure (and hence energy requirements) of CRF patients are similar to normal subjects [16], 35 kcal/kg per day is usually recommended [17].

More than 20 years ago, Kopple and Coburn [18] demonstrated that B_N was neutral or positive when CRF patients consumed about 0.6 g protein and 35 kcal/kg body weight per day. A VLPD supplemented with either EAAs or KAs also yields neutral B_N in CRF pa-

KARGER
E-Mail karger@karger.ch
Fax +41 61 306 12 34
http://www.karger.ch

Bradley J. Maroni, MD
Emory University School of Medicine
WMRB, Room 338, 1639 Pierce Drive
Atlanta, GA 30322 (USA)
Tel. +1 404 727 3959, fax +1 404 727 3425

tients [19, 20]. Further evidence that LPDs can maintain lean body mass in CRF patients is provided by studies indicating that normal values of serum proteins and anthropometrics and neutral B_N are maintained during long-term therapy [19, 21, 22]. In the full-scale Modification of Diet in Renal Disease Study, there were some minor, albeit statistically significant, changes in some nutritional parameters in the LPD groups (i.e., *increase* in serum albumin and *decrease* in serum transferrin and anthropometrics) [23]. Notably, abnormal values stabilized following the first few months of therapy and mean values for the LPD groups remained within the normal range. Most importantly, a lower protein intake was not associated with a higher rate of death, hospitalizations, or stop points.

These results suggest that patients with CRF must be able to invoke normal adaptive responses to dietary protein restriction. Indeed, Goodship et al. [24] found that when CRF patients were fed diets providing 0.6 or 1.0 g protein/kg per day, B_N and rates of whole-body turnover (i.e., protein synthesis and degradation) and leucine oxidation did not differ from control subjects. Similarly, we found that the B_N remained neutral and rates of whole-body protein turnover and leucine oxidation did not differ when measured around 3 weeks after initiation of the VLPD/KA regimen and following more than 1 year of therapy [22]. Neutral B_N was achieved by a marked suppression of amino acid oxidation and a postprandial inhibition of protein degradation [22].

Due in part to the association between hypoalbuminemia (and presumably nutritional status) and mortality in end-stage renal disease patients [9, 10], it has been suggested that LPDs should be avoided in patients with CRF (see below) [7, 8]. In fact, there is no evidence that the use of LPDs prior to the initiation of dialysis increases the risk of mortality during dialysis. Coresh et al. [25] found that when CRF patients were treated with a supplemented VLPD before starting dialysis, survival during the first 2 years on dialysis may actually have improved. These results lead to two conclusions; first, *nonacidotic* CRF patients can activate normal adaptive responses to protein restriction (fig. 1) and, second, LPDs can be used safely in this population.

Can LPDs be used safely in patients with the nephrotic syndrome? Until recently, it was recommended that nephrotic patients consume high-protein diets (HPDs) to compensate for urinary albumin losses and to promote anabolism [26]. This recommendation was recently questioned when Kaysen et al. [27] reported a distinct advantage of restricting dietary protein. They found that urinary albumin excretion *decreased* and serum albumin *increased* when nephrotic patients were fed a LPD providing 0.8 g protein/kg per day compared to a HPD containing 1.6 g protein/kg per day. We measured B_N and whole-body protein turnover in control subjects and nephrotic patients while they consumed diets providing 0.8 or 1.6 g protein (plus 1 g protein/g proteinuria) and 35 kcal/kg body weight per day [28]. The B_N was neutral or positive in both nephrotic and control subjects with both diets and anabolism was due to a postprandial suppression of whole-body protein degradation and stimulation of protein synthesis. The principal compensatory response to the LPD was a decrease in amino acid oxidation and this response was the same in both groups. With the LPD, leucine

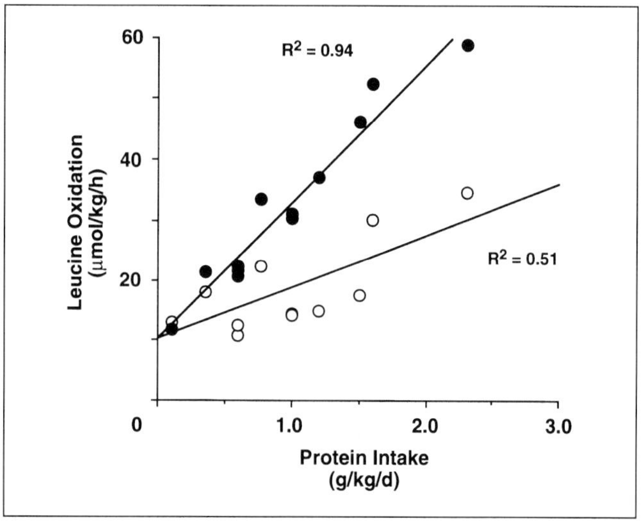

Fig. 1. Relationship between dietary protein intake during fasting (○) and feeding (●) and rates of leucine oxidation in normal subjects and CRF patients. Each symbol represents the average leucine oxidation rate for a group of subjects consuming the indicated level of dietary protein and are derived from Price et al. [39], Goodship et al. [24], Motil et al. [14], and Beaufrere et al. [40]. By linear regression there is a significant correlation between the level of dietary protein and leucine oxidation during fasting ($p = 0.009$) and feeding ($p < 0.001$). From Tom et al. [22] with permission.

oxidation during feeding correlated inversely with urinary protein losses suggesting the proteinuria is a stimulus to conserve dietary EAAs. Thus, nephrotic patients respond normally to dietary protein restriction and a diet providing 0.8 g protein/kg per day (plus 1 g protein/g proteinuria) is safe for nephrotic patients [28]. Moreover, when patients with the nephrotic syndrome were treated with a supplemented VLPD diet, serum albumin increased and serum cholesterol and proteinuria decreased: In four of six patients with an initial glomerular filtration rate (GFR) > 30 ml/min, this regimen induced a prolonged remission of nephrosis [29].

In conclusion, both nephrotic *and* nonnephrotic CRF patients can activate normal compensatory responses to dietary protein restriction allowing them to achieve a neutral B_N and maintain lean body mass during long-term therapy. These compensatory responses include a marked suppression of amino acid oxidation allowing more efficient utilization of dietary amino acids (fig. 1), and anabolism induced by a postprandial inhibition of protein degradation with or without a concomitant stimulation of protein synthesis [20, 22, 24, 28]. This inverse correlation between amino acid oxidation rates and urinary protein losses also suggests that proteinuria is a stimulus to suppress amino acid oxidation [28].

When prescribing LPDs, it is important to monitor dietary adequacy and compliance [30]. Dietary modification is challenging, but a simple method is available to estimate protein intake (and hence compliance) in outpatients [31]. A skilled nutritionist can facilitate menu planning and patient satisfaction while assisting in the monitoring of patients prescribed LPDs. Once patients are comfortable with the diet, we typically see outpatients every 3 months while (1) estimating protein intake from the 24-hour urinary urea nitrogen excretion [31], (2) estimating caloric intake from 3-day food diaries/recall, (3) monitoring serum albumin, transferrin, and anthropometrics, and (4) using these measurements to provide feedback to the patient.

Is a Spontaneous Reduction in Protein Intake of CRF Patients Consuming an Unrestricted Diet an Argument to Avoid LPDs?

It was recently reported that there is a spontaneous decrease in protein intake and some indices of nutritional status in patients with progressive CRF who were consuming unrestricted diets [7, 32]. These results led to the recommendation that dietary protein restriction should be used cautiously in patients with creatinine clearance (CCr) <25 ml/min and that early initiation of chronic dialysis therapy be considered if the patients spontaneous protein intake remains below 0.7–0.8 g/kg per day in spite of nutritional counseling [7, 8]. Although well intended, these recommendations are unfounded. First, the results just shown indicate that proper implementation of a LPD yields neutral B_N and maintenance of normal serum proteins and anthropometric indices during long-term therapy (see above). Second, hypoalbuminemia in hemodialysis patients appears to be more closely related to inflammation than to dietary adequacy [33, 34]. Third, use of a LPD in the predialysis period does not worsen and may actually improve survival during the first 2 years on dialysis [25].

It is unclear how the authors defined 'adequate nutritional counseling' since no dietary intervention was provided except to encourage patients to 'eat to satiety' and to limit their potassium intake if the serum potassium exceeded 6.0 mEq/l [7]. Without proper dietary education, CRF patients may inadvertently consume meals rich in protein, resulting in the accumulation of nitrogenous wastes and uremic symptoms for several days. In this case, satiety may actually represent anorexia, and an inadequate diet will increase the risk of malnutrition. Based on available data, it is clear that protein intake and nutritional status can deteriorate if uremic symptoms persist. If the diet is planned so that sufficient protein is prescribed to maintain neutral B_N while limiting the accumulation of nitrogenous wastes, then uremic symptoms will not appear and the nutritional status will remain normal.

Does Instituting Dialysis Improve Nutritional Intake and Reverse Malnutrition?

The rationale for initiating dialysis therapy when a CRF patient's spontaneous protein intake is below 0.7–0.8 g/kg per day warrants careful examination [7, 8]. The justification for this recommendation depends upon two unproven assumptions: (1) the decline in protein intake signifies the presence of irreversible uremic symptoms, and (2) initiation of dialysis will improve appetite and nutritional status. If one calculates the serum urea nitrogen (SUN) concentration that can be expected for different levels of renal function and protein intake (table 1), several conclusions are obvious. First, a clinically stable CRF patient would have a SUN < 70 mg/dl unless the protein intake is ≥1.2 g protein/kg per day or the CCr is <10/ml per min (table 1). In short, it is very unlikely that a CRF patient with a protein intake of 0.6–0.8 g/kg per day would be symptomatic unless there is acidosis or some other superimposed catabolic illness. The first step should be to carefully examine the patient to identify any reversible conditions contributing to malnutrition, e.g., metabolic acidosis, intercurrent illness, medication side effects.

Table 1. SUN (mg/dl) predicted from protein intake and renal function

CCr ml/min	Protein intake (g/kg per day)			
	0.6	0.8	1.0	1.2
10	53	79	105	130
15	35	52	70	87
20	26	39	52	65
25	21	31	42	52

SUN values were estimated for a 70-kg individual assuming $Cl_{urea} = 0.6 \times CCr$ using the formula:

$$SUN = \frac{I_N - 0.031 \text{ g N/kg body weight}}{Cl_{urea}},$$

where I_N = nitrogen intake in g N/day, 0.031 = nonurea nitrogen losses in g N/day [30].

Does dialysis increase dietary intake and/or ameliorate malnutrition? Although dialysis usually improves the appetite of uremic patients with terminal renal failure (e.g., CCr < 5 ml/min), the measured intake of maintenance dialysis patients is commonly below recommended levels and virtually every nutritional survey has revealed a high prevalence of protein-calorie malnutrition in this population [35–37]. Even in well-dialyzed populations (e.g., n = 585, equilibrated Kt/V = 1.30 ± 0.20; mean ± SD), the protein and energy intake of hemodialysis patients is frequently below recommended levels; e.g., 61% and 81% of these patients consumed <1 g protein and 30 kcal/kg per day, respectively [38]. Moreover, no correlation was seen between Kt/V and protein intake estimated from diet records (r = –0.01), indicating that more vigorous dialysis does not solve the problem.

Conclusion

In summary, both nephrotic *and* nonnephrotic CRF patients can activate normal compensatory responses to dietary protein restriction and maintain lean body mass during long-term therapy. There is no evidence that avoiding LPDs during the predialysis period improves the nutritional state after the initiation of dialysis. Evidence that the protein intake of CRF patients consuming unrestricted diets decreases as GFR declines is a persuasive argument in favor of the use of LPDs in patients with progressive CRF.

Acknowledgment

This work was supported in part by National Institute of Health grants DK40907, DK49243, and DK45215.

References

1 Aparicio HGM, Potaux L, Precigout V de, Bouchet JL, Aubertin J: Low protein and low phosphorus diet in patients with chronic renal failure: Influence on glucose tolerance and tissue insulin sensitivity. Metabolism 1987;36:1080–1085.

2 Barsotti G, Morelli E, Guiducci A, Ciardella F, Giannoni A, Lupetti S, Giovannetti S: Reversal of hyperparathyroidism in severe uremics following very low-protein and low-phosphorus diet. Nephron 1982;30:310–313.

3 Zeller KR, Whittaker E, Sullivan L, Raskin P, Jacobson HR: Effect of restricting dietary protein on the progression of renal failure in patients with insulin-dependent diabetes mellitus. N Engl J Med 1991;324:78–83.

4 Ihle BU, Becker GJ, Whitworth JA, Charlwood RA, Kincaid-Smith PS: The effect of protein restriction on the progression of renal insufficiency. N Engl J Med 1989;321:1773–1777.

5 Pedrini MT, Levey AS, Lau J, Chalmers TC, Wang PH: The effects of dietary protein restriction on the progression of diabetic and nondiabetic renal diseases: A meta-analysis. Ann Intern Med 1996; 124:627–632.

6 Levey AS, Adler S, Caggiula AW, England BK, Greene T, Hunsicker LG, Kusek JW, Rogers NL, Teschan PE, for the Modification of Diet in Renal Disease Study Group: Effects of dietary protein restriction on the progression of advanced renal disease in the Modification of Diet in Renal Disease Study. Am J Kidney Dis 1996;27:652–663.

7 Ikizler TA, Greene JH, Wingard RL, Parker RA, Hakim RM: Spontaneous dietary protein intake during progression of chronic renal failure. J Am Soc Nephrol 1995;6:1386–1391.

8 Hakim RM, Lazarus JM: Initiation of dialysis. J Am Soc Nephrol 1995;6:1319–1328.

9 Lowrie EG, Lew NL: Death risk in hemodialysis patients: The predictive value of commonly measured variables and an evaluation of death rate differences between facilities. Am J Kidney Dis 1990;15:458–482.

10 Owen WF Jr, Lew NL, Yan Liu SM, Lowrie EG, Lazarus JM: The urea reduction ratio and serum albumin concentration as predictors of mortality in patients undergoing hemodialysis. N Engl J Med 1993;329:1001–1006.

11 Felig P: Amino acid metabolism in man. Annu Rev Biochem 1975;44:935–955.

12 McNurlan MA, Garlick PJ: Influence of nutrient intake on protein turnover. Diabetes Metab Rev 1989;5:165–189.

13 Pacy PJ, Price GM, Halliday D, Quevedo MR, Millward DJ: Nitrogen homoeostasis in man: The diurnal responses of protein synthesis and degradation and amino acid oxidation to diets with increasing protein intakes. Clin Sci 1994;86:103–118.

14 Motil KJ, Matthews DE, Bier DM, Burke JF, Munro HN, Young VR: Whole-body leucine and lysine metabolism: Response to dietary protein intake in young men. Am J Physiol 1981;240:E712–E721.

15 Quevedo MR, Price GM, Halliday D, Pacy PJ, Millward DJ: Nitrogen homoeostasis in man: Diurnal changes in nitrogen excretion, leucine oxidation and whole body leucine kinetics during a reduction from a high to a moderate protein intake. Clin Sci 1994; 86:185–193.

16 Monteon FJ, Laidlaw SA, Shaib JK, Kopple JD: Energy expenditure in patients with chronic renal failure. Kidney Int 1986;30:741–747.

17 Kopple JD, Monteon FJ, Shaib JK: Effect of energy intake on nitrogen metabolism in nondialyzed patients with chronic renal failure. Kidney Int 1986; 29:734–742.

18 Kopple JD, Coburn JW: Metabolic studies of low protein diets in uremia. I. Nitrogen and potassium. Medicine 1973;52:583–595.

19 Mitch WE, Abras E, Walser M: Long-term effects of a new ketoacid-amino acid supplement in patients with chronic renal failure. Kidney Int 1982; 22:48–53.

20 Masud T, Young VR, Chapman T, Maroni B: Adaptive responses to very low protein diets: The first comparison of ketoacids to essential amino acids. Kidney Int 1994;45:1182–1192.

21 Walser M: Does prolonged protein restriction preceding dialysis lead to protein malnutrition at the onset of dialysis? Kidney Int 1993;44:1139–1144.

22 Tom K, Young VR, Chapman T, Masud T, Akpele L, Maroni BJ: Long-term adaptive responses to dietary protein restriction in chronic renal failure. Am J Physiol 1994;4:E668–E677.

23 Modification of Diet in Renal Disease Study Group (prepared by Kopple JD, Chumlea WC, Gassman JJ, Hollinger DL, Maroni BJ, Merrill D, Scherch LK, Schulman G, Zimmer GS): Nutritional response to diet prescription in the MDRD study (abstract). J Am Soc Nephrol 1994;5:335.

24 Goodship THJ, Mitch WE, Hoerr RA, Wagner DA, Steinman TI, Young VR: Adaptation to low-protein diets in renal failure: Leucine turnover and nitrogen balance. J Am Soc Nephrol 1990;1:66–75.

25 Coresh J, Walser M, Hill S: Survival on dialysis among chronic renal failure patients treated with a supplemented low-protein diet before dialysis. J Am Soc Nephrol 1996;6:1379–1385.

26 Blainey BJD: High protein diets in the treatment of the nephrotic syndrome. Clin Sci 1954;13:567–581.

27 Kaysen GA, Gambertoglio J, Jimenez I, Jones H, Hutchison FN: Effect of dietary protein intake on albumin homeostasis in nephrotic patients. Kidney Int 1986;29:572–577.

28 Maroni BJ, Staffeld C, Young VR, Manatunga A, Tom K: Mechanisms permitting nephrotic patients to achieve nitrogen equilibrium with a protein-restricted diet. J Clin Invest 1997;99:2479–2487.

29 Walser M, Hill S, Tomalis EA: Treatment of nephrotic adults with a supplemented, very low-protein diet. Am J Kidney Dis 1996;28:354–364.

30 Maroni BJ: Requirements for protein, calories, and fat in the predialysis patient; in Mitch WE, Klahr S (eds): Nutrition and the Kidney. Boston, Little, Brown, 1993, pp 185–212.

31 Maroni BJ, Steinman T, Mitch WE: A method for estimating nitrogen intake of patients with chronic renal failure. Kidney Int 1985;27:58–65.

32 Modification of Diet in Renal Disease Study Group (prepared by Kopple JD, Berg R, Houser H, Steinman TI, Teschan P): Relationship between GFR and nutritional status – Results from the MDRD study (abstract). J Am Soc Nephrol 1994; 5:335.

33 Kaysen GA, Rathore V, Shearer GC, Depner TA: Mechanisms of hypoalbuminemia in hemodialysis patients. Kidney Int 1995;48:510–516.

34 Bergstrom J, Heimburger O, Lindholm B, Qureshi AR: Elevated serum C-reactive protein is a strong predictor of increased mortality and low serum albumin in hemodialysis patients (abstract). J Am Soc Nephrol 1995;6:573.

35 Blumenkrantz MJ, Kopple JD, Gutman RA, Chan YK, Barbour GL, Roberts C, Shen FH, Gandhi VC, Tucker CT, Curtis FK, Coburn JW: Methods for assessing nutritional status of patients with renal failure. Am J Clin Nutr 1980;33:1567–1585.

36 Thuneberg BJ, Swamy AP, Cestero RVM: Cross-sectional and longitudinal nutritional measurements in maintenance hemodialysis patients. Am J Clin Nutr 1981;34:2005–2012.

37 Ahmed KR, Kopple JD: Nutrition in maintenance hemodialysis patients; in Kopple JD, Massary SG (eds): Nutritional Management of Renal Disease. Baltimore, William & Wilkins, 1996, pp 563–600.

38 HEMO Full-Scale Trial Study Group (prepared by Maroni BJ, Burkart J, Burrowes J, Cockram D, Drabik M, Dwyer J, Kusek J, Makoff R, Paranandi L, Power S, Rocco M): Baseline nutritional characteristics of the HEMO Study participants: Interim report (abstract). J Am Soc Nephrol 1996;7:1456.

39 Price GM, Halliday D, Pacy PJ, Quevedo MR, Millward DJ: Nitrogen homoeostasis in man: Influence of protein intake on the amplitude of diurnal cycling of body nitrogen. Clin Sci 1994;86:91–102.

40 Beaufrere B, Horber FF, Schwenk WF, Marsh HM, Matthews D, Gerich JE, Haymond MW: Glucocorticoids increase leucine oxidation and impair leucine balance in humans. Am J Physiol 1989;257: E712–E721.

Mineral and Electrolyte Metabolism

Miner Electrolyte Metab 1997;23:229–233

Giacomo Deferrari
Giacomo Garibotto
Cristina Robaudo
Stefano Saffioti
Rodolfo Russo
Antonella Sofia

Department of Internal Medicine,
Division of Nephrology,
University of Genoa, Italy

Protein and Amino Acid Metabolism in Splanchnic Organs in Metabolic Acidosis

Key Words
Amino Acids
Ammonia
Proteins
Splanchnic organs
Urea

Abstract
Metabolic acidosis causes a cooperative participation of different organs such as the liver, kidney, and muscle in restoring acid-base balance. In splanchnic organs, metabolic acidosis has repercussions on several nitrogen metabolism pathways. The decrease in urea synthesis due to reduced activity of urea cycle enzymes, ammonia uptake and amino acid transport, and changes in glutamine metabolism support renal ammoniagenesis thus offering a response to rid the body of excess protons. While some of the mechanisms are adaptive others may be harmful for the body. Metabolic acidosis may have effects on splanchnic protein turnover. In severe acidosis, synthesis of secreted liver proteins may be reduced. Acidosis may also modulate the response of the liver to growth hormone (GH) for insulin-like growth factor-I synthesis, thus inducing a state of GH resistance. Splanchnic abnormalities in acidosis might contribute to the malnutrition observed in uremia.

The body responds to metabolic acidosis with a variety of metabolic adaptations which contribute to restoring pH to normal values. There is general consensus that, while many of these adaptations are beneficial, the response of the body to excess acid ingestion or production may also be detrimental to patients, and somehow it imitates the 'trade-off' response leading to hyperparathyroidism [1–3]. The trade-off for the adaptation to chronic metabolic acidosis may include loss of lean body mass caused by stimulation of protein degradation when nutrient intake is reduced, bone mineral loss, progression of chronic renal failure (CRF) and generation and progression of kidney stones [1–3].

The effects of metabolic acidosis on the metabolism of amino acids (AAs) and proteins in splanchnic organs have received little attention. Metabolic acidosis may affect several aspects of nitrogen metabolism in splanchnic organs. Among these are reduced urea production and excretion, changes in the activity of ornithine cycle enzymes, inhibition of AA transport across the liver cells, and decreased synthesis of secretory proteins and insulin-like growth factor-I (IGF-I). Many of the changes in AA and protein metabolism occurring in splanchnic organs represent an integrated response to metabolic acidosis supporting the defense of the body against the acid load. However, some of these metabolic changes might also prove to be harmful on a chronic basis.

Chronic Metabolic Acidosis and Ureagenesis

Urea synthesis serves as a major mechanism for ammonium detoxification and removal of excess AAs, when the supply of protein or AAs is high because of changes in exogenous or endogenous supply to the liver [4].

Acute acidosis in the dog lowers urea excretion [5] and in the rat this decrease in urea excretion is equimolar with the increase in urine ammonium [6]. In humans, acidosis induced by the ingestion of hydrochloric acid for 4 days causes a shift of waste nitrogen: usually excreted predominantly as urea, it is eliminated as ammonium [7], indicating that metabolic acidosis has an inhibitory effect on hepatic urea formation.

Urea synthesis may be modulated by a rapid mechanism involving the synthesis of N-acetylglutamate or effects on AA transport [8]. A longer-term effect is mediated by modulation of the synthesis of urea cycle enzymes [8]. All these steps appear to be influenced by metabolic acidosis. Studies using perfused liver preparations have shown that the pH of the perfusate strongly influences the rate of urea production [8]. pH control of urea synthesis is partly mediated by an inhibition of ornithine cycle enzymes. It has been shown that the N-acetylglutamate synthesis rate decreases along with the decrease in pH, due to an increase in the K_m of the enzyme [9]. Moreover, the mitochondrial carbonic anhydrase activity is extremely sensitive to pH changes. An increase in pH from 6.9 to 7.8

KARGER
E-Mail karger@karger.ch
Fax +41 61 306 12 34
http://www.karger.ch

0378–0392/97/0236–0229 $12.00/0

Giacomo Deferrari, MD
Dipartimento di Medicina Interna
Viale Benedetto XV, 6, I-16132 Genova (Italy)
Tel. +39 10 3537972, fax +39 10 3538976

causes a sixfold increase in the activity of this enzyme [10], which is needed to produce HCO_3 for carbamoyl-phosphate synthase. Another possible control mechanism is through the protonation of ammonia. A decrease in pH causes a decrease in NH_3 and this might further diminish urea formation since carbamoyl-phosphate synthase requires NH_3 and not NH_4^+ [11].

Another possible mechanism causing an acidosis-mediated decrease in ureagenesis is an inhibition of glutamine degradation by the liver [8]. Liver glutaminase may adjust the urea cycle flux on the basis of acid-base balance requirements, due to the marked pH sensitivity of the flux through this enzyme [8]. A low pH reduces glutaminase activity and the rate of transport of glutamine across the cell membrane of hepatocytes [8, 12], thus reducing the ammonium supply for carbamoyl-phosphate synthetase and urea synthesis.

Based on the observation that ureagenesis is pH dependent, the idea that the liver is a primary organ of acid-base homeostasis has been proposed [13]. It has been emphasized that large quantities of HCO_3^- are formed every day as a product of protein metabolism [8, 13]. Liver urea synthesis disposes of excess HCO_3 and is therefore a bicarbonate-consuming process. However, it has also been pointed out that most of the ammonium produced by AA metabolism reacts with bicarbonate in the synthesis of urea without any effect on acid-base balance [14]. On the contrary, the metabolism of glutamine and other AAs in the kidney for ammonia formation constitutes a net systemic bicarbonate-generating process when the ammonium produced is excreted in urine [14]. If ammonium returns to the liver, the urea synthesis consumes the bicarbonate produced during the metabolism of the carboxylate anion in the kidney and no net gain of bicarbonate is obtained. Accordingly, in metabolic acidosis, the decrease in liver uptake of AA and the changes in glutamine metabolism shift nitrogen from ureagenesis to renal ammoniagenesis, thus inducing a net bicarbonate gain.

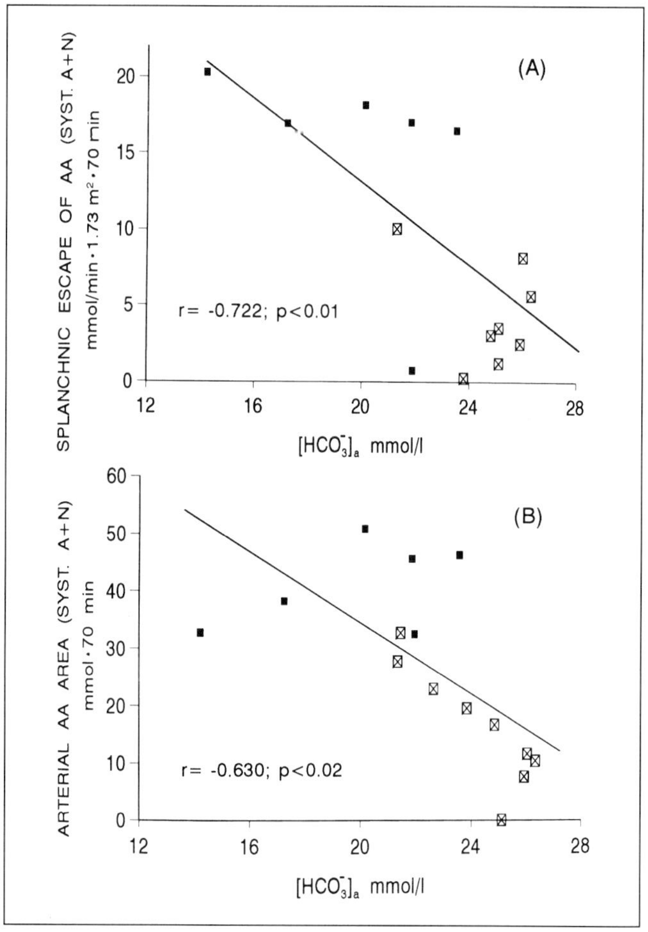

Fig. 1. Relationships between splanchnic escape of AAs (systems A and N) (**A**) and their arterial increments (**B**) to $[HCO_3]_a$ in patients with CRF (■) and controls (☒) after AA ingestion. Adapted from Deferrari et al. [21].

Metabolic Acidosis and Liver AA Transport

Amino acid transport by liver cells is another pH-controlled step of ureagenesis. In vitro studies have shown that AA transport through systems A and N is inhibited by low pH [8]. In liver perfusion studies and in isolated hepatocytes, a decrease in pH reduces alanine consumption [15, 16]. In vivo acid loading in fed rats causes an increase in arterial levels of many nonessential amino acids (NEAAs), which are transported across the liver by systems A and N [17]. The hepatic uptake of alanine and glycine, as well as total AA uptake, is reduced in rats with hydrochloric acid acute acidosis as compared to alkalosis [18].

Defective liver transport of AAs in metabolic acidosis might explain the altered postprandial AA pattern observed in rats with NH_4Cl acidosis [17] and in patients with CRF [19, 20]. In patients with CRF, a protein [20] or an AA [19] meal is followed by high increases of NE AAs in arterial blood, pointing to an 'intolerance' to NE AAs. This alteration rises from a defective splanchnic organ AA metabolism which allows many AAs, especially glutamine and other AA transported by A and N systems, to increase in arterial blood more than branched-chain amino acids (BCAAs), which under normal conditions predominate in the protein-fed AA profile [21]. Recently, we looked into possible relationships between acidosis and the altered splanchnic AA metabolism. Figure 1 shows that after AA ingestion, the splanchnic escape and the arterial increments of AAs which are transported by systems A and N are inversely related to arterial bicarbonate levels, both in patients with CRF and in controls. The lower the bicarbonate levels, the higher the splanchnic escape of AAs and their arterial increments. It is well known that in the protein-fed state, splanchnic organs play a major role in regulating the pattern of absorbed AAs subsequently presented to the peripheral circulation [22]. The above-mentioned observations suggest that in CRF, metabolic acidosis modifies both liver AA uptake and the delivery of AAs to peripheral tissues in the protein-fed state.

The observation of an altered splanchnic handling of AAs in the protein-fed state in CRF patients with metabolic acidosis is consistent with the concept of the metabolic zonation in the liver acinus. It is now generally accepted that AA degradation, ureagenesis, and gluconeogenesis all occur in the periportal zone of the acinus [8]. Catabolic stimuli, such as sepsis and inflammation have been shown to affect specifically the metabolism of periportal hepatocytes [23]. Interestingly, in studies performed in the postabsorptive state, we failed to find significant differences in the splanchnic uptake of AA, both in acidotic CRF patients [21] and in a preliminary study performed in postabsorptive humans with metabolic acidosis [24]. Accordingly, after oral AA ingestion, the associated increase in portal AA levels and AA supply to periportal hepatocytes may show a liver transport defect undetectable in the postabsorptive state.

Complementing the altered splanchnic and arterial AA profile, an altered pattern of AA repletion in peripheral tissues has been observed in acidotic patients with CRF [25]. A large percentage of the AAs taken up by the leg during the absorptive period is no longer comprised of BCAAs but NEAAs [25]. It is not known if the altered AA exchange in protein-fed renal patients has repercussions on splanchnic or muscle protein turnover.

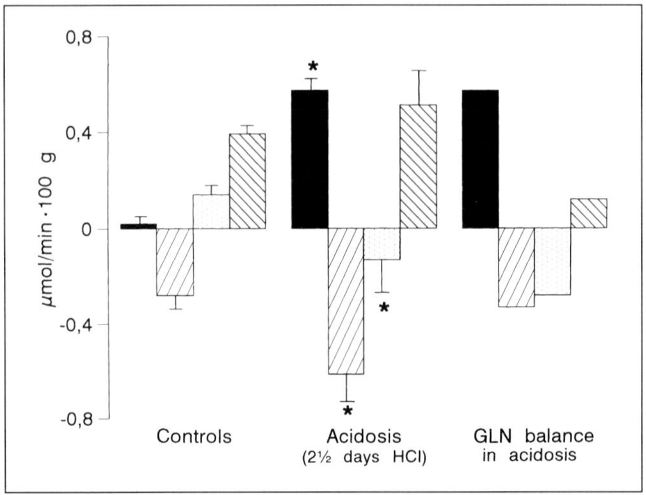

Fig. 2. Interorgan glutamine (GLN) flow in rats with HCl metabolic acidosis. ■ = Kidney; ▨ = muscle; ▢ = liver; ▧ = gut. Adapted from Schrock and Goldstein [28].

Interorgan Exchange of Glutamine in Metabolic Acidosis

Glutamine is a major substrate for urea synthesis and renal ammonia production [8]. Under normal conditions, glutamine is produced by peripheral tissues and is taken up by the splanchnic organs, thus providing a fuel for the intestine and a substrate for ureagenesis.

Most of the ammonia produced by the kidney and released into the renal veins during chronic metabolic acidosis, is derived from glutamine [26]. Glutamine extracted by the kidney during metabolic acidosis increases greatly and large amounts of glutamine are removed from the blood over a short period. Therefore, blood glutamine pools would be rapidly depleted if it were not for a compensatory reaction in other organs. It has been stressed that base generation may be limited by the supply of glutamine, its availability, and the ease in transferring protein nitrogen to glutamine [27].

Changes in the interorgan glutamine flow have recently been reviewed [27]. Glutamine metabolism in splanchnic organs is strikingly modified by metabolic acidosis. Schrock and Goldstein [28] studied the effects of chronic hydrochloric-acid-induced acidosis on the interorgan flow of glutamine in rats. During acidosis, the increased use of glutamine by the kidney was associated with an increased muscle release of the same AA. At the same time, hepatic uptake of glutamine was reversed into a release, and gut metabolism was unmodified (fig. 2). The changes in splanchnic glutamine metabolism are explained by augmented liver synthesis and/or reduced liver degradation or transport of glutamine [27, 28].

As for humans with chronic metabolic acidosis, investigations have provided insights into renal exchange of glutamine and other substrates [26], whereas the extrarenal metabolism of ammoniagenic precursors is poorly known. Moreover, in most studies carried out so far, metabolic acidosis has been induced by ammonium chloride, which might affect AA metabolism by providing a nitrogen load to the organs. In healthy volunteers with 6-day ammonium chloride metabolic acidosis, the arterial–hepatic difference for glutamine was found to be reduced (fig. 3). In this way, despite the increased renal uptake, the arterial levels of glutamine are only slightly reduced [29].

Therefore, the available evidence indicates that the augmented renal demand for glutamine in animals with metabolic acidosis is met not only by an augmented release by peripheral tissues but also by decreased liver uptake of glutamine, thus making increased ammoniagenesis possible. However, whether such metabolic modifications are observed in humans with metabolic acidosis is not completely understood.

Acidosis and Protein Turnover in Splanchnic Organs

The general picture of the effects metabolic acidosis has on ureagenesis and liver AA transport suggests that metabolic acidosis may affect protein turnover in the liver and/or intestine. Over the last few years, several studies have shed light on the catabolic effects of metabolic acidosis on whole-body protein metabolism. Increased protein and AA degradation in muscle have been reported to play an important role in the rise in acidosis-induced whole-body protein catabolism [2]. However, the effects, if any, of metabolic acidosis on protein and AA metabolism in splanchnic organs in humans are poorly known. Protein synthesis in splanchnic organs accounts for about 27% of whole-body protein synthesis, nearly matching the amount contributed by total skeletal muscle (about 33%) [30]. The average daily synthesis of albumin, the major secretory protein of the liver, is about 15 g, and is thus an important component of splanchnic protein synthesis (about 70 g/day) [22].

The trade-off hypothesis suggests that metabolic acidosis blunts adaptive protein metabolism responses when a low-protein diet is prescribed or nutrient intake is reduced in CRF [2]. This may also occur for splanchnic protein metabolism. Splanchnic organs parti-

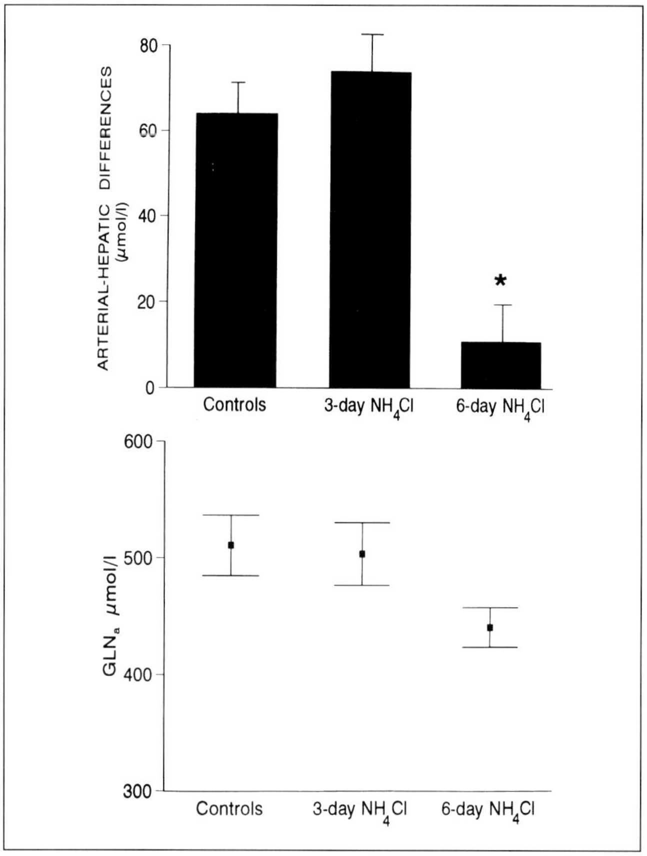

Fig. 3. Splanchnic uptake of glutamine (GLN) in humans with NH_4Cl metabolic acidosis. Adapted from Tizianello et al. [29].

cipate in adapting to low nutrient intake by means of various mechanisms. In the postabsorptive state, the splanchnic protein balance is neutral since protein degradation is no different from synthesis [30]. Thus, under conditions of mild substrate deprivation such as overnight fasting, a net loss of visceral proteins does not occur, most likely because AAs are supplied to the splanchnic area by peripheral tissues. However, animal studies have shown that the protein liver content depends on dietary factors such as the exogenous AA supply. When rats are given protein-free diets, the liver loses protein without changes in fractional protein synthesis [31]. On the other hand, when rats are starved, liver protein synthesis declines abruptly, with a parallel fall in liver protein concentration [32]. During low nutrient intake, intracellular protein degradation provides AAs for gluconeogenesis and the synthesis of new export proteins. The reduction in liver protein content contrasts with the slight (only 8%) decrease in muscle protein content during fasting [33]. It has been pointed out that the rapid loss of liver protein by the rat in early starvation corresponds to the rate of protein export that is maintained at the expense of liver structural proteins [32]. It is interesting that clinical investigations in humans are in keeping with this idea. Ballmer et al. [34] found that ammonium chloride metabolic acidosis caused both a reduction in fractional synthesis of albumin and a negative nitrogen balance in normal adults. However, while a negative nitrogen balance had already been observed in the presence of mild metabolic acidosis (mean HCO_3 19.1 mmol/l), the decrease in fractional albumin synthesis was observed only in subjects receiving larger doses of ammonium chloride and having more severe acidosis (mean HCO_3 15.1 mmol/l). Thus, body protein degradation in humans is affected early by metabolic acidosis, while synthesis of albumin in the liver is relatively spared and is depressed only when metabolic acidosis is severe. Observations in animals suggest that metabolic acidosis may have a negative effect on the metabolism of structural splanchnic proteins as well. Besides finding evidence of defective rat liver AA transport, Boon et al. [18] observed a decrease in the liver protein/DNA ratio, suggesting increased protein degradation in metabolic acidosis. The mechanism(s) for splanchnic protein catabolism may also involve increased oxidation of essential AAs, such as BCAAs. BCAAs are essential AAs which limit protein synthesis; moreover they may have regulating effects on protein metabolism [22]. In humans, splanchnic organs express a high rate of leucine oxidation, the total amount of leucine oxidized by the splanchnic organs being about one-half that oxidized by skeletal muscle [26]. Chronic metabolic acidosis increases whole body leucine oxidation and branched-chain ketoacid (BCKA) dehydrogenase activity in rat muscle and liver [35]. In contrast, renal BCKA decreases in acidosis. Interestingly, such an observation suggests that increased AA catabolism may be a response to metabolic acidosis which is common to muscle and liver, but not to kidney. Whether humans undergo changes in interorgan protein turnover and AA oxidation in acidosis remains to be studied.

Resistance to Growth Hormone (GH) and IGF-I

Recent studies indicate that metabolic acidosis causes a resistance to the action of GH and/or IGF-I. IGF-I is produced locally in most tissues, but the liver is the main source for circulating IGF-I [36]. Like insulin, IGF-I promotes whole-body anabolism by inhibiting protein degradation and/or increasing protein synthesis [37]. Plasma levels of IGF-I are low in ammonium chloride metabolic acidosis [34]. Challa et al. [38] found that metabolic acidosis significantly reduces the liver GH receptor mRNA in rats, thereby inducing a state of GH resistance. In Sprague-Dawley rats made acidemic with ammonium chloride, the expression of liver IGF-I is also reduced and this is associated with impaired weight gain. However, the effects of acidosis on the expression of liver IGF-I may be partly mediated by malnutrition [38]. The changes in the GH/IGF-I axis induced by acidosis combined with inhibition of GH secretion in acidosis [36] may be an important cause for growth failure in children with renal tubular acidosis.

Acknowledgements

We are grateful to Francesca Tincani for her assistance in the preparation of the manuscript and figures. The studies were supported by grants from the Ministero dell'Università e della Ricerca Scientifica e Tecnologica (Assegnazione per la ricerca scientifica 40%), and Progetti di Ricerca di Ateneo-Università di Genova.

References

1 Halperin ML, Etheir JH, Kamel KS: Ammonium excretion in chronic metabolic acidosis: Benefits and risks. Am J Kidney Dis 1989;14:267–271.
2 Reaich D, Price SR, England BK, Mitch WE: Mechanisms causing muscle loss in chronic renal failure. Am J Kidney Dis 1995;26:242–247.
3 Alpern RJ: Trade-off in the adaptation to acidosis. Kidney Int 1995;47:1205–1215.
4 Cohen PP: Regulation of the ornithine-urea cycle enzymes; in Waterlow JC, Stephen JML (eds): Nitrogen Metabolism in Man. London, Applied Science, 1981, pp 215–228.
5 Keeton RW: Ammonia excretion following experimental administration of acids via the stomach and peripheral vein. J Biol Chem 1921;49:411–427.
6 Oliver J, Bourke E: Adaptations in urea and ammonium excretion in metabolic acidosis in the rat: A reinterpretation. Clin Sci Mol Med 1975;48:515–520.
7 Fine A, Carlyle JE, Bourke E. The effects of administrations of HCl, NH_4Cl and NH_4HCO_3 on the excretion of urea and ammonium in man. Eur J Clin Invest 1977;7:587–589.
8 Meijer AJ, Lamers WH, Chamuleau RAFM: Nitrogen metabolism and ornithine cycle function. Physiol Rev 1990;70:701–748.
9 Kamemoto ES, Atkinson DE: Modulation of the activity of rat liver acetylglutamate synthase by pH and arginine concentration. Arch Biochem Biophys 1985;243:100–107.
10 Haussinger D, Gerok W: Hepatic urea synthesis and pH regulation: Role of CO_2, HCO_3, pH and the activity of carbonic anhydrase. Eur J Biochem 1985;152:381–386.
11 Rémésy C, Demigné C, Fafournoux P: Control of ammonia distribution ratio across the liver cell membrane and of ureagenesis by extracellular pH. Eur J Biochem 1986;158:283–288.
12 Sobol S, Lenzen C, Haussinger A: Regulation of glutamine transport in isolated perfused rat liver. J Hepatol 1987;4(suppl 1):44.
13 Atkinson DE, Bourke E: The role of ureagenesis in pH homeostasis. Trends Biochem Sci 1984;9:297–300.
14 Dubose TD Jr, Cogan MG, Rector FC Jr: Acid-base disorders; in Brenner BM (ed): The Kidney. Philadelphia, Saunders, 1996, vol 22, pp 929–998.
15 Kilberg MS, Handlogten ME, Christensen HN: Characteristics of an amino acid transport system in rate liver for glutamine, asparagine, histidine, and closely related analogs. J Biol Chem 1980;225:4011–4019.
16 Sestoft LP, Bartels PO, Folke M: Pathophysiology of metabolic acidosis: Effect of low pH on the hepatic uptake of lactate, pyruvate and alanine. Clin Physiol 1982;2:51–58.
17 Cupisti A, Baker F, Brown J, Lock C, Bevington A, Harris KPG, Walls J: Effects of acid loading on serum amino acid profiles and muscle composition in normal fed rats. Clin Sci 1993;85:445–449.
18 Boon L, Blommaart PJE, Meijer AJ, Lamers WH, Schoolwerth AC: Effect of chronic acidosis on hepatic amino acid uptake and gene regulation: Implication for control of acid-base balance. Contrib Nephrol, Basel, Karger, vol 110, pp 138–143.
19 Garibotto G, Deferrari G, Robaudo C, Saffioti S, Salvidio G, Paoletti E, Tizianello A: Effect of amino acid ingestion on blood amino acid levels in patients with chronic renal failure. Am J Clin Nutr 1987; 46:949–954.
20 Garibotto G, Deferrari G, Robaudo C, Saffioti S, Paoletti E, Pontremoli R, Tizianello A: Effects of a protein meal on blood amino acid profile in patients with chronic renal failure. Nephron 1993;64:216–225.
21 Deferrari G, Garibotto G, Robaudo C, Sala MR, Tizianello A: Splanchnic exchange of amino acid after amino acid ingestion in patients with chronic renal insufficiency. Am J Clin Nutr 1988;48:72–83.
22 Young VR, Yu Y: Protein and amino acid metabolism; in Fischer JE (ed): Nutritional and Metabolism in the Surgical Patients. Boston, Little, Brown, 1996, pp 159–201.
23 Ballmer PE, Ballmer-Hofer K, Repond F, Kohler H, Studer H: Acute suppression of albumin synthesis in systemic inflammatory disease: An individual graded response of rat hepatocytes. J Histochem Cytochem 1992;40:201.
24 Deferrari G, Garibotto G, Robaudo C, Saffioti S, Russo R, Sala MR, Bruzzone M, Tizianello A: Renal ammoniagenesis and interorgan flow of glutamine in chronic metabolic acidosis. Contrib Nephrol. Basel, Karger, vol 110, pp 144–149.
25 Garibotto G, Deferrari G, Robaudo C, Saffioti S, Sofia A, Russo R, Tizianello A: Disposal of exogenous amino acids by muscle in patients with chronic renal failure. Am J Clin Nutr 1995;62:136–142.
26 Tizianello A, Deferrari G, Garibotto G, Robaudo C, Bruzzone M, Passerone GC: Renal ammoniagenesis during the adaptation to metabolic acidosis in man; in Tannen RL, Goldstein L, Lemieux G, Simpson D, Vinay P (eds): Renal Ammonia Metabolism. Contrib Nephrol. Basel, Karger, vol 31, pp 40–46.
27 Welbourne TC, Joshi S: Interorgan glutamine metabolism during acidosis. J Parent Enter Nutr 1990; 14:778–858.
28 Schrock H, Goldstein L: Interorgan relationships for glutamine metabolism in normal and acidotic rats. Am J Physiol 1981;240:E519–E525.
29 Tizianello A, Deferrari G, Garibotto G, Gurreri G: Effects of chronic renal insufficiency and metabolic acidosis on glutamine metabolism in man. Clin Sci Mol Med 1978;55:391–397.
30 Tessari P, Garibotto G, Inchiostro S, Robaudo C, Saffioti S, Vettore M, Zanetti M, Russo R, Deferrari G: Kidney, splanchnic, and leg protein turnover in humans: Insight from leucine and phenylalanine kinetics. J Clin Invest 1996;98:1481–1492.
31 Garlick PJ: The effect of protein deprivation and starvation on the rate of protein synthesis in tissues of the rat. Biochem Biophys Acta 1975;414:71–79.
32 Levenson SM, Imegwu OJ, Cang TH: Starvation: Metabolic and physiologic responses; in Fischer JE (ed): Nutrition and Metabolism in the Surgical Patient. Boston, Little, Brown, 1996, pp 523–576.
33 Addis T, Poo LJ, Lew W: The quantities of protein lost by the various organs and tissues of the body during a fast. J Biol Chem 1936;115:111–116.
34 Ballmer PE, McNurlan MA, Hulter HN, Anderson SE, Garlick PJ, Krapf R: Chronic metabolic acidosis decreases albumin synthesis and induces negative nitrogen balance in humans. J Clin Invest 1995; 95:39–45.
35 May RC, Masud T, Logue B, Bailey J, England B: Chronic metabolic acidosis accelerates whole body proteolysis and oxidation in awake rats. Kidney Int 1992;41:1535–1542.
36 Krieg RJ Jr, Santos F, Chan JCM: Growth hormone, insulin-like growth factor and the kidney. Kidney Int 1995;48:321–336.
37 Fryburg DA: Insulin-like growth factor-1 exerts growth hormone- and insulin-like actions on human muscle protein metabolism. Am J Physiol 1994;267:E331–E336.
38 Challa A, Chan W, Krieg RJ Jr, Thabet MA, Liu F, Hintz RL, Chan JCM: Effect of metabolic acidosis on the expression of insulin-like growth factor and growth hormone receptor. Kidney Int 1993;44: 1224–1227.

Mineral and Electrolyte Metabolism

Miner Electrolyte Metab 1997;23:234–236

John Walls

Department of Nephrology,
Leicester General Hospital,
Leicester, UK

Effect of Correction of Acidosis on Nutritional Status in Dialysis Patients

Key Words
Acidosis
Buffer
Haemodialysis
CAPD

Abstract
Malnutrition is a well-recognised feature of end-stage renal failure and contributes to the continuing high morbidity and mortality in this group of patients. One of the aetiological factors is metabolic acidosis which has been shown to increase protein degradation in both experimental models of chronic renal failure and in humans with uraemia. Many patients currently receiving haemodialysis have subnormal values of plasma bicarbonate. However, the values can be normalised by using a dialysate bicarbonate concentration of 35–40 mmol/l and in continuous ambulatory peritoneal dialysis (CAPD), a similar increment in plasma bicarbonate can be achieved using a dialysate lactate content of 35–40 mmol/l. In short-term studies in haemodialysis patients there is evidence of an increase in body weight and other anthroprometric parameters when the plasma bicarbonate has been normalised by increasing the dialysate bicarbonate content. A long-term study in CAPD patients has demonstrated increased body weight, tricep skinfold thickness and midarm muscle circumference in those patients with a plasma bicarbonate of 27.2 ± 0.3 mmol/l, compared to those with a value of 23.0 ± 0.3 mmol/l. These studies strongly suggest that correction of acidosis by increased dialysate buffering capacity will improve nutritional status for patients with end-stage renal failure.

Introduction

'Life is a struggle, not against sin, not against the money power, not against malicious animal magnetism, but against hydrogen ions'.
H.L. Mencken, *About the Meaning of Life and Death* [1919]

During the pioneering days of dialysis therapy it was recognised that it was necessary to add a buffer to the dialysate in order to replete the reduced plasma bicarbonate levels and correct metabolic acidosis occurring as a result of uraemia. Initially, the dialysate contained sodium bicarbonate, and to prevent precipitation of calcium carbonate from the calcium chloride added to the dialysate to maintain calcium balance, the pCO_2 of the dialysate was increased by bubbling CO_2 gas through the dialysate. An alternative strategy was to make the pH of the dialysate acidic. Neither of these techniques was entirely satisfactory and in 1964 Mion et al. [1] described the use of acetate as a buffer for haemodialysis. Acetate was metabolized by the liver to bicarbonate and did not precipitate calcium carbonate in the dialysis bath. The next decade witnessed several technical developments in haemodialysis, including the use of large-surface-area dialysers with a shortening of dialysis times. However, many patients, especially the elderly, suffered from cardiovascular instability as a result of the 'acetate infusion', and to overcome these problems bicarbonate was reintroduced as the major dialysate buffer in the late 1970s [2] and has now become the standard or preferred dialysate buffer.

In continuous ambulatory peritoneal dialysis (CAPD), lactate has become the standard buffer as acetate-containing peritoneal dialysate caused sclerosis of the peritoneal membrane. Recently, preliminary results with carbonate-containing peritoneal dialysate have been encouraging, maintaining adequate plasma bicarbonate levels, and it is possible that by using a more physiological solution, peritoneal membrane viability will be maintained [3].

In addition to the buffering capacity of the dialysate, the use of oral phosphate binders and calcium carbonate have increased the plasma bicarbonate levels. Adequate doses of calcium carbonate i.e. 500 mg three times a day will increase the plasma bicarbonate levels by 3–4 mmol/l [4, 5] whilst also controlling the serum phosphate. A recent experience with calcium α-ketogluterate has shown similar results [6].

Degree of Correction of Acidosis

It is a common clinical observation that haemodialysis patients have lower plasma bicarbonate levels, especially predialysis, than CAPD patients. Data from the Leicester dialysis programme are shown in table 1. As can be seen, the interdialytic plasma bicarbonate value in 100 haemodialysis patients using acetate dialysate was 23.8 ± 0.2 mmol/l compared to 25.6 ± 0.3 mmol/l for CAPD patients. The latter were using a peritoneal dialysate with a lactate

KARGER
E-Mail karger@karger.ch
Fax +41 61 306 12 34
http://www.karger.ch

0378–0392/97/0236–0234 $12.00/0

John Walls, MD
Department of Nephrology
Leicester General Hospital
Leicester LE4 5PW (UK)

Table 1. Plasma bicarbonate levels in dialysis patients

Haemodialysis (n = 100)	predialysis	22.6 ± 0.3 mmol/l
	postdialysis	25.1 ± 0.3 mmol/l
	interdialytic	23.8 ± 0.2 mmol/l
CAPD (n = 100)	Clinic visit	25.6 ± 0.3 mmol/l*

Haemodialysis patients used acetate dialysate. Both groups used $CaCO_3$ as a phosphate binder. * $p < 0.001$.

content of 35 mmol/l. A similar range of plasma bicarbonate values in haemodialysis and CAPD patients has been reported by others using a dialysate bicarbonate concentration of 34 mmol/l [7, 8]. More recently, Williams et al. [9] have demonstrated that using a high-bicarbonate dialysate solution containing 40 mmol/l of bicarbonate that the predialysis plasma bicarbonate can be maintained in the normal range, i.e. between 25 and 27 mmol/l. When the lactate content of the peritoneal dialysate is increased to 40 mmol/l, the plasma bicarbonate level will increase to the upper range of the normal values and may in fact exceed it (31.5 ± 0.5 mmol/l) [10] without any apparent adverse effects.

Nutritional Status

Haemodialysis Patients

Over the past decade there have been several studies in haemodialysis patients demonstrating an improvement in nutritional status with an increase in plasma bicarbonate levels. These studies have used a variety of parameters to assess nutritional state. In 1985, Papadoyannakis et al. [11] demonstrated in a group of maintenance haemodialysis patients, on a fixed protein intake, that supplementation with calcium carbonate increased the plasma bicarbonate and significantly reduced the protein catabolic rate (PCR) from 1.2 g/kg per 24 h to 1 g/kg per 24 h. An increase in body weight was noted in a group of haemodialysis patients transfered from one haemodialysis unit using a dialysate bicarbonate of 32 mmol/l to a second unit where the dialysate bicarbonate was 36 mmol/l [12]. Recently, the study of Williams et al. [9] demonstrated improved nutritional status with an increase in triceps skinfold thickness on changing from the low (30 mmol/l) to high (40 mmol/l) bicarbonate dialysate. These findings reversed when the dialysate bicarbonate concentration was changed in the opposite direction. In 1990, Bergström et al. [13] demonstrated a significant positive correlation between the predialysis plasma bicarbonate value and muscle valine content. Studies by Moran et al. [14] in 1993 demonstrated a decrease in urea generation from 5.0 ± 1.26 to 4.83 ± 1.63 mg/min, which was negatively correlated with the increase in plasma bicarbonate in a group of haemodialysis patients.

CAPD

In a short-term study (2 weeks), Stein et al. [15] demonstrated that by increasing the plasma bicarbonate with sodium bicarbonate supplementation in CAPD patients, the PCR significantly decreased, a finding not unlike that of Papadoyannakis et al. [11] in haemodialysis patients. Using the leucine turnover technique to measure protein synthesis and degradation, Graham et al. [16] demonstrated a significant decrease in protein degradation and protein synthesis but no change in amino acid oxidation in CAPD patients using a similar protocol to that of Stein et al. [15].

It is important to determine whether the results of short-term studies as described above can be translated into long-term treatment strategies. In a prospective randomised trial in 200 CAPD patients, Graham et al. [16] demonstrated that a peritoneal dialysate with a lactate content of 40 mmol/l and oral calcium carbonate produce significant increases in body weight, tricep skinfold thickness, midarm circumference and midarm muscle circumference when compared at 1 year to a control group receiving a dialysate with a lactate content of 35 mmol/l. There was a significant difference in the plasma bicarbonate of the two groups (23.0 ± 0.3 vs. 27.2 ± 0.3 mmol/l, $p < 0.001$). In this study, serum albumin did not change, being maintained within the normal range in both groups. However, those patients with the higher plasma bicarbonate level had a lower hospital admission rate and a lower overall hospital stay, suggesting that a better correction of acidosis reduces morbidity in CAPD patients.

Conclusion

In the frequently quoted report of Lowrie and Lew [17], in addition to the adverse effect of reduced serum albumin, a reduction in the plasma bicarbonate level was also associated with an increased risk of mortality in haemodialysis patients. As it is now well recognised that metabolic acidosis increases protein degradation in uraemia [18–20], and that this may be a contributory factor in the malnutrition seen in both haemodialysis and CAPD patients, attempts must be made to maintain the plasma bicarbonate level within the normal range at all times. During the past decade this issue has been addressed by gradually increasing dialysate buffering capacity, whether it be bicarbonate in haemodialysis patients or lactate in CAPD patients. There are now some encouraging results to suggest that if the plasma bicarbonate is in the upper range of normal or even slightly above the upper range, there are long-term beneficial effects.

References

1 Mion CM, Hegstrom RM, Boen ST, Scribner BH: Substitution of sodium acetate for sodium bicarbonate in the bath fluid for haemodialysis. Trans ASAIO 1964;10:110–113.

2 Graefe U, Milutinovich J, Follette WC, Vizzo JE, Babb AL, Scribner BH: Less dialysis-induced morbidity and vascular instability with bicarbonate in dialysate. Ann Intern Med 1978;88:332–336.

3 Feriani M, Biasoli S, Barbacini S, Ronco C, Chiaramonte S, Bragantini L, Dell'Aquila R, Milan M, Brendolan A, La Greca G: Acid-base correction in bicarbonate CAPD patients. Adv Periton Dial 1989;5:191–194.

4 Fournier A, Moriniere P, Sebert JL, Dkhissi H, Atik A, Leflon P, Renaud H, Gueris J, Gregoite I, Garabedian M: Calcium carbonate, an aluminum free agent for control of hyperphosphataemia, hypocalcaemia and hyperparathyroidism in uraemia. Kidney Int 1986;29(suppl):S114–S119.

5 Williams B, Vennegoor M, O'Nunan T, Walls J: The use of calcium carbonate to treat the hyperphosphataemia of chronic renal failure. Nephrol Dial Transplant 1989;4:725–729.

6 Kardasz S, Johnstone J, Walls J: Calcium ketoglutanate (CaKg) in chronic renal failure. Nephrol Dial Transplant 1996;11:A133.

7 Hakim RM, Pontzer MA, Tilton D, Lazarus JM, Cottlieb MN: Effects of acetate and bicarbonate dialysis in stable chronic dialysis patients. Kidney Int 1985;28:535–540.

8 Bergström J, Lindholm B: Nutrition and adequacy of dialysis: How do haemodialysis and CAPD compare? Kidney Int 1993,43(suppl 40):539–550.

9 Williams AJ, Dittmer ID, McArley A, Clarke J: High bicarbonate dialysate in haemodialysis patients: Long term effects on acid-base status and body composition. Nephrol Dial Transplant, in press.

10 Stein A, Feehally J, Walls J: Does low calcium dialysate improve the nutritional status of CAPD patients? Periton Dial Int 1993;13:69.

11 Papadoyannakis NJ, Stefanides CJ, Patricarea A, Kotsovasilis K, Georgiopoulou A: The effect of calcium carbonate administration on nitrogen metabolism in patients on haemodialysis. Proc EDTA-ERA 1985;22:83–87.

12 Seyffart G, Ensminger A, Scholz R: Increase in body mass during long term bicarbonate dialysis. Kidney Int 1987;31(suppl 22):S174–S177.

13 Bergström J, Alvestrand A, Furst P: Plasma and muscle free amino acids in maintenance haemodialysis patients without protein malnutrition. Kidney Int 1990;38:108–114.

14 Moran JE, Schmidt B, Blumenstein M, Gurland HJ: Reduction of urea generation rate by ameliorating acidosis in haemodialysis. ERA-EDTA abstracts 1993;150.

15 Stein A, Bennett S, Larratt C, Baker F, Feehally J, Walls J: The effect of correction of acidosis on protein catabolic rate (PCR) in peritoneal dialysis (PD) patients. Clin Sci 1993;85:4.

16 Graham KA, Reaich D, Channon SM, Downie S, Gilmour E, Passlick-Ddetjen J, Goodship THJ: Correction of acidosis in CAPD decreases whole body protein degradation. Kidney Int 1996;49:1396–1400.

17 Lowrie EG, Lew NL: Death risk in haemodialysis patients: The predictive value of commonly measured variables and an evaluation of death rate differences between facilities. Am J Kidney Dis 1990;15:458–482.

18 Mitch WE: Cellular mechanisms of catabolism activated by metabolic acidosis. Blood Purif 1995;13:368–374.

19 Williams B, Layward E, Walls J: Skeletal muscle degradation and nitrogen wasting in rats with chronic metabolic acidosis. Clin Sci 1991;80:457–462.

20 Williams B, Hattersley J, Layward E, Walls J: Metabolic acidosis and skeletal muscle adaptation to low protein diets in chronic uraemia. Kidney Int 1991;40:779–786.

Mineral and Electrolyte Metabolism

Miner Electrolyte Metab 1997;23:237–242

Cytokines and the Malnutrition of Chronic Renal Failure

M.S. Abdullah[a, c]
G. Wild[b]
V. Jacob[a]
A. Milford-Ward[b]
R. Ryad[c]
M. Zanaty[c]
M.H. Ali[c]
A.M. El Nahas[a]

[a] Sheffield Kidney Institute and
[b] Department of Immunology, Northern General Hospital Trust, Sheffield, UK;
[c] Departments of Medicine and Nephrology, Zagazig University, Egypt

Key Words
Chronic renal failure
Insulin-like growth factor-I
Interleukin-1β
Interleukin-6
Malnutrition
Tumor necrosis factor-α

Abstract
Malnutrition in uremic patients remains one of the major causes of morbidity and mortality. Its mediators remain largely unknown. Uremia is characterized by changes in circulating levels of catabolic cytokines and anabolic growth factors. The aim of this study was to investigate whether these changes are associated with the malnutrition of patients with chronic renal failure (CRF). We have studied the prevalence of malnutrition in a small group of patients ($n = 20$) with CRF (serum creatinine $= 551 \pm 105$ μmol/l, mean ± SD) and 25 age-matched controls. Nutritional status was assessed by dietary diaries, subjective global assessment (SGA), and by measurement of anthropometric parameters. Regression analysis was applied to examine the relationship between biochemical and anthropometric parameters. Simultaneously, we have investigated changes in the circulating levels of catabolic cytokines [tumor necrosis factor-α (TNF-α), interleukin (IL)-1β and IL-6] and an anabolic growth factor [insulin-like growth factor-I (IGF-I)]. We observed a high prevalence of malnutrition as judged initially by SGA: 50% moderately malnourished and 15% severely malnourished. This was confirmed by anthropometric measurements. We noted a significant reduction in both triceps skinfold thickness (TST; 35% of patients <25th centile) and midarm muscle circumference (MAMC; 65% of patients <25th centile). We also noted a reduction in serum IGF-I in malnourished patients (IGF-I in well-nourished patients = 207 ± 48 μg/l, in malnourished patients = 133 ± 33 μg/l, $p < 0.01$). IGF-I correlated with TST ($r = 0.71$, $p < 0.001$) and MAMC ($r = 0.47$, $p < 0.05$). IGF-I had a high predictive value for TST ($R^2 = 51\%$, $p < 0.001$). In contrast, TNF-α levels were higher in malnourished patients: 19.5 ± 30 pg/ml compared to 3.9 ± 8 pg/ml in healthy patients ($p < 0.001$) and TNF-α showed a negative correlation with MAMC ($r = -0.69$, $p < 0.01$; $R^2 = 47\%$, $p < 0.01$). IL-1β levels were higher in CRF than in controls but did not correlate with nutritional parameters. No significant changes could be detected in serum IL-6. A significant percentage of predialysis patients with CRF suffer from some degree of malnutrition. This may be attributed in part to a fall in circulating anabolic growth factors and an increase in catabolic cytokines.

Introduction

Malnutrition in chronic uremia is associated with considerable morbidity and mortality [1–3]. It is thought to be multifactorial in etiology with changes in the endocrine and metabolic milieu playing an important role [4]. However, little attention has been paid to the potential role of changes in anabolic growth factors and catabolic cytokines in the pathogenesis of uremic malnutrition.

It is known that the circulating levels of some of these growth factors/cytokines are altered in chronic uremia. For instance, an increase in circulating levels of insulin-like growth factor-I (IGF-I) has been described [5]. Similarly, high circulating plasma levels of proinflammatory cytokines such as tumor necrosis factor-α (TNF-α) and interleukin (IL)-1 and IL-6 have been described in dialysis patients and nondialyzed chronic renal failure (CRF) patients [6, 7]. However, the possible contribution of these factors to the malnutrition of uremic patients has not been fully investigated. Some reports have suggested that malnourished patients on hemodialysis (HD) [8] and on continuous ambulatory peritoneal dialysis (CAPD) [9] have lower serum IGF-I levels. Rises in circulating levels of TNF-α and IL-1β have been associated with nonuremic catabolic states [10, 11] and are thought to be associated with a poor prognosis [12].

KARGER
E-Mail karger@karger.ch
Fax +41 61 306 12 34
http://www.karger.ch

0378–0392/97/0236–0237 $12.00/0

Professor A.M. El Nahas
Sheffield Kidney Institute
Northern General Hospital Trust
Sheffield S5 7AU (UK)
Tel. +44 114 2714018, fax +44 114 2562514

Table 1. Clinical and demographic characteristics of the studied population

Parameters	Controls (n = 25)	CRF (n = 20)	p
Age, years	49.6 ± 14.3	51.8 ± 18.7	NS
Female/male	9/25	6/20	NS
Etiology of renal failure			
Chronic glomerulonephritis		7 (35%)	
Diabetic nephropathy		3 (15%)	
Polycystic kidney disease		3 (15%)	
Chronic pyelonephritis		2 (10%)	
Hypertensive nephrosclerosis		2 (10%)	
Obstructive nephropathy		3 (15%)	
Others		4 (20%)	
Estimated dietary intake			
Protein, g/kg per day		0.62 ± 0.12	
Energy, kcal/kg per day		32.05 ± 4.45	
Anthropometric measurements			
BMI, kg/m^2	28.6 ± 4.4	25.8 ± 5.1	NS
TST, mm	20.5 ± 7.2	16.1 ± 8.7	< 0.01
MAMC, cm	26.1 ± 3.1	23.6 ± 3.2	< 0.01
Serum biochemical parameters			
Urea, mmol/l	5.3 ± 1.1	24.9 ± 5.9	< 0.001
Creatinine, μmol/l	91.5 ± 14.9	551 ± 105	< 0.001
Bicarbonate, mmol/l	25.1 ± 0.41	23.1 ± 2.1	NS
Protein, g/l	75.5 ± 4.3	69.9 ± 4.4	< 0.01
Albumin, g/l	41.9 ± 2.8	37.1 ± 3.3	< 0.001
Prealbumin, g/l	0.29 ± 0.05	0.38 ± 5.9	< 0.001
C-reactive protein, mg/l	< 5.1	6.3 ± 2.8	NS

NS = Not significant; BMI = body mass index; TST = triceps skinfold thickness; MAMC = midarm muscle circumference.

We have investigated in this study a small group of patients with moderate to severe CRF in order to determine the prevalence of malnutrition. We have also studied the changes in circulating levels of catabolic cytokines (TNF-α, IL-1β, and IL-6) and IGF-I and attempted to determine whether they correlated with nutritional parameters.

Subjects and Material

Subjects

We studied 20 patients (15 males, 5 females). At the time of the study, their renal function was significantly compromised as judged by their serum creatinine values (551 ± 105 μmol/l, mean ± SD). The etiology of their renal failure and clinical characteristics are listed in table 1. Patients were studied when they attended the outpatient clinic of the Sheffield Kidney Institute. Their clinical evaluation was undertaken by the same observer (M.S.A.). They were compared to a population of 25 age-matched healthy volunteers from the research and medical staff and from the anticoagulation clinic. This study was approved by the Ethics Committee of the Northern General Hospital NHS Trust.

Dietary Assessments

Dietary intake of protein and calories was calculated from 4-day dietary diaries. Nutrient intake was calculated with a computer program (Microdiet, Salford University). In general, we do not prescribe dietary protein restriction to our patients with CRF. They are all encouraged to consume an adequate amount of calories (around 35 kcal/kg per day). All the dietary assessments were undertaken and analyzed by one of the authors (V.J.).

Subjective Global Assessment

Subjective global nutritional assessment (SGA) was undertaken by one of the authors (M.S.A.). Criteria from the medical history included four major parameters: (1) weight changes, especially over the previous month, (2) dietary intake, (3) gastrointestinal symptoms such as loss of appetite, anorexia, vomiting, and diarrhea, and (4) functional capacity of the patient and underlying diseases, particularly as they may affect the nutritional requirements. Criteria from physical examination included loss of subcutaneous fat mass, particularly in the triceps area and chest, and muscle wasting, especially in the quadriceps area and shoulder girdle. Each medical history and physical examination parameter was rated subjectively as A (well nourished), B (mild or moderately malnourished) or C (severely malnourished). The final score was assigned according to the investigator's subjective opinion of the patient's nutritional status. The severely malnourished rating (C) was given whenever a patient had physical signs of malnutrition such as loss of subcutaneous fat and severe muscle wasting, in the presence of a medical history suggestive of risk such as continuing weight loss with a net loss of 10% or more or a decline in dietary intake. If the severely malnourished (C) or the well-nourished (A) rating was not clearly defined, the patient was assigned to the moderately malnourished (B) classification [13].

Anthropometric Measurements

The following anthropometric indices were evaluated: body weight (Wt), height (Ht), midarm circumference (MAC), and triceps skinfold thickness (TST). Body mass index (BMI) was calculated as Wt (kg)/Ht^2 (m^2). MAC was also calculated as MAMC (cm) = MAC − [3.14 × triceps skinfold thickness (TST; cm)]. MAC was measured with a tape to the nearest centimeter at the midpoint between the tip of the acromial and the olecocranon process. TST was measured to the nearest millimeter by a Lange skinfold caliper. We compared the measurements to centiles of an age-matched population [14, 15]. Values equal or less than the 5th percentile indicated severe malnutrition (depletion); those between the 5th and 25th percentiles indicated mild to moderate malnutrition (undernutrition). Anthropometric measures were also taken by the same clinical investigator (M.S.A.).

Biochemical Indices
Blood was drawn and collected without anticoagulation from nonfasting patients on the day of nutritional assessment and stored at −20 °C until measurements of the following biochemical indices were made: urea, creatinine, protein, albumin (A), prealbumin (PA), and IGF-I. Samples for TNF-α, IL-1β, and IL-6 assays were collected in tubes containing EDTA and immediately centrifuged at 4 °C for 10 min. Plasma was removed without disturbing the buffy coat and centrifuged at 4 °C for a further 15 min to remove platelets. Aliquots were stored in polypropylene tubes at −70 °C until the day of measurement.

Laboratory Assays
Serum protein, albumin, urea, and creatinine were measured with a parallel analyzer (American Monitor, Indianapolis, Ind., USA). PA was measured by a single radial immunodiffusion technique. IGF-I was measured by a two-site immunometric assay using commercially available kits (Octeia IGF-I IEMA, Immunodiagnostics, Bolton, UK). The method incorporates a sample pretreatment to avoid interference from IGF-binding proteins. The sensitivity of the test is 8 μg/l, and the coefficient of variation (CV) is 20.5%, with a mean recovery of 99%. There is no interference from IGF-II, insulin, or proinsulin. Results are expressed as μg/l.

Plasma levels of TNF-α, IL-β, and IL-6 were measured by enzyme-linked immunosorbant assay using commercially available kits (Predicta, Genzyme, USA). Quality control was ensured using the control samples with each assay. Then TNF-α assay does not cross-react with TNF-β, IL-1β, IL-2, IL-4, or IL-6. Intra-assay and inter-assay reproducibility (CV) are 6.06% and 7.39%, respectively. The detection limit of the TNF-α assay is 10 pg/ml. The detection limits of IL-1β and IL-6 assays are 3 pg/ml and 18 pg/ml, respectively. Results are expressed as pg/ml and samples below the detection limit of the assays were considered zero.

Statistical Evaluation
Results are expressed as the mean ± SD. Nonparametric tests and χ^2 analyses were applied when appropriate. Correlation between variables was estimated by the Pearson product-moment correlation coefficient with two-tailed probabilities. Regression analysis using a stepwise variable-selection method was applied to examine the relationships between biochemical and anthropometric parameters and the power of the former in predicting the latter. A probability of less than 5% was taken as significant. Statistical analysis was performed using standard computer software (Statistical Program for Social Sciences SPSS, 1993).

Results

Table 1 summarizes the clinical and demographic characteristics of our study population.

Dietary Intake
The majority of the CRF patients had a low protein and calorie intakes (0.62 ± 0.12 g/kg per day and 32.05 ± 4.45 kcal/kg per day, respectively). There was no correlation between the patients' nutrient intake and their serum creatinine levels. On the other hand, a strong positive correlation was observed between patients' protein and energy intake and MAMC ($r = 0.60$, $p < 0.01$ and $r = 0.59$, $p < 0.01$, respectively).

SGA
According to the SGA, the patients studied could be divided into those who were well nourished (A: 7 of 20, 35%), those who had mild–moderate malnutrition (B: 10 of 20, 50%), and those with severe malnurition (C: 3 of 20, 15%). This SGA score correlated with the parameters of nutrition based on objective anthropometric measurements including TST ($r = 0.69$, $p < 0.01$) and MAMC ($r = 0.59$, $p < 0.01$). Moreover, the SGA score predicted TST ($R^2 = 44\%$, $F = 16.47$, $p < 0.001$) and MAMC ($R^2 = 32\%$, $F = 10.07$, $p < 0.01$). The SGA score also correlated with both protein and energy intake ($r = 0.48$, $p < 0.05$ and $r = 0.45$, $p < 0.05$, respectively).

Anthropometric Indices
Body weight and BMI were decreased in 15% of patients, but 65% showed some depletion of muscle/protein stores (MAMC < 25th percentile), and 35% showed depletion of fat stores (TST < 25th percentile). Some (30%) had depletion of both. Anthropometric parameters correlated with protein intake and SGA score (table 2).

Serum Biochemical Indices
IGF-I. IGF-I levels were higher in patients with CRF (166 ± 55 μg/l) when compared to the healthy controls (142 ± 26 μg/l, $p < 0.05$). In contrast, IGF-I was significantly lower in malnourished patients with CRF (SGA = B or C and TST or MAMC < 25th percentile) compared to their robust counterparts (CRF malnourished 133 ± 33 μg/l, CRF healthy 207 ± 48 μg/l, $p < 0.01$) (table 3). There was a strong negative correlation between the degree of malnutrition judged by SGA score and serum IGF-I levels ($r = -0.77$, $p < 0.001$). IGF-I also correlated closely with anthropometric parameters including BMI, TST and, to a lesser extent, MAMC (table 2) (fig. 1a). It also had a high predictive power for fat depletion (low TST; < 25th perentile) in the population with CRF ($R^2 = 51\%$, $p < 0.001$). However, the predictive value of IGF-I for protein depletion (MAMC; < 25th percentile) was low ($R^2 = 9\%$).

TNF-α. TNF-α levels were higher in patients with CRF (12.5 ± 24 pg/ml) than in healthy controls (3.4 ± 9 pg/ml, $p < 0.05$). Plasma levels of TNF-α were above the detection limit of the assay (10 pg/ml) in 4 of the 25 controls (16%) and in 8 of the 20 patients (40%) with CRF ($p < 0.05$). In CRF, circulating TNF-α levels correlated with the severity of the renal insufficiency ($r = 0.65$, $p < 0.01$). TNF-α levels were significantly higher in malnourished compared to the well-nourished CRF patients (19.5 ± 29.8 pg/ml vs. 3.9 ± 7.8 pg/ml, respectively, $p < 0.001$). Plasma TNF-α levels showed a significant negative correlation with protein intake ($r = -0.53$, $p < 0.02$, table 2). Moreover, TNF-α levels correlated negatively with both BMI and MAMC (table 2, fig. 1b). TNF-α predicted MAMC in patients with CRF ($R^2 = 47\%$, $p < 0.01$). However, the predictive value of TNF-α for TST was poor ($R^2 = 7\%$).

IL-1β and IL-6. Among the healthy controls, plasma IL-1β levels were consistently below the detection level of the assay we used (3 pg/ml). In contrast, IL-1β levels were above the detection level in 5 of the 20 (25%) patients with CRF. The plasma level of the CRF group was 2.95 ± 5.1 pg/ml. IL-6 levels were low in all the

Table 2. Correlation matrix of nutritional and biochemical parameters in CRF patients

	SGA	DPI	DEI	BMI	TST	MAMC	S. urea	S. creatinine	S. A	S. PA	IGF-1	TNF
DPI	0.48*											
DEI	0.45*	0.77***										
BMI	0.61**	0.31	0.39									
TST	0.69**	0.16	0.17	0.62**								
MAMC	0.59**	0.60**	0.59**	0.66**	0.22							
S. urea	−0.41	−0.37	−0.40	−0.38	−0.25	−0.51*						
S. creatinine	−0.21	−0.11	−0.28	−0.27	−0.22	−0.27	−0.51*					
S. A	−0.03	−0.19	0.01	0.10	0.10	0.02	0.05	0.14				
S. PA	−0.07	−0.34	−0.26	0.03	0.07	−0.18	−0.07*	−0.49*	−0.18			
S. IGF-1	0.77***	0.35	0.37	0.68**	0.71**	0.49*	−0.44	−0.43	−0.24	0.11		
S. TNF-α	−0.41	−0.53**	−0.40	−0.49*	−0.03	−0.69**	−0.65**	−0.01	0.05	0.36	−0.35	
S. IL-1β	0.18	−0.08	−0.01	−0.29	0.09	−0.16	0.06	−0.07	0.09	0.12	0.05	0.47*

DPI = Dietary protein intake; DEI = dietary energy intake; S. = serum; * $p < 0.05$, ** $p < 0.01$, *** $p < 0.001$.

Table 3. Nutritional and biochemical parameters in well-nourished (WN) compared to malnourished (MN) CRF patients

Parameter	CRF-WN	CRF-MN	p
Age, years	55.1±14.7	49.1±21.8	NS
BMI, kg/m²	29.0±3.8	23.1±4.5	<0.05
TST, mm	21.2±9.3	11.7±4.7	<0.01
MAMC, cm	26.1±1.3	21.6±2.9	<0.05
S. creatinine, μmol/l	503±63	590±118	NS
S. A, g/l	37.2±4.0	36.9±2.2	NS
S. PA, g/l	0.37±0.07	0.39±0.07	NS
S. IGF-I, μg/l	166.4±54.6	133.1±33.3	<0.01
S. TNF-α, pg/ml	3.9±7.8	19.5±29.8	<0.001
S. IL-1β, pg/ml	3.4±5.5	2.5±5.7	NS

NS = Not significant; S. = serum.

patients studied and did not differ between healthy individuals and patients with CRF. Neither IL-1β or IL-6 correlated with any of the nutritional parameters studied.

Other biochemical parameters such as serum protein, A, and PA did not discriminate between healthy and malnourished CRF patients.

Discussion

Considerable emphasis has been put in recent years on the effects of malnutrition on the morbidity and mortality of patients treated with long-term dialysis replacement therapy [1–3]. Studies have demonstrated increased risk ratios for survival on dialysis in patients with biochemical indications of malnutrition such as low serum A, creatinine, phosphorus or cholesterol [1–3]. This is the case for patients treated by HD as well as those treated by CAPD [1–3]. We have also described a high predictive value for IGF-I for malnutrition in HD [8] and CAPD [9] patients. In this study, we extend these observations by describing a high prevalence of malnutrition in predialysis patients with CRF and also attribute a useful discriminatory value for serum IGF-I and TNF-α in this population. At present we can only speculate on the relevance of the changes observed in these growth factors/cytokines to the pathogenesis of malnutrition in patients with CRF.

The observation that up to 6% of predialysis patients with CRF have some degree of malnutrition confirms previous observations [16]. The finding that 15% of the patients suffer, according to SGA, from severe malnutrition is a cause for considerable concern. This concern relates to the subsequent poor prognosis of these patients on dialysis. While some may improve nutritionally after the initiation of dialysis replacement therapy, it is likely that the majority will remain malnourished. The other cause of concern relates to the correlations observed between their nutritional status and their dietary nutrient intake. Those with decreased protein intake had lower nutritional parameters. Clearly, advanced uremia is itself associated with a reduction in dietary protein intake which could in turn exacerbate malnutrition [17]. However, to make matters worse, dietary protein restriction continues to be prescribed in many renal units, a practice we deplore as it is likely to further compromise the nutritional status of patients with advanced CRF [18] and has proved ineffective in slowing the progression of their renal disease [19].

Our study also highlights the clinical value of SGA in predialysis patients with CRF. In these patients, we noted that such assessment is valuable and accurate in predicting malnutrition. This simple clinical evaluation is therefore a useful tool to identify malnourished patients with CRF and possibly monitor their response to nutritional rehabilitation. It warrants broader clinical application in view of the widespread malnutrition in this population.

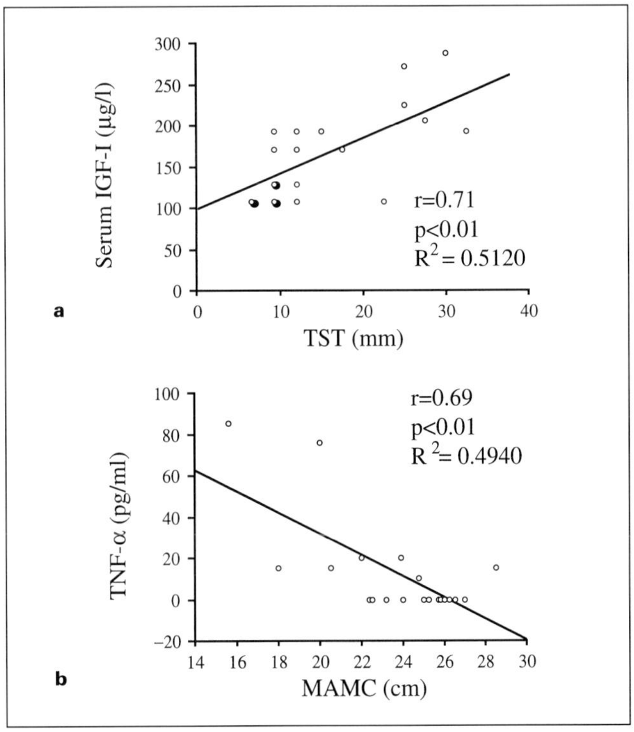

Fig. 1. Correlation between serum IGF-I levels and TST (**a**) and between TNF-α and MAMC (**b**) in patients with CRF.

Another finding of our current study is the value of circulating levels of IGF-I in reflecting the patient's nutritional status. This confirms our previous observations in HD and CAPD patients where this test was found to be the best discriminatory serological marker of malnutrition [8, 9]. In patients with CRF, this is particularly true of the predictive value of IGF-I for TST, a reflection of body fat store depletion. IGF-I is also well known to correlate with malnutrition in nonrenal patients [20]. This growth factor has major anabolic properties similar to those of insulin [21, 22]. Its short half-life makes it a valuable reflection of the patient's recent nutritional status and offers a guide for the response to nutritional treatment [23]. More recently, interest has focused on the potential therapeutic role of IGF-I in the correction of malnutrition in renal [24] and nonrenal [25] patients. The administration of IGF-I to patients with advanced renal insufficiency and malnutrition may have the added therapeutic advantage of improving in their renal function. It was recently demonstrated that the short-term administration of recombinant human IGF-I to patients with severely impaired renal failure led to a marked, albeit transient, increase in their glomerular filtration rate [26].

We also noted in this study an increase in the circulating levels of catabolic cytokines, in particular TNF-α and IL-1β, in CRF. Elevated plasma levels of TNF-α, IL-1β and IL-6 have been found in both nondialyzed patients with CRF and in dialysis patients [6, 7]. The detection of circulating proinflammatory cytokines in nondialyzed patients with CRF suggests that factors related to uremia could be sufficient to activate mononuclear cells and trigger cytokine production. However, an elevated plasma level of a cytokine or its inhibitor in patients with renal insufficiency could also be due to decreased clearance [6, 7].

In this study, TNF-α levels were found to correlate negatively with patients' protein intake and protein stores (MAMC). This is to our knowledge the first description of such changes in this population and may have important nutritional implications. The release of proinflammatory cytokines is clearly associated with significant metabolic aberrations that can compromise the patient's nutritional status [27]. In animals and humans, TNF-α induces a net release of amino acids from peripheral tissues after myofibrillar protein degradation [28]. Both TNF-α and IL-1β may act individually or in concert to increase the catabolism of muscle protein, an effect that may be mediated by prostaglandin E_2 [28], the stimulation of branched-chain ketoacid dehydrogenase [29], and increased plasma corticosterone concentration [30]. The observation that plasma TNF-α levels correlated negatively with protein intake suggests that proinflammatory cytokines may play a role, at least in part, in the anorexia in this patient population. It has been noted that IL-1β and TNF-α alter firing rates of glucose-sensitive neurons of the feeding centers of the lateral hypothalamus [31], and that both have been shown to reduce food intake and gastric emptying in healthy animals [32]. In addition to the direct effect on central satiety centers, anorexia associated with TNF-α and IL-1β may be mediated by the release of various neuroendocrine hormones and neuropeptides, including prostaglandins [33] and cholecystokinin [34]. These effects attributed to TNF-α, along with other uremic factors, may contribute to the observed spontaneous reduction in dietary protein intake in our predialysis patients. Others have described a similar decrease in protein intake in patients with CRF and a creatinine clearance of less than 10 ml/min [17].

In conclusion, our study identified a major nutritional problem and deficiency in predialysis patients. Clinical evaluation of these patients with SGA and the monitoring of their serum IGF-I and TNF-α levels may allow the early detection of malnutrition. This should in turn lead to appropriate nutritional rehabilitation and if need be, the early initiation of dialysis replacement therapy. Our suspicion is that if these patients start their dialysis treatment malnourished, they may remain so or even deteriorate and therefore have a poor prognosis with high morbidity and mortality rates. Finally, it is tempting to speculate that the progressive increase in plasma levels of proinflammatory cytokines, especially TNF-α and IL-1β, caused by CRF, contributes at least in part to the observed muscle wasting and poor nutrient intake. This would be compounded by the fall in circulating levels of anabolic hormones such as IGF-I as well as the resistance to their peripheral action [22]. The prospect of therapy based on the manipulation of growth factors and cytokines and aimed at the correction of uremic malnutrition is an exciting one.

Acknowledgments

The authors are grateful to the Northern General Hospital Trust for its support. M.S.A. is a visiting fellow supported by the Egyptian Government.

References

1 Lowrie EG, Lew NL: Death risks in hemodialysis patients: The predictive value of commonly measured variables and evaluation of death rate differences between facilities. Am J Kidney Dis 1990;15:458–482.

2 Avram MM, Mittman N, Bonomini L, Chattopadhyay J, Fein P: Markers for survival in dialysis: A seven-year prospective study. Am J Kidney Dis 1995;26:209–219.

3 Young GA, Kopple JD, Lindholm B, Vonesh EF, De Vecchi A, Scalamogna A, Castelnova C, Oreopoulos DG, Anderson GH, Bergstrom J: Nutritional assessment of continuous ambulatory peritoneal dialysis patients: An international study. Am J Kidney Dis 1991;17:462–471.

4 Ikizler TA, Hakim RM: Nutrition in end-stage renal disease. Kidney Int 1996;50:343–357.

5 Zumkeller W, Schofield PN: The role of insulin-like growth factors and IGF-binding proteins in the physiological and pathological processes of the kidney. Virchows Arch B Cell Pathol 1992;62:207–220.

6 Herbelin A, Nguyen AT, Zingraff J, Urena P, Descamps-Latscha B: Influence of uraemia and hemodialysis on circulating interleukin-1 and tumour necrosis factor-α. Kidney Int 1990;37:116–125.

7 Pereira BJ, Shapiro L, King AJ, Falagas ME, Strom JA, Dinarello CA: Plasma levels of IL-1β, TNF-α and their specific inhibitors in undialyzed chronic renal failure, CAPD, and haemodialysis patients. Kidney Int 1994;45:890–896.

8 Jacob V, Le Carpentier JE, Salazano S, Naylor V, Wild G, Brown CB, El Nahas AM: IGF-I, a marker of undernutrition in haemodialysis patients. Am J Clin Nutr 1990;52:39–44.

9 Jacob V, Marchant PR, Wild G, Brown CB, Moorhead PJ, El Nahas AM: Nutritional profile of continuous ambulatory peritoneal dialysis patients. Nephron 1995;71:16–22.

10 McMurray J, Abdullah I, Dargie H, Shapiro D: Increased concentrations of tumour necrosis factor in 'cachectic' patients with severe chronic heart failure. Br Heart J 1991;66:356–358.

11 Mooradian AD, Reed RL, Meredith KE, Scuderi P: Serum levels of tumour necrosis factor-α and IL-1β in diabetic patients. Diabetes Care 1991;14:63–65.

12 Mooradian A, Reed R, Osterweel D, Scuderi P: Detectable serum levels of tumour necrosis factor-α may predict early mortality in elderly institutionalized patients. J Am Geriatr Soc 1991;39:891–894.

13 Detsky AS, Mc Laughlin JR, Baker JP, Johston N, Jeejeebhoy KN: What is Subjective Global Assessment of nutritional status? J Parent Enter Nutr 1987;11:8–13.

14 Metropolitan Height and Weight Table. Stat Bull 1983;64:Jan-Jun.

15 Heymsfield SB, Mc Manus CB, Seitz SB, Nixon DW, Andrews JS: Anthropometric assessment of adult protein-energy malnutrition; in Wright RA, Heymsfield SB (eds): Nutritional Assessment. Cambridge, Blackwell, 1984, pp 27–77.

16 Campher C: Nutrition assessment in chronic renal failure. Nutr Support Serv 1985;7:18–21.

17 Ikizler TA, Greene J, Wingard RL, Parker RA, hakim RM: Spontaneous dietary protein intake during progression of chronic renal failure. J Am Soc Nephrol 1995;6:1386–1391.

18 El Nahas AM, Coles GA: Dietary treatment of chronic renal failure: Ten unanswered questions. Lancet 1986,i:597–600.

19 Klahr S, Levey AS, Beck GJ, Caggiula AW, Hunsicker L, Kusek JW, Striker G: The effects of dietary protein restriction and blood pressure control on the progression of chronic renal disease. N Engl J Med 1994;330:877–884.

20 Underwood LE, Clemmons DR, Maes M, D'Ercole AJ, Ketelsegers E: Regulation of somatomedin-C/insulin-like growth factor-I by nutrients. N Engl J Med 1991;325:695.

21 El Nahas AM, Sayed-Ahmed N: IGF-1 and the kidney: Friend or foe? Exp Nephrol 1993;1:205–217.

22 Ikizler TA, Wingard RL, Hakim RM: Interventions to treat malnutrition in dialysis patients: The role of the dose of dialysis, intradialytic parenteral nutrition and growth hormone. Am J Kidney Dis 1995;26:256–265.

23 Isley WL, Underwood LE, Clemmons DR: Dietary components that regulate serum somatomedin C concentration in humans. J Clin Invest 1983;71:175–182.

24 Peng S, Fouque D, Kopple J: Insulin-like growth factor-I causes anabolism in malnourished CAPD patients. J Am Soc Nephrol 1993;4:414.

25 Wilmore DW: Catabolic illness: Strategies for enhancing recovery. N Engl J Med 1991;325:695.

26 Mille SB, Moulton M, O'Shea M, Hammermann MR: Effects of IGF-I on renal function in end-stage chronic renal failure. Kidney Int 1994;46:201–207.

27 McNamara MJ, Alexander HR, Norton JA: Cytokines and their role in the pathophysiology of cancer cachexia. J Parent Enter Nutr 1990;16(suppl):S50–S55.

28 Flores EA, Bistrian BR, Pomposelli JJ, Dinarello CA, Blackburn GL, Istfan NW: Infusion of tumour necrosis factor/cachectin promotes muscle catabolism in the rat: A synergistic effect with interleukin-1. J Clin Invest 1989;83:1614–1622.

29 Nawabi MD, Block KP, Chakrabarti MC: Administration of endotoxin, tumour necrosis factor or interleukin-1 to rats activates skeletal muscle branched alpha-ketoacid dehydrogenase. J Clin Invest 1990;85:256–263.

30 Mealy K, Lonschot JB van, Robinson BG, Rounds J, Wilmore DW: Are the catabolic effects of tumour necrosis factor-α mediated by glucocorticoids? Arch Surg 1990;125:42–48.

31 Plata-Salaman CR, O'omura Y, Kai Y: Tumour necrosis factor-α and interleukin-1β: Suppression of food intake by direct action in the central nervous system. Brain Res 1988;448:106–114.

32 Robert A, Olafsson AS, Lancaster C, Zhang W: IL-1 is cytoprotective, antisecretory, stimulates prostaglandin E2 synthesis by stomach, and retards gastric emptying. Life Sci 1991;48:123–134.

33 Otterness IG, Golden HW, Seymour PA, Eskra JD, Daumy GO: Role of prostaglandins in the behavioral changes induced by murine IL-1 in the rat. Cytokine 1991;3:333–338.

34 Daun JM, Mc Carthy DO: The role of cholecystokinin in IL-1-induced anorexia. Physiol Behav 1993;54:237–241.

Mineral and Electrolyte Metabolism

Miner Electrolyte Metab 1997;23:243–248

Update on Renal Acidification: A Physiological View

Giovambattista Capasso [a]
Caterina Saviano [a]
Maria Rizzo [a]
Angelo Pica [a]
Dorotea Capodicasa [a]
Nicola Mascolo [b]
Natale G. De Santo [a]

[a] Chair of Nephrology,
[b] Experimental Pharmacology,
Second University of Naples, Italy

Key Words
H^+-ATPase
Intracellular pH
Na^+/H^+ exchanger

Abstract
The kidney plays a major role in the regulation of acid-base balance. This process is mainly dependent on H^+ secretion in the tubular lumen. Two acid extruder proteins are involved: the Na^+/H^+ exchanger and H^+-ATPase. Studies using in vivo and in vitro microperfusion and isolated membrane vesicles have clearly demonstrated that the Na^+/H^+ exchanger is the main mechanism regulating H^+ secretion/HCO_3^- reabsorption along the proximal nephron. Moreover, several reports indicate that this protein is involved in intracellular pH (pH_i) regulation. Newer studies using molecular biology techniques have identified at least five isoforms of the Na^+/H^+ exchanger: NHE-1 is the housekeeping isoform, while NHE-3 seems to be implicated in transepithelial acid-base transport, although other isoforms could be involved too. H^+-ATPase is the major acid extruder protein along the distal nephron, but it is also expressed along the proximal tubule, where a Na^+-independent bicarbonate reabsorption has been described. There are a few studies indicating that the proton pump participates in pH_i regulation, particularly in the presence of a large acid load. Its absence along the distal nephron may be one of the causes of distal tubular acidosis.

One of the major functions of the kidney is the regulation of acid-base balance. This is achieved by at least three mechanisms: (1) reabsorption of almost all the filtered bicarbonate in exchange for secreted H^+ [1, 2]; (2) excretion of titratable acidity [3], and (3) generation and excretion of NH_4^+ [4]. All of these processes are dependent on H^+ secretion by tubular cells into the lumen. Several acid-extruder proteins participate in this action. In this paper, we will analyze the properties, functions and significance of two of these proteins: the Na^+/H^+ exchanger and H^+-ATPase. In addition, recent data on their molecular biology characteristics will be pointed out. We will also present evidence of the importance of these acid-extruders in selected pathophysiological conditions.

Na^+/H^+ Exchanger

The Na^+/H^+ exchanger has been found in the plasma membranes of many cell types [5]. It is a secondary active transport system, deriving its energy from the electrochemical gradient of Na^+, established in turn by a primary active transporter, the Na^+/K^+ pump. Its normal mode of action is to mediate the uphill extrusion of H^+ coupled to the downhill flux of Na^+ into the cell. Under physiological conditions, the cellular gradient for the movements of these two ions is far from equilibrium. In most animal cells, due to the active extrusion of Na^+ via the Na^+/K^+ pump, there is a large, inwardly directed, Na^+ gradient. On the other hand, the concentration of H^+ is generally higher in the intracellular than in the extracellular compartments. Therefore, there is a large thermodynamic driving force favoring the exit of internal H^+ in exchange for the entry of external Na^+. The Na^+/H^+ exchanger has been implicated in a variety of very important physiological functions, such as the induction of cell growth and proliferation [6], the metabolic response to hormones [7], the control of cell volume [8], the regulation of transepithelial bicarbonate transport and of intracellular pH (pH_i). In the following, we will analyze more careful experimental evidence related to the last two points.

Control of Transepithelial Bicarbonate Transport
Following the hypothesis of Pitts [1] that proton secretion by the tubular cells is coupled to bicarbonate reabsorption, Ullrich et al. [9] demonstrated that the Na^+/H^+ exchanger was indeed implicated in H^+ secretion/HCO_3^- reabsorption along the rat proximal tubule. These authors, using the stop flow microperfusion technique with simultaneous capillary perfusion, evaluated the secretory rate

0378–0392/97/0236–0243 $12.00/0

Giovambattista Capasso, MD
Chair of Nephrology, Second University of Napoli
Padiglione 17, Policlinico Nuovo, Via Pansini 5, I-80131 Napoli (Italy)
Tel. +39 81 5666652, fax +39 81 5666655
E-Mail gcapasso@cds.unina.it

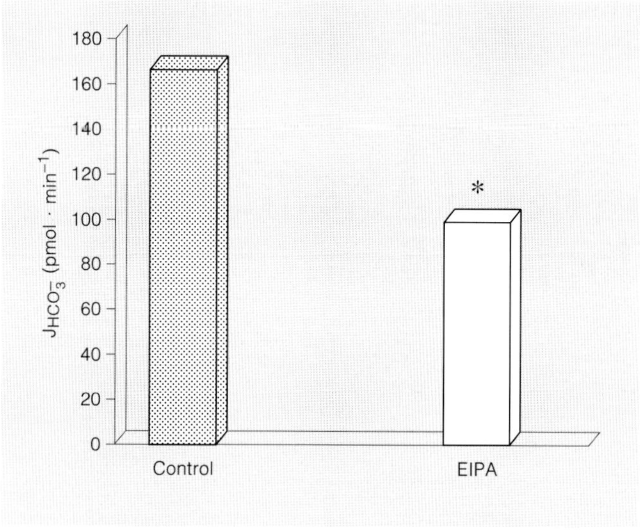

Fig. 1. Effect of EIPA on net bicarbonate reabsorption (J_{HCO_3}) along in vivo microperfused loop of Henle [from ref. 10].

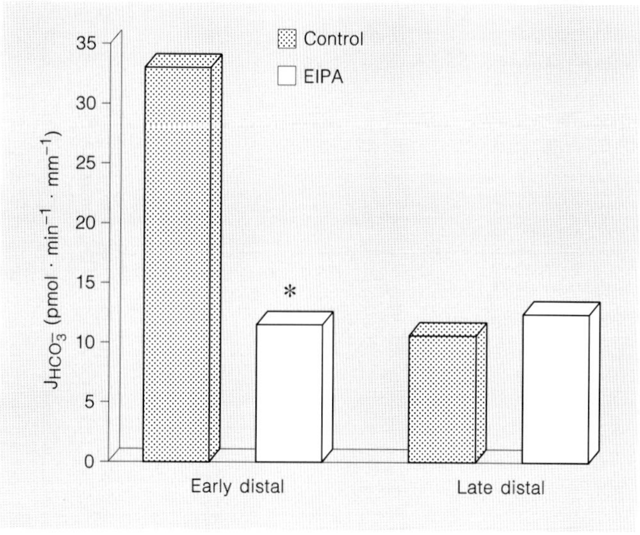

Fig. 2. Effect of EIPA on net bicarbonate reabsorption (J_{HCO_3}) along in vivo microperfused early and late distal tubules [from ref. 19].

of H^+ in the proximal tubule by measuring the reabsorption of glycodiazine, a bicarbonate-like buffer. They showed that when Na^+ was replaced in the ambient perfusion solutions by Li^+ or choline$^+$, glycodiazine transport was significantly reduced. These results strongly supported the hypothesis that, by analogy with other Na^+-dependent transport systems like sugars and amino acids, there could be a direct coupling between Na^+ and H^+ flux across the brush border membrane. A clear proof of this link appeared from the study of Murer et al. [2]: using renal brush border membrane vesicles, these authors evaluated proton transport from the intravesicular space to the incubation medium by monitoring pH changes, in the membrane suspension, induced by sudden Na^+ replacement. The addition of sodium salts to the membrane suspension caused an acidification of the incubation medium. When Na^+ uptake by brush border microvilli was studied, it was found that, in the presence of a proton gradient directed from the intravesicular space to the incubation medium, there was a transient accumulation of Na^+ inside the vesicles. These studies clearly demonstrated that in renal brush border membranes isolated, mostly, from the proximal tubule, there is a Na^+/H^+ exchanger that can generate a proton gradient, energized by the Na^+ concentration difference.

More recently, the importance and the contribution of this acid-extruder to renal acid-base transport has been extended to segments beyond the proximal tubule. Among the distal nephron segments, the loop of Henle has a very important role since, under normal conditions, it reabsorbs approximately 10–15% of the filtered bicarbonate [10]. The importance of the loop of Henle in acid-base balance was first established by micropuncture experiments which showed that a significant fraction of filtered bicarbonate was reabsorbed along this nephron segment [11, 12]. Subsequent studies of isolated thick ascending limbs (TALs) from rat kidneys confirmed directly the ability of this tubule segment to reabsorb bicarbonate by a carbonic-anhydrase-sensitive mechanism of Na^+/H^+ exchange [13]. Such investigations also provided evidence for a possible physiological role of the TAL in bicarbonate reabsorption by demonstrating a reduction of reabsorptive transport in chloride depletion alkalosis and stimulation in acidosis [14, 15]. Perfusion studies of the loop of Henle in vivo have confirmed the contribution of the Na^+/H^+ exchanger to bicarbonate absorption [10, 16, 17]. In particular, using a specific inhibitor of the exchanger, the ethylisopropyl-amiloride (EIPA), it has been possible to demonstrate that the contribution of the exchanger to the total net bicarbonate reabsorption could be in the range of 60% (fig. 1). The situation along the distal tubule is more complicated. The heterogeneity of this segment, as defined in micropuncture experiments, has created some problems in identifying the contribution of the various subsegments to the process of bicarbonate transport. According to Kriz and Bankir [18], the cortical distal tubule of the rat includes the distal convoluted tubule, the connecting tubule and the initial collecting tubule; these subsegments are made up of several cell types, located either in the early or in the late part of the cortical distal tubule. Recent in vivo experiments [19], performed to allow perfusion of individual early and late distal tubule segments, have demonstrated that at the level of the early distal tubule, the bulk of bicarbonate reabsorption is mediated by the Na^+/H^+ exchanger; along the late part of the distal tubule, the situation looks different, because H^+ secretion seems to be most mediated by an amiloride-insensitive mechanism (fig. 2).

Regulation of pH_i

We have monitored pH_i in the mouse convoluted proximal tubule (S_2 segment) using a video-imaging microscopy apparatus that allows pH_i measurements at the single-cell level. To this end, on the television monitor, in addition to a pseudocolor image of the tubular cells, areas of interest could be identified from the image. The computer program allowed multiple areas to be displayed simulta-

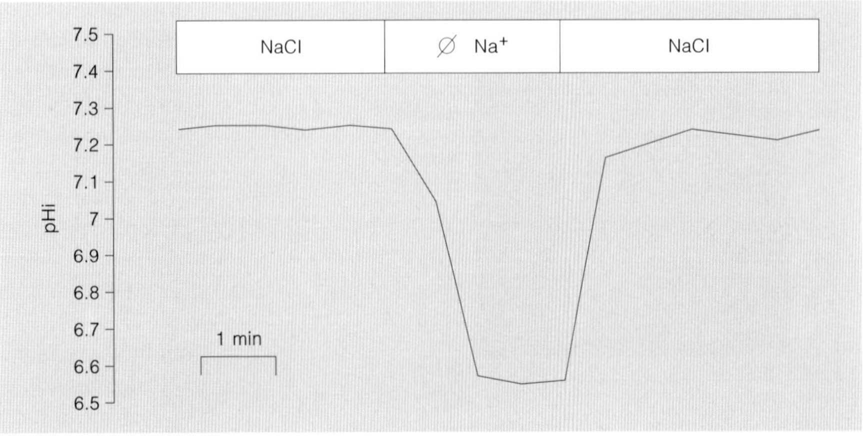

Fig. 3. Effect of removing Na^+ from the perfusion solution on intracellular pH (pH_i) of a single cell of mouse proximal tubule.

neously. Using a stereo microscope, the convoluted portion of the proximal tubule (S_2 segment) of superficial nephrons was identified and isolated. The dissected tubules were transferred to a solution containing 10 μM of the fluorescent pH-sensitive dye 2′,7′-bis-2-carboxyethyl-5-6-carboxyfluorescein acetoxymethyl ester. The dye-loaded tubule cells were alternatively excited at 490 and 450 nm, while the fluorescence emission was recorded at 530 nm. The emission measured while exciting at 490 nm was divided by that measured while exciting at 450 nm, producing a fluorescence excitation ratio. At the end of each experiment, a calibrated ratio was obtained using the nigericin/high $[K^+]$ method. Using this approach, we performed experiments, in the nominal absence of HCO_3, where the tubule was initially exposed to a Na^+-containing Ringer's solution (fig. 3); the recorded pH_i was around 7.24. Following removal of Na^+ (substituted with choline), there was a rapid fall in pH_i that was completely reversed after the readdition of Na^+ to the superfused solution. This experiment, together with other reports, indicates that the Na^+/H^+ exchanger is an important determinant of pH_i in a wide variety of animal cells [8, 20, 21].

Na^+/H^+ exchanger activity can be modulated by several stimuli, including pH_i. The first solid evidence for stimulation of the extruder by low pH_i came from experiments on membrane vesicles showing that the reduction in pH_i stimulates amiloride-sensitive sodium efflux [22]. Since such stimulation could not be explained by a first-order reaction mechanism involving a single proton binding site, it has been proposed that intracellular $[H^+]$, independent of its role as transportable substrate, can act as an allosteric activator of the Na^+/H^+ exchanger.

Molecular Biology

A very important breakthrough in the field of the Na^+/H^+ exchanger was the cloning by Sardet et al. [23] of the complementary DNA (cDNA) for the human fibroblast amiloride-sensitive Na^+/H^+ exchanger. Using the highly selective 'H^+-killing' secreening test, these authors isolated mutant mouse fibroblasts deprived of Na^+/H^+ activity. The cells were then transfected with human DNA. Those where the activity of the acid extruder was restored were used as a probe to clone the exchanger from a cDNA library obtained from a mouse transformant in which the transfected human gene was amplified. The amino acid sequence predicted a protein of molecular weight 99,354. The validity of these results was confirmed by the observation that stable expression of the transfected cDNA in Na^+/H^+-exchanger-deficient cells restored most of the property of the exchanger, such as pH_i regulation and proton-dependent Na^+ influx. Subsequent studies have unveiled the existence of a large Na^+/H^+ exchanger multigene family; to date at least five isoforms have been cloned and named NHE-1 to NHE-5 [24–27]. NHE-1 is present in most cells. In epithelial cells it is located on the basal lateral side. It is very sensitive to amiloride and its main function is the maintenance of a correct pH_i; therefore, it is considered the housekeeping isoform. NHE-2 is expressed in the liver, kidney and intestine and a similar distribution exists for NHE-4. So far there are no solid data about their role and function. NHE-3 has been identified as the main isoform present along the luminal membrane of the tubular cells. It is inhibited by high concentrations of amiloride and it is recognized as the exchanger involved in acid-base transepithelial transport. Based on these data we have concentrated our interest on this last isoform. Our aim was to detect the presence and to quantify the expression of NHE-3 mRNA along identified tubular segments using a RT-PCR method. To this end we have isolated proximal tubules and TALs of rat kidneys after perfusion with a collagenase-enriched solution. Using aligo(dT)-cellulose, the mRNA was purified and then transcribed into cDNA by incubation with a reverse transcriptase. The PCR reaction was performed using NHE-3-specific primers. β-Actin was used as an internal control. The PCR products were size fractionated by gel electrophoresis on 1% agarose. The predicted size for PCR products was 649 base pairs (bp) for β-actin and 223 bp for NHE-3. The results of the gel electrophoresis are shown in figure 4: in lane 1, the molecular weight standard is displayed (123 DNA ladder). At the bottom of the figure are the bands for NHE-3 obtained from renal cortex, outer and inner medullar (lanes 2, 4, and 6, respectively), while in lanes 8 and

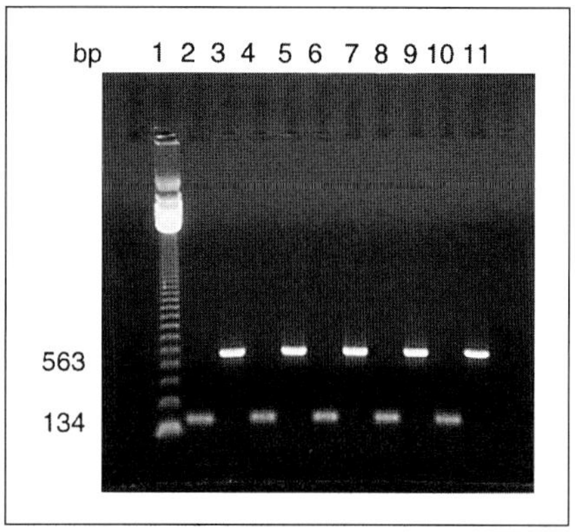

Fig. 4. Gel electrophoresis on 1% agarose of NHE-3 isoform and β-actin from different parts of the kidney, including late proximal tubules and TAL. For details see text.

10 are the bands collected from proximal (S_1 and S_2) and TAL tubules, respectively. In odd lanes there are the bands for the corresponding β-actin. These results indicate that NHE-3 is present in the cortex, along the proximal S_1 and S_2 segments, and in renal medulla, along the TAL. They confirm a recent report from Amemiya et al. [28] where, using polyclonal antisera against cytoplasmic epitopes of rat NHE-3, it has been demonstrated that this isoform is strongly expressed along the apical membrane of S_1 and S_2 segments, but is absent in S_3. There was intense staining along the TAL that disappeared at the conjuction with the distal tubule. This report seems to indicate that the acid-base transport along the S_3 and the distal tubule could be mediated by other isoforms of the NHE superfamily.

Use of these new molecular biology tools has provided important information about several aspects of renal physiology and pathophysiology. For example, it is known that neonates have a lower threshold for bicarbonate, due to a lower rate of bicarbonate reabsorption by the neonatal proximal tubule [29]. Baum et al. [30] have demonstrated that the maturation of tubular bicarbonate reabsorption is parallelel by an increase in renal cortical NHE-3, but not NHE-1, mRNA and protein abundance; moreover, this selective maturation can be accelerated by administration of glucocorticoids. Further important information is related to the molecular mechanism of the endothelin-induced increase in blood pressure. It is possible that this effect is mediated, beside vasoconstrictive actions, through renal sodium retention. Chu et al. [31] have demonstrated that the renal effect of endothelin is by NHE-3 activation, stimulated in part through increases in intracellular Ca^{2+} and Ca-calmodulin kinase.

H^+-ATPase

This vacuolar proton pump was first described in lysosomes and chromaffin granules [32, 33]. It is a primary active acid extruder that derives its energy from the hydrolysis of ATP. This process is adequate to create a large H^+ gradient across the cell membrane (up to three pH units). However, the proton pump works at a very low rate compared, for example, with the Na^+/H^+ exchanger. This transporter is, therefore, appropriate for epithelia, like the distal nephron, where a high-gradient, low-capacity H^+ extruder is needed.

The first evidence that the H^+-ATPase were a new class of proton pump came from a study on the properties of H^+-ATPase responsible for electrogenic H^+ secretion by the turtle urinary bladder [34]. It was found that acidification was stimulated by addition of ATP, inhibited by dicyclohexylcarbodiimide (DCCD), and insensitive to oligomycin, azide and vanadate. At the kidney level, Ullrich et al. [35], using in vivo microperfusion of proximal tubules of golden hamster, demonstrated that luminal H^+ secretion, measured as glycodiazine reabsorption, could be Na^+ independent and they hypothesized that such ion movement could be driven by metabolic energy, i.e., ATP. Later, Kinne et al. [36] showed that an ATPase activity was present in brush border membrane vesicles; such activity was stimulated by protonophores and inhibited by DCCD and could therefore be indicative for the presence of a H^+-ATPase in the membrane. However, the first solid evidence for the presence of a vacuolar proton pump in renal structures came from immunocytochemical staining of rat kidney with anti-H^+-ATPase antibodies, which demonstrated extensive staining in the apical membrane of the proximal tubule, TAL and distal convoluted tubules. Along the collecting ducts, the situation was more complex: principal cells were virtually unlabeled, while intercalated cells had intense staining with an apical, basal lateral or diffuse pattern in the cortex, and exclusively apical staining in the medulla [37]. This study also suggested a role for the H^+-ATPase in acid-base transport along the tubules and in pH_i regulation of tubular cells.

Acid-Base Transport

Although the major mechanism responsible for luminal H^+ secretion along the proximal tubule is the Na^+/H^+ exchanger, there are data indicating that some H^+ secretion is preserved even after Na^+ transport has been greatly reduced [35, 38]. These results were confirmed by Bank et al. [39], who, using in vivo microperfusion experiments of the late proximal convoluted tubules and applying the microcalorimetric method to measure total CO_2 in collected tubular fluid samples, demonstrated that at least 50% of total bicarbonate reabsorption along the proximal tubule is mediated by sodium-independent H^+ secretion [39]. A similar situation was also found along the loop of Henle. In this study [10], to estimate the contribution of the proton pump to the total bicarbonate reabsorption, we used bafilomycin A_1, a macrolide antibiotic shown to be a potent and quite specific (at low concentration) inhibitor of vacuolar H^+-ATPase [40]. When we perfused the loop of Henle with bafilomycin A_1, it produced a 20% fall in net bicarbonate reabsorption (fig. 5). This result is consistent, on the one hand, with the view that a significant fraction of loop bicarbonate transport is not medi-

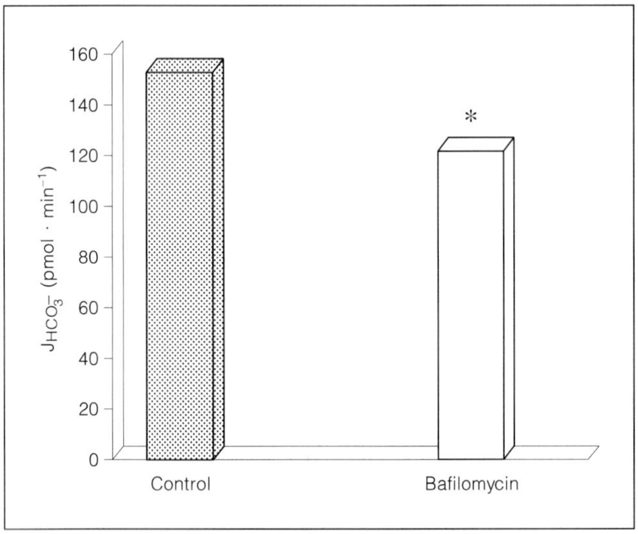

Fig. 5. Effect of bafilomycin on net bicarbonate reabsorption (J_{HCO_3}) along in vivo microperfused loop of Henle [data from ref. 10].

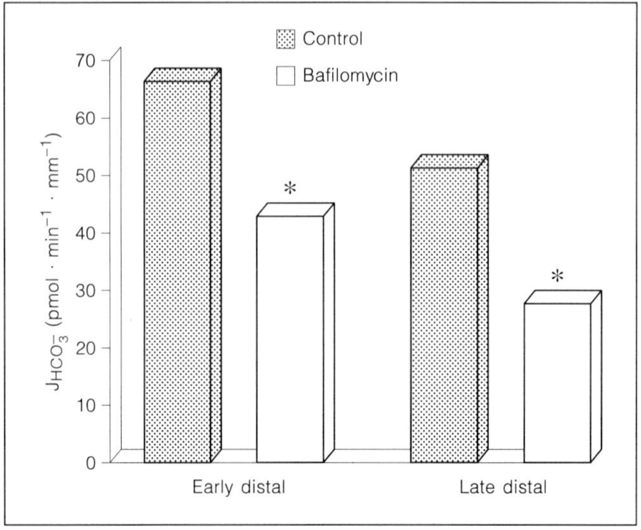

Fig. 6. Effect of bafilomycin on net bicarbonate reabsorption (J_{HCO_3}) along in vivo microperfused early and late distal tubules [data from ref. 19].

ated by the Na^+/H^+ exchanger and, on the other hand, with the immunohistochemical demonstrations of proton pump localization along the loop of Henle [37]. The role of the H^+-ATPase in mediating bicarbonate reabsorption is even more important along the distal tubule. As shown in figure 6 that summarizes results of microperfusion studies in cortical distal tubules [19, 41], it is apparent that bafilomycin inhibits bicarbonate reabsorption both in the early and late distal tubule, suggesting that vesicular H^+-ATPase is important in H^+ secretion at this site. There is also biochemical evidence for the presence of H^+-ATPase in the mammalian cortical distal tubule and collecting duct [42] and immunocytochemical localization to the apical membrane of distal tubule cells and α-intercalated cells is firmly established [37].

Control of pH$_i$

Following the work of Selvaggio et al. [43] on collecting duct cells, cultured from inner medulla, suggesting the contribution of the H^+-ATPase to the recovery of pH_i after acute intracellular acid loading, Geibel [20] investigated this issue in more detail using in vitro microperfusion of rabbit proximal tubules. When these cells were exposed to extreme acid loads, their pH_i could spontaneously recover even in the continued absence of Na^+. Moreover, this acid extrusion was blocked by the addition to the luminal perfusate of N-ethylmaleimide, an inhibitor of H^+-ATPase activity. Very recently, we have monitored pH_i of single proximal tubular cells of mouse kidney [44] using a video-imaging microscopy apparatus. The experiments were performed in the nominal absence of HCO_3^-. The lack of Na^+ in the superfusion solution caused a rapid fall in pH_i; however, when we prolonged the time of superfusion with Na^+-free solution, pH_i showed slow, but constant spontaneous recovery. This was significantly accelerated by readdition of Na^+ to the superfusion solution, resulting in full recovery of pH_i. In a second set of experiments we tested the effect of bafilomycin A_1 on spontaneous pH_i recovery in an Na^+-free solution. The addition of the specific H^+-ATPase inhibitor blocked the Na^+-independent pH_i recovery. In a third set of experiments, pH_i was lowered by an NH_4Cl prepulse technique. In the absence of Na^+ in the superfusion solution, pH_i recovery was completely inhibited by bafilomycin A_1, while the readdition of Na^+ induced its prompt and full recovery. Taken together, all these data indicate that, in several cells of different tubules and species, there is an Na^+-independent H^+ extruder (H^+-ATPase) that contributes to the maintenance of pH_i.

Clinical Importance of H$^+$-ATPase

Distal tubular acidosis (type I) is a clinical syndrome characterized by metabolic acidosis associated with the incapacity of the kidney to produce an appropriate acid urine. Among other explanations, the disorder may be related to a defect in H^+ secretion along the collecting duct. Cohen et al. [45] have investigated this hypothesis in a patient with Sjogren's syndrome and distal tubular acidosis whose kidney biopsy was examined by fluorescent immunocytochemistry with an antibody to a subunit of the vacuolar H^+-ATPase. On the biopsy sample, staining was absent in the intercalated cells but present in the subvillous invaginations of proximal tubular cells. This study demonstrates that the defect in proton secretion in some patients with distal tubular acidosis may reside in the absence of H^+-ATPase in the interaclated cells. Other reports have appeared to confirm this hypothesis [46]. This is one among many examples of the importance of the studies on H^+-extruder proteins for understanding physiological and pathophysiological states.

Acknowledgements

This study was supported by EC contract No. CHRX-CT94-0595 and by a grant from the National Research Council No. 96.03311. CT04 to GC.

References

1 Pitts RF: Renal excretion of acid. Fed Proc 1948; 7:418–426.
2 Murer H, Hopfer U, Kinne R: Sodium/proton antiport in brush-border-membrane vesicles isolated from rat small intestine and kidney. Biochem J 1976; 154:597–604.
3 Costa Silva VL, Campiglia SS, De Mello Aires M, Malnic G, Giebisch G: Role of luminal buffers in renal tubular acidification. J Membrane Biol 1981; 63:13–24.
4 Pitts RF, Lotspeich WD, Schiess WA, Ayer JL: The renal regulation of acid-base balance in man. I: The nature of the mechanism for acidifying the urine. J Clin Invest 1948;27:48–56.
5 Aronson PS: Kinetic properties of the plasma membrane Na^+-H^+ exchanger. Annu Rev Physiol 1985; 47:545–560.
6 Pouysségur J: The growth factor-activable Na^+/H^+ exchanger system: A genetic approach. Trends Biochem Sci 1985;10:453–455.
7 Moore RD: Stimulation of Na:H exchange by insulin. Biophys J 1981;33:203–210.
8 Grinstein S, Rothstein A: Mechanisms of regulation of the Na^+/H^+ exchanger. J Membrane Biol 1986; 90:1–12.
9 Ullrich KJ, Rumrich G, Bauman K: Renal proximal tubular buffer-(glycodiazine) transport. Pfluegers Arch 1975;357:149–163.
10 Capasso G, Unwin R, Agulian S, Giebisch G: Bicarbonate transport along the loop of Henle. I: Microperfusion studies of load and inhibitor sensitivity. J Clin Invest 1991;88:430–437.
11 Gottschalk CW, Lassiter WE, Mylle M: Localization of urine acidification in the mammalian kidney. Am J Physiol 1960;198:581–585.
12 Malnic G, De Mello Aires M, Giebisch G: Micropuncture study of renal tubular hydrogen ion transport in the rat. Am J Physiol 1972;222:147–158.
13 Good DW: Sodium-dependent bicarbonate absorption by cortical thick ascending limb of rat kidney. Am J Physiol 1985;248:F821–F829.
14 Good DW: Bicarbonate absorption by the thick ascending limb of Henle's loop. Semin Nephrol 1990;10:132–138.
15 Good DW: Adaptation of HCO_3 and NH_4^+ in rat MTAL: Effects of chronic metabolic acidosis and Na^+ intake. Am J Physiol 1990;258:F1345–F1353.
16 Capasso G, Unwin R, Ciani F, De Tommaso G, Vinciguerra M, Russo F, De Santo NG: The effect of acute metabolic alkalosis on bicarbonate transport along the loop of Henle: The role of active transport processes and passive paracellular backflux. Pfluegers Arch 1994;429:44–49.
17 Capasso G, Unwin R, Ciani F, De Santo NG, De Tommaso G, Russo F, Giebisch G: Bicarbonate transport along the loop of Henle. II: Effects of acid-base, dietary and neurohumoral determinants. J Clin Invest 1994;94:830–838.
18 Kriz W, Bankir L: A standard nomenclature for structures of the kidney. Kidney Int 1988;33:1–7.
19 Wang T, Malnic G, Giebisch G, Chan YL: Renal bicarbonate reabsorption in the rat. 4: Bicarbonate transport mechanisms in the early and late distal tubule. J Clin Invest 1993;91:2776–2784.
20 Geibel J: Apical H^+-ATPase activity in the rabbit proximal tubule. Cell Physiol Biochem 1993;3:34–41.
21 Boron WF: Cell pH regulation and acidification mechanisms in the proximal nephron. Annu Rev Physiol 1982;44:336–342.
22 Aronson PS, Nee J, Suhm MA: Modifier role of internal H^+ in activating the Na^+-H^+ exchanger in renal microvillus membrane vesicles. Nature 1982; 299:161–163.
23 Sardet C, Franchi A, Pouyssegur J: Molecular cloning, primary structure and expression of the human growth factor activatable Na^+-H^+ antiporter. Cell 1989;56:271–280.
24 Orlowsky J, Kandasamy R, Shull G: Molecular cloning of putative members of the Na^+/H^+ exchanger gene family. J Biol Chem 1992;267:9331–9339.
25 Collins JF, Honda T, Knobel S, Bulus NM, Conary J, Dubois M, Ghishan FK: Molecular cloning, sequencing, tissue distribution, and functional expression of Na^+/H^+ exchanger (NHE-2). Proc Natl Acad Sci USA 1993;90:3938–3942.
26 Orlowski J, Kandasamy RA, Shull GE: Molecular cloning of putative members of the Na/H exchanger gene family: cDNA cloning, deduced amino acid sequence, and mRNA tissue expression of the rat Na/H exchanger NHE-1 and two structurally related protein. J Biol Chem 1992;267:9331–9339.
27 Wang Z, Orlowski J, Shull GE: Primary structure and functional expression of a novel gastrointestinal isoform of the rat Na/H exchanger. J Biol Chem 1993;268:11925–11928.
28 Amemiya M, Loffing J, Lotscher M, Kaissling B, Alpern RJ, Moe OW: Expression of NHE-3 in the apical membrane of rat renal proximal tubule and thick ascending limb. Kidney Int 1995;48:1206–1215.
29 Schwartz GJ, Evan AP: Development of solute transport in rabbit proximal tubule. I: HCO_3 and glucose absorption. Am J Physiol 1983;245:F382–F390.
30 Baum M, Biemesderfer D, Gentry D, Aronson PS: Ontogeny of rabbit renal cortical NHE3 and NHE1: Effect of glucocorticoids. Am J Physiol 1995;268: F815–F820.
31 Chu T, Peng Y, Cano A, Yanagisawa M, Alpern RJ: $Endothelin_B$ receptor acitivates NHE-3 by a Ca^{2+}-dependent pathway in OPK cells. J Clin Invest 1996;97:1454–1462.
32 Mergo JL, Farb RM, Barnes J: An adenosine triphosphate-dependent stabilization of proteolytic activity in heterolysosomes: Evidence for a proton pump. Biochem J 1972;128:763–769.
33 Philips JH, Allison VP: Proton translocation of the bovine chromaffin-granule membrane. Biochem J 1978;170:661–672.
34 Gluck S, Kelly S, Al-Awqati Q: The proton translocating ATPase responsible for urinary acidification. J Biol Chem 1982;257:9230–9233.
35 Ullrich KJ, Capasso G, Rumrich G, Papavassiliou F, Kloss S: Coupling between proximal tubular transport processes: Studies with ouabain, SITS and HCO_3-free solutions. Pfluegers Arch 1977;368: 245–252.
36 Kinne R, Kinne-Saffran E, Echütz H, Schölermann B: Ammonium transport in medullary thick ascending limb of rabbit kidney: Involvment of the Na^+, K^+, Cl^--cotransporter. J Membrane Biol 1986;94:279–284.
37 Brown D, Hirsch S, Gluck S: Localization of a proton-pumping ATPase in the rat kidney. J Clin Invest 1988;82:2114–2126.
38 Chan YL, Biagi B, Giebisch G: Control mechanisms of bicarbonate transport across the rat proximal convoluted tubule. Am J Physiol 1982;242: F532–F543.
39 Bank N, Aynedjian HS, Mutz BF: Proximal bicarbonate absorption independent of Na^+-H^+ exchange: Effect of bicarbonate load. Am J Physiol 1989;256:F577–F582.
40 Bowman EJ, Siebers A, Altendorf K: Bafilomycins: A class of inhibitors of membrane ATPases from microorganisms, animal cells, and plant cells. Proc Natl Acad Sci USA 1988;85:7972–7976.
41 Capasso G, Malnic G, Wang T, Giebisch G: Acidification in mammalian cortical distal tubule. Kidney Int 1994;45:1543–1554.
42 Garg LC, Narang N: Changes in H-ATPase activity in the distal nephron segments of the rat during metabolic acidosis and alkalosis; in Morii (ed): Calcium-Regulating Hormones. II: Calcium Transport, Bone Metabolism and New Drugs. Contrib Nephrol. Basel, Karger, 1991, vol 92, pp 39–45.
43 Selvaggio AM, Schwartz JH, Bengele HH, Gordon FD, Alexander EA: Mechanisms of H^+ secretion by inner medullary collecting duct cells. Am J Physiol 1988;254:F391–F400.
44 Capasso G, Saviano C, Rizzo M, De Santo NG: The effect of bafilomycin A_1 on intracellular pH of single cells in the mouse proximal tubule (abstract). J Am Soc Nephrol 1996;7:1253.
45 Cohen EP, Bastani B, Cohen MR, Kolner S, Hemken P, Gluck SL: Absence of H^+-ATPase in cortical collecting tubules of a patient with Sjogren's syndrome and distal renal tubular acidosis. J Am Soc Nephrol 1992;3:264–271.
46 Bastani B, Haragsim L, Gluck S, Siamopoulos KC: Lack of H-ATPase in distal nephron causing hypokalaemic distal RTA in a patient with Sjogren's syndrome. Nephrol Dial Transplant 1995;10:908–909.

Mineral and Electrolyte Metabolism

Miner Electrolyte Metab 1997;23:249–252

Dieter Häussinger

Medizinische Universitätsklinik Freiburg, Deutschland

Liver Regulation of Acid-Base Balance

Key Words
Acid-base balance
Acidosis
Liver
Urea synthesis

Abstract
Traditionally, lungs and kidneys have been viewed as the sole and principal organs involved in systemic acid-base homeostasis in mammals, but this view is not entirely compatible with basic principles of chemistry. Recent conceptual developments point to a role for the liver in pH homeostasis in addition to the well-established role of lungs and kidneys. Hepatic and renal nitrogen metabolism are linked by an interorgan glutamine flux, which couples both renal ammoniagenesis and hepatic ureogenesis to systemic acid-base regulation. Hepatic urea synthesis is a major pathway for the removal of metabolically generated bicarbonate. A structural-functional organization in the liver acinus uncouples urea cycle flux control from the vital need to maintain ammonium homeostasis. There is a sensitive and complex control of bicarbonate disposal via hepatic ureogenesis by the extracellular acid-base status, suggestive of a feedback control loop between acid-base status and the rate of bicarbonate elimination, i.e., a hepatic bicarbonate-homeostatic response. Some pathophysiological implications arising from the pH-stat function of the liver are discussed.

Introduction

Recent conceptual development point to a role of the liver in pH homeostasis in addition to the well-established role of lungs and kidneys. Hepatic and renal nitrogen metabolism are linked by an interorgan glutamine flux, coupling both renal ammoniagenesis and hepatic ureagenesis to systemic acid-base regulation. A hepatic role in this interorgan effort is based upon several features. (1) The presence of a quantitatively important and liver-specific pathway for irreversible removal of metabolically generated bicarbonate, i.e., urea synthesis. (2) A structural-functional organization, which uncouples urea cycle flux control from the vital need to maintain ammonium homeostasis. (3) A sensitive and complex control of bicarbonate disposal via hepatic ureagenesis by the extracellular acid-base status, suggestive of a feedback control loop between the acid-base status and the rate of bicarbonate elimination, i.e., a hepatic bicarbonate homeostatic response. (4) Inhibition of amino acid uptake into the liver during acidosis at the level of plasma membrane transport, which may result in a shift of amino acid catabolism to extrahepatic tissues.

General Considerations

The complete oxidation of proteins generates large amounts of HCO_3^- and thus a threat of alkalosis. This is because oxidation of bipolar amino acids yields the strong base HCO_3^- together with the weak acid NH_4^+ in almost equimolar amounts. For elimination of both NH_4^+ and HCO_3^-, mammals have developed a pathway catalyzing an irreversible proton transfer from NH_4^+ to HCO_3^-, namely urea synthesis. In chemical terms, urea synthesis is an energy-driven neutralization of the strong base HCO_3^- by the weak acid NH_4^+ [1–4]:

$$2\,HCO_3^- + 2\,NH_4^+ \rightarrow \text{Urea} + CO_2 + 3H_2O$$

Thus, the daily formation and excretion of 30 g urea in humans is equivalent to the disposal of 1 mol HCO_3^- and NH_4^+ each, i.e., amounts corresponding to the complete oxidation of about 100 g protein. An alternative pathway for disposal of NH_4^+ is the direct excretion into urine (renal ammoniagenesis); under these conditions, the bicarbonate, which is produced together with NH_4^+ during protein breakdown, remains in the body. Thus, the factors governing the partitioning between hepatic urea synthesis and renal ammoniagenesis exert control on bicarbonate homeostasis [4]. In other words, increases or decreases of urea cycle flux relative to the rate of protein catabolism will diminish or expand the bicarbonate pool in the body, respectively. The mechanisms which adjust renal ammoniagenesis, which might be seen as a primary ammonium homeostatic response, to the needs of the acid-base status are beyond the scope of this article.

Two major mechanisms may contribute to the adjustment of bicarbonate-consuming urea synthesis to the needs of acid-base homeostasis. Amino acid uptake by the liver is decreased in acidosis at the level of membrane transport [5–7]. This is expected to shift the site of amino acid catabolism to nonhepatic tissues such as the kidney. The other mechanism involves a pH-sensitive regulation

KARGER
E-Mail karger@karger.ch
Fax +41 61 306 12 34
http://www.karger.ch

0378–0392/97/0236–0249 $12.00/0

Prof. Dr. Dieter Häussinger
Abteilung für Gastroenterologie und Infektiologie
Heinrich-Heine Universität
Moorenstr. 5
D–40225 Düsseldorf (Germany)

of ammonia and glutamine metabolism inside the liver, which leads to a pH-dependent shift of hepatic ammonium detoxication from urea to glutamine synthesis in acidosis [3]. When the acid-base status is in equilibrium, the rate of bicarbonate removal (urea synthesis) from the organism must match the rate of bicarbonate production (protein catabolism) and the regulation of urea cycle flux appears to be controlled by the rate of protein breakdown.

In acidosis, hepatic urea synthesis decreases relative to the rate of protein catabolism, resulting in a retention of HCO_3^- as a pH homeostatic response by the liver. Such a sparing of HCO_3^- by diminishing urea cycle flux in acidosis occurs without the threat of hyperammonemia, despite the ultimate link of HCO_3^- and NH_4^+ generation during protein catabolism. It is the structural and functional organization of nitrogen-metabolizing pathways in the liver acini which uncouples urea synthesis from the need to dispose of ammonium and provides the basis for an independent control of ammonium and bicarbonate homeostasis in the liver. The pH sensitivity of renal ammoniagenesis finally serves to eliminate the ammonium ions which cannot be used for urea synthesis and represents an ammonium homeostatic response.

Structural and Functional Organization of Hepatic Nitrogen Metabolism

The structural and functional organization of ammonium-metabolizing pathways in the liver acini is one prerequisite for the understanding of a hepatic role in acid-base regulation. At the acinar level, there is remarkable functional hepatocyte heterogeneity with respect to nitrogen metabolism [8–11]. Urea synthesis and glutaminase are present in periportal hepatocytes, whereas glutamine synthetase is found only in perivenous hepatocytes. The borderline between the periportal urea-synthesizing and the perivenous glutamine-synthesizing compartment is very strict: glutamine synthetase is exclusively found in a small hepatocyte population (about 7% of all hepatocytes of an acinus) surrounding the terminal hepatic venule. These cells are virtually free of urea cycle enzymes. Accordingly, when one follows the sinusoidal blood stream, the two major hepatic ammonia detoxification systems, urea and glutamine synthesis, are anatomically located behind each other. Functionally, this represents the sequence of a periportal low-affinity, but high-capacity system (ureagenesis) and a perivenous high-affinity system for ammonium detoxication (glutamine synthesis) [12]. The difference in ammonium affinity between urea and glutamine synthesis results from the higher K_m (1–2 mM) for NH_4^+ of carbamoyl phosphate synthetase, the rate-controlling enzyme of the urea cycle, compared to that for isolated glutamine synthetase ($K_m = 0.3$ mM). Hepatic clearance of physiologically low portal ammonia concentrations (0.2–0.3 mM) requires intact glutamine synthetase activity ('high-affinity system'): about one-third of a physiological portal ammonium load is converted by perivenous cells into glutamine, although portally delivered ammonium first comes into contact with cells capable of urea synthesis. Here, perivenous glutamine synthetase acts as a high-affinity scavenger for the ammonium which escaped periportal urea synthesis [12]. The remarkable specialization of perivenous scavenger cells for high-affinity removal of ammonium, which escaped more upstream (periportal) urea synthesis, is also underlined by an almost exclusive uptake of vascular glutamate, aspartate, α-oxoglutarate, and malate by perivenous glutamine-synthetase-containing hepatocytes, but not by the much larger periportal cell population. The important scavenger role of perivenous glutamine synthesis for the maintenance of physiologically low ammonium concentrations in the hepatic vein becomes rapidly evident after inhibition of glutamine synthetase by methionine sulfoximine or after destruction of perivenous cells by CCl_4 treatment. In the latter case, hyperammonemia ensues, due to an almost complete failure to synthesize glutamine, although periportal urea synthesis is not affected. This sophisticated structural and functional organization of hepatic ammonium metabolism uncouples urea synthesis from the vital need to dispose of potentially toxic ammonium, because perivenous scavenger cells act as a 'backup system' for ammonia detoxication, even when urea synthesis is inhibited, as occurs in acidosis.

Whereas glutamine synthetase is perivenous, glutaminase is localized in periportal hepatocytes [12] and has a joint mitochondrial localization together with carbamoyl phosphate synthetase. Recent in situ hybridization studies suggest that glutaminase is even restricted to a small periportal cell subpopulation at the inflow of the acinar bed [13]. Accordingly, in the middle of the acinus there is a cell population that contains urea cycle enzymes, but that does not contain either glutaminase or glutamine synthetase (fig. 1). Liver glutaminase is immunologically and kinetically different from glutaminases in other tissues [for a review see ref. 14] and requires ammonium as an essential activator. In the intact rat liver, half-maximal activation of glutaminase occurs at the physiological portal ammonium concentration of 0.2–0.3 mM. Thus, physiological fluctuations of the portal ammonium concentrations are paralleled by activity changes of hepatic glutaminase, and the function of glutaminase is now to amplify ammonium inside the mitochondria in parallel to that delivered via the portal vein or that which arises during amino acid breakdown [3, 8]. Because urea synthesis normally is controlled by flux through carbamoyl phosphate synthetase, which largely depends on the actual ammonium concentration inside the mitochondria, amplification of mitochondrial ammonium via glutaminase flux becomes an impor- tant determinant of urea cycle flux, especially at physiologically low ammonium concentrations, which are about one order of magnitude below the K_m (ammonia) of carbamoyl phosphate synthetase.

Acid-Base Regulation of Urea Synthesis

Acid-base control of urea synthesis occurs primarily at the level of amino acid uptake and ammonium and bicarbonate delivery to carbamoyl phosphate synthetase, but not in the urea cycle itself. Flux through liver glutaminase is very sensitive to small pH changes; when the extracellular pH drops from 7.4 to 7.3, flux through

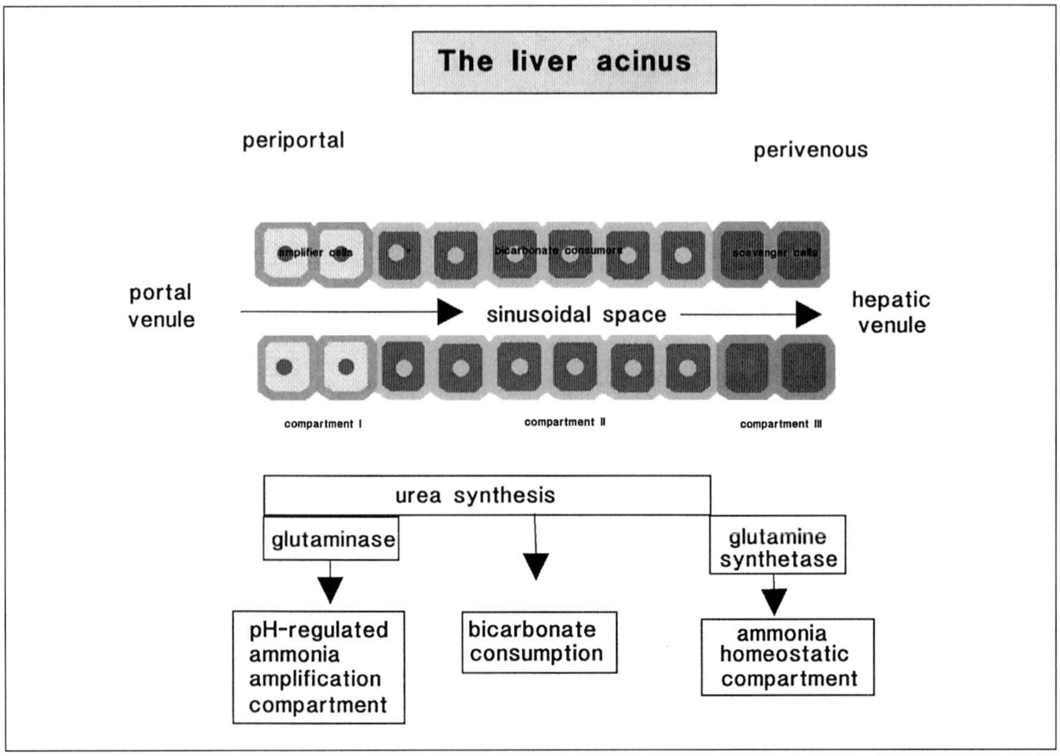

Fig. 1. Acinar organization of hepatic nitrogen metabolism. The acinus represents the functional unit of the liver; three different compartments are distinguished which are sequentially perfused by the blood circulation. Ammonium-rich portal blood first reaches periportal cells containing glutaminase (compartment I). This enzyme acts as a pH-regulated ammonium amplifier and is inhibited in acidosis. Ammonia amplification in this compartment involves the consumption of glutamine, but increases the ammonium concentration and counteracts the low ammonium affinity of the urea cycle. The extent of pH-governed ammonium amplification is thus an important determinant of urea cycle flux (in compartments I and II), which is equivalent to hepatic bicarbonate consumption. Compartment III consists of perivenous scavenger cells, which are free of glutaminase and urea cycle enzymes. These cells contain glutamine synthetase, which acts as a high-affinity scavenger for the ammonium which was not used for urea synthesis. When the acid-base status is in equilibrium, glutamine breakdown in compartment I is matched by resynthesis in compartment III (the so-called intercellular glutamine cycle). In acidosis, the liver becomes a net glutamine producer due to inhibition of glutaminase in compartment I, decreased urea synthesis (in compartments I and II), and increased glutamine synthesis in compartment III due to an increased delivery of ammonium to scavenger cells.

glutaminase decreases by about 70%, due to an inhibition of glutamine transport across the plasma and mitochondrial membrane and a decreased ammonia and hormone activation of glutaminase at low pH [8]. Accordingly, acidosis turns off the mitochondrial ammonium amplifier glutaminase in periportal hepatocytes and consequently urea cycle flux. The input of bicarbonate into the carbamoyl phosphate synthetase reaction is also controlled by the acid-base status at the level of mitochondrial carbonic anhydrase V [for a review see ref. 15] and discrimination of respiratory versus metabolic acidosis occurs because the spontaneous CO_2 conversion into bicarbonate is pH insensitive, whereas the carbonic-anhydrase-catalyzed conversion is inhibited at low pH. Urea synthesis is also controlled by extracellular CO_2 and HCO_3^- concentrations, independent of the pH. Even at a normal extracellular pH in vitro, the liver seems to respond to a decrease in buffer capacity of the extracellular HCO_3^-/CO_2 system with an inhibition of urea synthesis and, accordingly, a sparing of bicarbonate [15].

In summary, the complex and sensitive control of ureagenesis by pH, $[HCO_3^-]$ and $[CO_2]$ in experimental systems, such as perfused rat liver, seems well designed for an important contribution of the liver to maintenance of acid-base homeostasis in vivo: whenever the pH and/or the HCO_3^- and CO_2 concentration in the extracellular space fall, the liver responds with an inhibition of bicarbonate-consuming urea synthesis. It is the ultimate link between HCO_3^- and NH_4^+ formation during protein catabolism, which requires not only acid-base control of ureagenesis, but also mechanisms which adjust urea cycle flux to the rate of protein catabolism. These mechanisms, such as N-acetylglutamate activation of carbamoyl phosphate synthetase and ornithine availability, have been reviewed in detail elsewhere [15].

Perivenous scavenger cells maintain ammonium homeostasis, even when urea synthesis is switched off, as occurs in metabolic acidosis. Renal glutamine hydrolysis and subsequent urinary NH_4^+ excretion (so-called renal ammoniagenesis) provide the final sink for surplus ammonium; both processes underlie complex and sensitive short- and long-term regulation and are stimulated in acidosis [for a review see ref. 11]. Thus, renal ammoniagenesis may be seen to act as a finely tuned spillover for the ammonium requiring elimination without concomitant HCO_3^- consumption and may represent a primary ammonium homeostatic response rather than a bicarbonate homeostatic one.

Pathophysiological Aspects

The coordinated action of liver and kidney in the maintenance of ammonium and bicarbonate homeostasis predicts that disturbances of function in either organ should lead to acid-base derangements. In chronic renal failure, renal ammoniagenesis and urinary NH_4^+ excretion are impaired. Failure of this renal spillover function for NH_4^+ may lead to inadequate ammonium stimulation of urea synthesis and consequently to bicarbonate wasting: hepatic urea synthesis prevents hyperammonemia at the price of hypobicarbonatemia [8, 11]. On the other hand, impairment of hepatic urea synthesis in chronic liver disease may diminish hepatic bicarbonate disposal and give rise to metabolic alkalosis [8, 16, 17]. There is an increase in plasma bicarbonate concentration with a progressive loss of urea cycle capacity in humans, as no other causes for the development of hyperbicarbonatemia having been identified [16, 17]. Because alkalosis, in turn, must be viewed as a potent stimulus of urea cycle flux by activating the ammonium amplifier glutaminase and increasing mitochondrial availability of HCO_3^- for carbamoyl phosphate synthetase, it seems that the development of a certain degree of hyperbicarbonatemia and alkalosis in liver cirrhosis produces a new, albeit more alkalotic, metabolic steady-state maintaining life-compatible urea cycle flux, despite a reduction in urea cycle enzyme activities. Interestingly, the progressive loss of urea cycle capacity is paralleled by an increase in renal ammonium excretion, despite co-existing metabolic alkalosis, indicating that the kidney assumes the task of eliminating ammonia when urea synthesis fails, despite the presence of alkalosis [17].

References

1 Atkinson DE, Camien M: The role of urea synthesis in the removal of metabolic bicarbonate and the regulation of blood pH. Curr Top Cell Regul 1982; 21:261–302.
2 Atkinson DE, Bourke E: The role of ureogenesis in pH homeostasis. Trends Biochem Sci 1984;9:297–300.
3 Häussinger D, Gerok W, Sies H: Hepatic role in pH regulation: Role of the intercellular glutamine cycle. Trends Biochem Sci 1984;9:300–302.
4 Häussinger D, Gerok W, Sies H: The effect of urea synthesis on extracellular pH in isolated perfused rat liver. Biochem J 1986;236:261–265.
5 Häussinger D, Lenzen C, Sies H, Soboll S: pH control of hepatic glutamine degradation: Role of transport. Eur Biochem 1987;146:483–488.
6 Boon L, Blommaert PJE, Meijer AJ, Lamers WH, Schoolwerth AC: Acute acidosis inhibits liver amino acid transport: No primary role for the urea cycle in acid base balance. Am J Physiol 1994;267: F1015–F1020.
7 Christensen HN, Kilberg MS: Hepatic amino acid transport primary to the urea cycle in regulation of biological neutrality. Nutr Rev 1995;53:74–76.
8 Häussinger D: Nitrogen metabolism in liver: Structural and functional organization and physiological relevance. Biochem J 1990;267:281–290.
9 Jungermann K, Katz N: Functional specialization of different hepatocyte populations. Physiol Rev 1989;69:708–764.
10 Traber PG, Chianale J, Gumucio JJ: Physiologic significance and regulation of hepatocellular heterogeneity. Gastroenterology 1988;95:30–45.
11 Häussinger D, Lamers WH, Moorman AFM: Metabolism of amino acids and ammonia. Enzyme 1993;46:72–93.
12 Häussinger D: Hepatocyte heterogeneity in glutamine and ammonia metabolism and the role of an intercellular glutamine cycle during ureogenesis in perfused rat liver. Eur J Biochem 1983;133:269–274.
13 Moorman AFM, de Boer PAJ, Watford M, Dingemanse MA, Lamers WH: Hepatic glutaminase is confined to part of the urea cycle domain in the adult rodent liver lobule. FEBS Lett 1994;356:76–80.
14 Curthoys NP, Watford M: Regulation of glutaminase and glutamine transport. Ann Rev Nutr 1995; 15:133–159.
15 Häussinger D (ed): pH homeostasis. London, Academic Press, 1988.
16 Häussinger D, Steeb R, Gerok W: Ammonium and bicarbonate homeostasis in chronic liver disease. Klin Wschr 1991;68:175–182.
17 Häussinger D, Steeb R, Gerok W: Metabolic alkalosis as a driving force for urea synthesis in liver disease: Pathogenetic model and therapeutic implications. Clin Investig 1992;70:411–415.

Mineral and Electrolyte Metabolism

Miner Electrolyte Metab 1997;23:253–260

Shaul G. Massry
Miroslaw Smogorzewski

Division of Nephrology and the Department of Medicine, University of Southern California, School of Medicine, Los Angeles, Calif., USA

Role of Elevated Cytosolic Calcium in the Pathogenesis of Complications in Diabetes mellitus

Key Words
Cell dysfunction in diabetes
Calcium channel blockers
Polymorphonuclear leucocytes
mRNA of hormone receptors

Abstract
Both type I and type II diabetes mellitus are associated with derangements in the regulation of intracellular calcium. Hyperglycemia causes an acute rise in cytosolic calcium ($[Ca^{2+}]_i$) due to increased calcium influx and in certain cells to mobilization of intracellular calcium stores as well. The increase in calcium entry is secondary to the activation of calcium channels inhibitable by verapamil, nifedipine, or amlodipine. The stimulation of these calcium channels is mediated by the activation of G protein(s), leading to stimulation of various cellular pathways. Chronic hyperglycemia is also associated with decreased calcium exit from cells. The combination of increased calcium influx and decreased calcium efflux leads to sustained elevation in basal levels of $[Ca^{2+}]_i$. The latter abnormality may adversely affect cell function. Treatment of diabetic animals with calcium channel blockers normalizes cell $[Ca^{2+}]_i$ and prevents and/or reverses the derangements in cellular function.

Diabetes mellitus is associated with complications affecting almost every organ, including vasculopathy, neuropathy, central nervous system dysfunction, retinopathy, cataracts, nephropathy, cardiomyopathy in the absence of appreciable coronary artery disease, atherosclerosis, and impaired function of polymorphonuclear leucocytes (PMNLs) and B lymphocytes. The mechanisms responsible for this wide range of organ dysfunctions are not fully understood. Many factors may be involved, such as hyperglycemia, genetic predisposition, environment, hypertension, insulin deficiency (insulin-dependent diabetes mellitus, IDDM), and elevated blood levels of insulin (non-insulin-dependent diabetes mellitus, NIDDM). Thus dissection of the role of the entire spectrum of these factors in the genesis of diabetic complications is a very difficult and taxing endeavor.

Support for the 'glucose hypothesis', implying an important role for hyperglycemia in the pathogenesis of diabetic complications, is growing. Epidemiological studies [1], investigations with experiment diabetes [2, 3], and the results of the Diabetes Control and Complication Trials (DCCT) [4] provide strong evidence that hyperglycemia is a major culprit in the genesis of diabetic complications. The availability of the assay of glycosylated hemoglobin (HbA_{1c}) [5], as a measure of glycemic control, has permitted better and more reliable epidemiologic studies exploring the relationship between hyperglycemia and diabetic complications. Data are now available supporting a relationship between hyperglycemia and retinopathy in both NIDDM [7] and IDDM [8]. In addition, Singer et al. [9] reported an association between prevalance of cardiovascular disease and HbA_{1c} levels. In addition, data from animal research showed that stringent control of hyperglycemia in animals with diabetes mellitus was associated with amelioration of the microvascular disease [2] or glomerulopathy [3]. Similar observations were reported in humans with diabetes mellitus. Indeed, the DCCT clearly demonstrated that intensive therapy for the control of hyperglycemia in patients with IDDM was associated with significant reductions in retinopathy, neuropathy, and nephropathy [4]. It stands to reason, therefore, that strict control of hyperglycemia is necessary for the control of the diabetic complications. However, a very large segment of the population of diabetic patients does not have good hyperglycemic control, and this is especially true for those with NIDDM. In fact, in a large survey of a population of such patients in the Denver, Colorado area, the average fasting blood glucose was found to be about 200 mg/dl and their HbA_{1c} 11.5–12.0%, independent of their treatment with insulin or oral hypoglycemic agents [R.W. Schrier, University of Colorado, personal communication]. It is apparent, therefore, that understanding the mechanisms through which hyperglycemia exerts its deleterious effects is critical for designing therapeutic approaches to ameliorate the complications of diabetes mellitus. Many issues need to be considered. These include, but are not limited to, four aspects which we mention because we believe that they provide the initial investigative steps in researching the deleterious effects of hyperglycemia.

KARGER
E-Mail karger@karger.ch
Fax +41 61 306 12 34
http://www.karger.ch

Shaul G. Massry, MD
Chief, Division of Nephrology
University of Southern California, School of Medicine
2025 Zonal Avenue
Los Angeles, CA 90033 (USA)

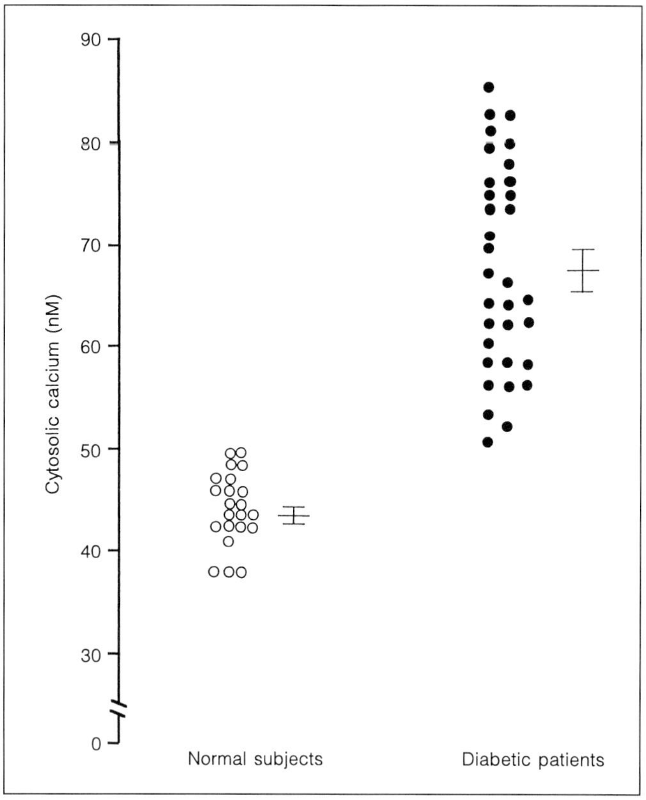

Fig. 1. Basal levels of cytosolic calcium of PMNLs from normal subjects and patients with type II diabetes mellitus. Each point represents one subject and the horizontal line denotes the mean (±1 SE). Modified from Alexiewicz et al. [43].

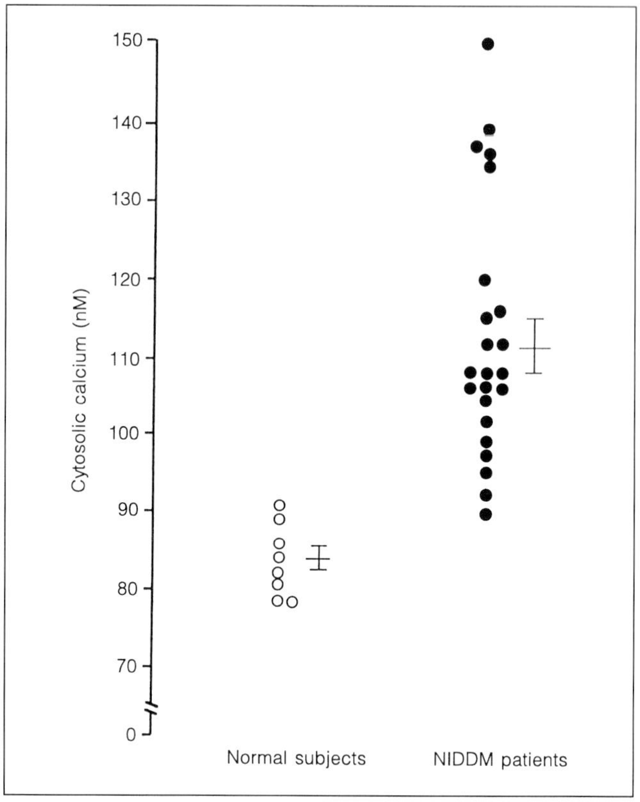

Fig. 2. Basal levels of cytosolic calcium of B cells from normal subjects and patients with type II diabetes mellitus. Each point represents one subject and the horizontal line denotes the mean (±1 SE). Reproduced by permission from Alexiewicz et al. [44].

(1) How does hyperglycemia mediate its deleterious effects? (2) Are the mechanisms underlying the hyperglycemia-induced complications uniform and applicable to all organs? Are there differences in the response of various organs to hyperglycemia? (3) Are the pathways through which hyperglycemia exerts its deleterious effects preventable even if the hyperglycemia is not adequately controlled? (4) Are there medical therapeutic approaches to treat the complications in patients with poorly controlled diabetes mellitus?

Certain data are available regarding these issues. Two consequences of hyperglycemia have received a great deal of attention. First, the aldose reductase-sorbitol pathway which is activated by hyperglycemia leading to the accumulation of sorbital in cells [10, 11]. The increase in sorbitol content may be harmful to cell function. Indeed, sorbitol accumulation is associated with an increased osmotic gradient and swelling of cells, reduced myoinositol content and deranged energy metabolism [11–13]. Available data show that sorbitol accumulation is harmful to the eye lens and may lead to cataracts, and can damage the pericytes and nerve tissue and may contribute to neuropathy. These data have prompted the use of aldose reductase inhibitors in the management of the neuropathy of diabetes mellitus with encouraging results in animal models [14, 15] but unconvincing outcomes in humans [16–18]. It must be mentioned that the activation of aldose reductase occurs only in cells into which glucose can enter independently of insulin. Thus, the stimulation of this pathway may not occur uniformly in all cells of animals or humans with hyperglycemia. This may explain why sorbitol accumulation is implicated in the genesis of cataracts and neuropathy but not in other complications of diabetes mellitus. Thus, even if the use of aldose reductase inhibitors is helfpul, their usefulness may be limited to only some diabetic complications.

The second consequence of hyperglycemia, which has received much attention and may play an important role in the genesis of the complications of diabetes, is the production of advanced glycosylation end products (AGEs) [19]. The initial step of this proces is the Maillard reaction [20] which involves the nonenzymatic glycosylation of amino acids by glucose with the early products being the Schiff base and the Amadori adducts [21–23]. The production and breakdown of these two compounds are fixed, and an equilibrium between them and a protein of sufficient life span is reached within hours or days; once this has happened,the amount of Amadori adducts bound to the protein does not change over time [24]. These products are the precursors of AGEs that form slowly by rearrangement, oxidation, dehydration, and polymerization processes of the Amadori products [25]. The latter may undergo oxidative cleavage and the resulting products do not propagate advanced glycosylation [26, 27]. Therefore, the formation of these

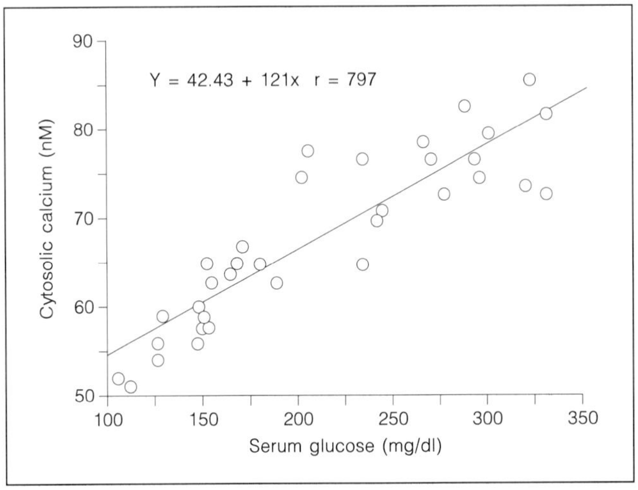

Fig. 3. Relationship between cytosolic calcium of PMNLs and blood levels of glucose in 34 patients with type II diabetes mellitus. Reproduced with permission from Alexiewicz et al. [43].

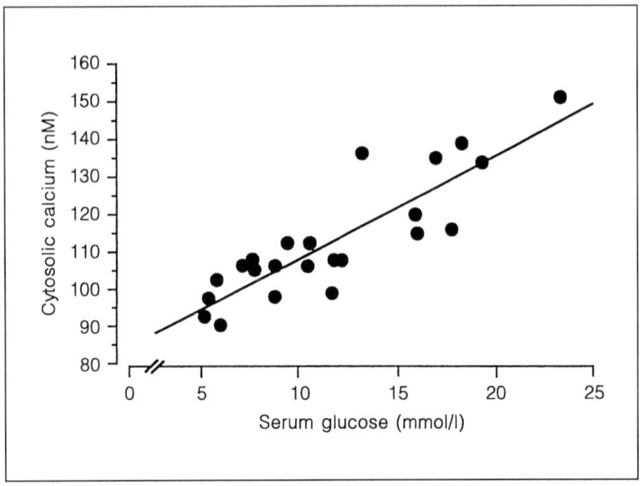

Fig. 4. Relationship between cytosolic calcium of B cells and blood levels of glucose in 23 patients wtih type II diabetes mellitus. Reproduced with permission from Alexiewicz et al. [44].

cleavage products may limit the progressive accumulation of AGEs. These compounds are irreversibly attached to proteins for the entire life span of the protein [24, 25], but their formation in vivo may be retarded by the action of reductase enzymes that reduce 3-deoxy-*D*-glucosone to less-reactive 3-deoxy-*D*-fructose [26, 27]. Inhibition of reductase enzymes may lead to an increase in AGE accumulation. Furthermore, the net accumulation of AGE proteins in tissue matrix reflects a balance between glucose-accelerated accumulation and macrophage-mediated removal [25]. Inhibition of marcrophage activity would result in an increase in AGEs. Since AGEs are cleared by the kidney [28], renal failure is associated with high AGE levels [29, 30]. The circulating levels of AGEs and their tissue content are elevated in diabetes mellitus [24, 25, 29], and a large body of evidence suggests that the accumulation of AGEs in diabetic patients plays an important role in the pathogenesis of many of the complications in these patients [24, 25, 29]. Studies on the usefulness of AGE inhibitors by aminoguanidine in the management of diabetic complication are currently being conducted.

A third potential mechanism that may participate in the pathogenesis of diabetic complications that has received less attention is a derangement in the cellular homeostasis of calcium and elevation in the basal levels of cytosolic calcium ($[Ca^{2+}]_i$). Indeed, a review of the literature indicates that the basal level of $[Ca^{2+}]_i$ is elevated in erythrocytes [31], platelets [32, 33] and adipocytes [34, 35] of diabetic patients as well as in many cells of animals with experimental diabetes mellitus [36–39]. This phenomenon led Levy et al. [40] to propose that many of the complications of diabetes mellitus are, at least in part, related to the elevation in $[Ca^{2+}]_i$. This is a plausible proposition because other conditions in which the basal levels of $[Ca^{2+}]_i$ are elevated, such as chronic renal failure [41] and phosphate depletion [42], are associated with organ dysfunctions similar to those seen in diabetes mellitus. Demonstration of a relationship between the elevation in $[Ca^{2+}]_i$ in diabetes mellitus and a specific cell dysfunction is limited. We have reported that the basal levels of $[Ca^{2+}]_i$ of both PMNLs [43] (fig. 1) and B lymphocytes [44] (fig. 2) from NIDDM patients with poorly controlled hyperglycemia are elevated and there were significant correlations between the $[Ca^{2+}]_i$ of these cells and the degree of the hyperglycemia (fig. 3, 4). The elevation in $[Ca^{2+}]_i$ was associated with impaired phagocytosis of PMNLs [43] and with impaired proliferation of B cells [44]; normalization of blood glucose of these patients with the oral hypoglycemic agent, glyburide, was associated with reversal of the $[Ca^{2+}]_i$ derangements in these cells and their abnormal functions. These observations suggested that hyperglycemia causes a rise in $[Ca^{2+}]_i$ of these cells and the high $[Ca^{2+}]_i$ interferes with their function.

The mechanisms responsible for the hyperglycemia-induced elevation in basal levels of $[Ca^{2+}]_i$ are not well understood. Hyperglycemia may be associated with increased calcium influx into cells with a consequent acute rise in $[Ca^{2+}]_i$. Indeed, certain studies have shown that a high glucose concentration induces an acute rise in $[Ca^{2+}]_i$ of insulinoma cells [45], pancreatic islets [46], cultured rat tail artery vascular smooth muscle cells [47], PMNLs [48], and human red blood cells [49].

The pathways that mediate the acute rise in $[Ca^{2+}]_i$ of cells exposed to a high glucose concentration are not fully elucidated. It has been reported that hyperglycemia may increase the activity [50] or the translocation of protein kinase C [51], or both [52]. Activation of protein kinase C causes a rise in $[Ca^{2+}]_i$ of many cells including thymocytes [53], pancreatic islets [54], hepatocytes [55], adipocytes [56], and renal proximal tubular cells [57]. It is, therefore, possible that activation of protein kinase C by glucose participates in the glucose-induced rise in $[Ca^{2+}]_i$ of PMNLs. In an effort to gain insight into the cellular pathways responsible for the glucose-induced acute rise in $[Ca^{2+}]_i$, we carried out studies in PMNLs of rats exposed to high glucose concentrations. We found that hyperglycemia, through its osmotic effect and cell shrinkage, stimulates a

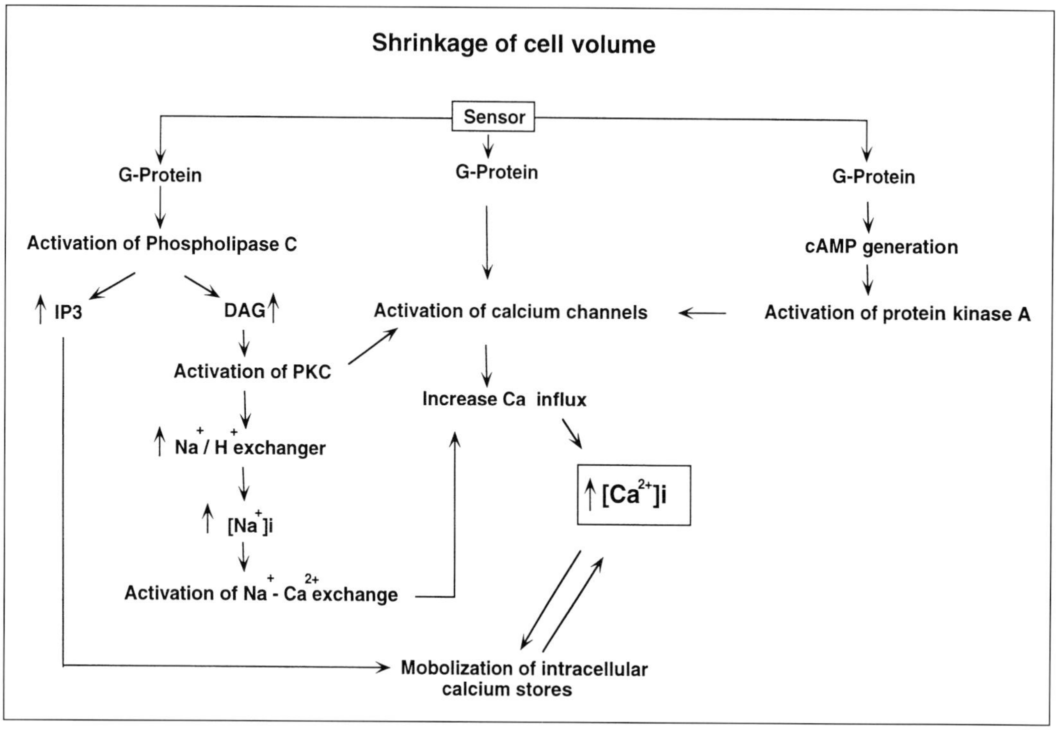

Fig. 5. Schematic representation of the potential cellular events that lead to an elevation in cytosolic calcium ($[Ca^{2+}]_i$) of PMNLs exposed to a high glucose concentration (30 m*M*) in vitro. IP3 = inositol triphosphate; DAG = diacylglycerol; PKC = protein kinase C.

myriad of intracellular pathways including G proteins, the adenylate cyclase-cAMP protein kinase A system, and the phospholipase C pathway with consequent activation of protein kinase C. These cellular events activate calcium channels which permit calcium influx into cells, and mobilize calcium from intracellular stores (fig. 5). Indeed, the calcium channel blockers, verapamil, nifedipine, or amlodipine, markedly inhibited the magnitude of the glucose-induced rise in $[Ca^{2+}]_i$. These effects of hyperglycemia are mimicked by equimolar concentrations of mannitol or choline chloride, demonstrating that the effects described above are not related to glucose per se but to its osmotic activity. At present it is not known whether similar events occur in every cell exposed to high glucose concentration and whether various cells respond differently under such circumstances.

It must be mentioned, however, that an increase in calcium influx into cells by itself is not adequate to cause an increase in the basal levels of $[Ca^{2+}]_i$ because cells are endowed with powerful mechanisms that directly (Ca^{2+}-ATPase, Na^+/Ca^{2+} exchanger) and indirectly (Na^+-K^+-ATPase) extrude calcium from cells [58]. Therefore, for the basal levels of $[Ca^{2+}]_i$ to increase, the function of these pumps must be impaired. Indeed, it has been found that in other settings where basal levels of $[Ca^{2+}]_i$ are elevated, the continued increase in calcium influx sets into motion cellular events including decreased ATP content, impaired activity of Ca^{2+}-ATPase and Na^+/K^+-ATPase and reduced Na^+/Ca^{2+} exchange [41]. These changes lead to decreased extrusion of calcium from cells.

Certain data indicate that Ca^{2+}-ATPase activity is reduced in sarcolemmal preparations of hearts from rats with streptozotocin-induced diabetes [59] and in erythrocytes of Zucker diabetic rats [60]. Furthermore, Na^+-K^+-ATPase is reduced in the heart of rats with NIDDM [37], in the sciatic nerve of rats with streptozotocin-induced diabetes [61], and in the platelets of patients with NIDDM [62]. In addition, Kuriyama et al. [63] found that vascular smooth muscle cells cultured in media containing high glucose had elevated levels of $[Ca^{2+}]_i$ and reduced Na^+-K^+-ATPase activity. Finally, the Na^+/Ca^{2+} exchanger was found to be reduced in sarcolemmal preparations of heart from diabetic rats [37, 59]. It is not known, however, whether derangements in these pumps that may be associated with diabetes mellitus are present in other cells.

Thus, the limited available data permit the proposition that the initial event leading to the rise in the basal levels of $[Ca^{2+}]_i$ in diabetes is the hyperglycemia-induced calcium influx into cells. This is supported by the observation that the control of hyperglycemia is associated with normalization of the basal levels of $[Ca^{2+}]_i$ at least in two cell types, i.e., PMNLs [43] and B cells [44]. Chronic and sustained entry of calcium into cells, that occurs with uncontrolled hyperglycemia, may result in a decrease in cellular ATP content. Indeed, it has been found that sustained entry of calcium into cells in other settings is associated with inhibition of mitochondrial oxidation and phosphorylation and hence reduced ATP production [64, 65] and a fall in ATP content. We reported that the ATP content of PMNLs of patients with NIDDM is indeed reduced, returning to normal after the control of hyperglycemia and normalization of PMNL $[Ca^{2+}]_i$ [43]. Since ATP is required for the normal function of Ca^{2+}-ATPase and Na^+-K^+-ATPase [58], a decrease in ATP content may be followed by a decrease in the activity of these enzymes. Understanding the mechanisms responsible for the elevation in the basal levels of $[Ca^{2+}]_i$ would permit the development of therapeutic approaches to prevent the rise in $[Ca^{2+}]_i$ even if the hyperglycemia is uncontrolled.

If the initial event leading to the rise in the basal levels of $[Ca^{2+}]_i$ in diabetes mellitus is the hyperglycemia-induced rise in calcium influx, blocking this effect may prevent the rise in the basal levels of $[Ca^{2+}]_i$ despite the hyperglycemia. Further, if the rise in the basal levels of $[Ca^{2+}]_i$ plays an important role in the genesis of the diabetic complications, preventing the rise in $[Ca^{2+}]_i$ may ameliorate or reverse cell dysfunction. We examined these issues in PMNLs of rats with streptozotocin-induced diabetes mellitus. We found that the basal levels of PMNL $[Ca^{2+}]_i$ are increased and phagocytosis is impaired after the induction of diabetes mellitus. Treatment of these rats with the calcium channel blocker, amlodipine, prevented and reversed the rise in $[Ca^{2+}]_i$ basal levels and the impaired phagocytosis of PMNLs despite sustained and marked hyperglycemia [66] (fig. 6, 7). It is of interest that Pierce et al. [67] found that treatment of rats with streptozotocin-induced diabetes with the calcium channel blockers, verapamil or diltiazem, markedly and significantly reduced the incidence of cataract, and partially protected against the development of retinopathy despite sustained hyperglycemia. The lenses of the diabetic rats had a high calcium content which correlated significantly with cataract formation. Verapamil treatment caused significant reduction in lenticular calcium. These in vivo observations again show that hyperglycemia is associated with an increased calcium content of the lens, that this derangement plays an important role in cataract formation, and both abnormalities could be mitigated by treatment with calcium channel blockers.

The mechanisms through which an elevation in the basal levels of $[Ca^{2+}]_i$ in diabetes mellitus adversely affect cell function are not known. An elevation in $[Ca^{2+}]_i$ in other conditions such as chronic renal failure or phosphate depletion is associated with (1) reduced cellular ATP content [68–79], (2) an inhibitory effect on many enzymes such as Ca^{2+}-ATPase [70, 71, 73, 77–79], Na^+-K^+-ATPase [71, 72, 77–81], synaptosomal tyrosine hydroxylase, monoamine oxidase [82], and choline kinase [83], hepatic lipase [84], lipoprotein lipase [85, 86] and phosphofructokinase 1 [73, 87], (3) an inhibitory effect on transport systems such as the Na^+/Ca^{2+} exchanger [77–79] and Na^+-H^+ antiport [88], (4) reduced phospholipid contents in e.g., brain synaptosomes [89, 90] and myocardium [91], and (5) alterations in the molecular machinery of the cells [92–94]. Indeed, an elevation in the basal levels of $[Ca^{2+}]_i$ in chronic renal failure or in phosphate depletion is associated with down-regulation of the mRNA of many proteins. These include the mRNA of the PTH-PTHrP receptors of the kidney [92, 93], heart [94] and liver [92, 95], the mRNA of the angiotensin II and vasopressin receptors of the liver [94], and the mRNA of hepatic lipase [84]. Normalization of $[Ca^{2+}]_i$ of the various cells by treatment of animals with chronic renal failure or phosphate depletion with a calcium channel blocker normalized $[Ca^{2+}]_i$ and significantly mitigated or completely reversed the down-regulation of the mRNA of these proteins [92–94].

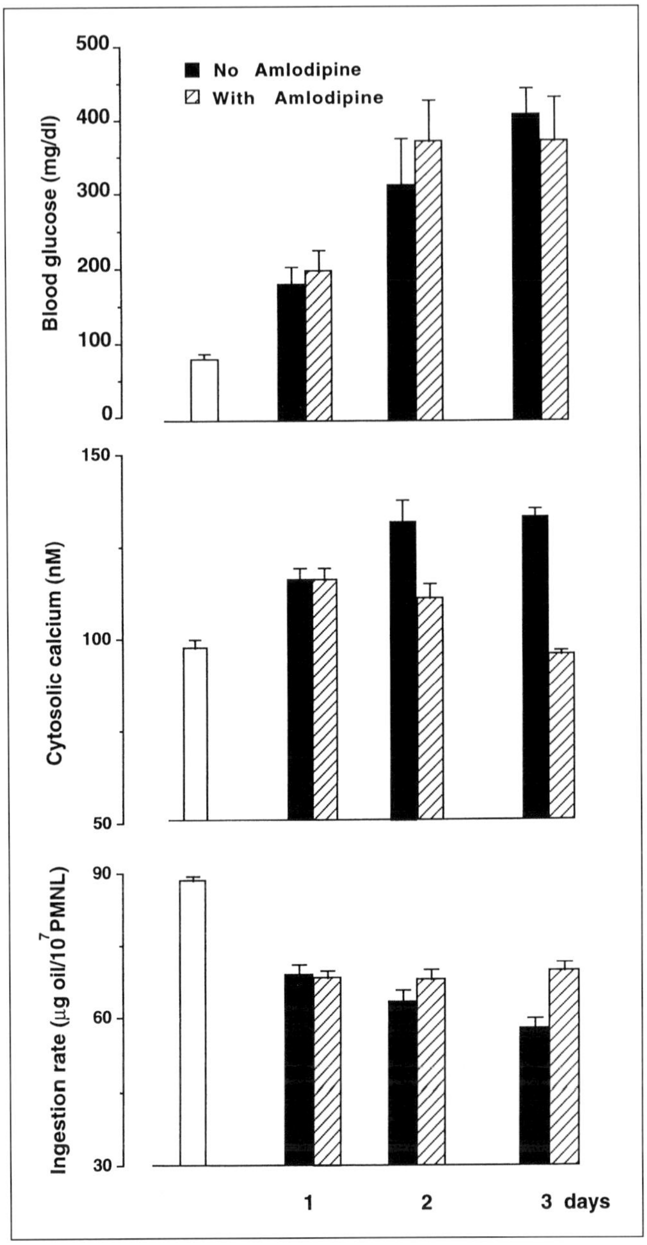

Fig. 6. Blood glucose concentration, and basal levels of $[Ca^{2+}]_i$ and phagocytosis of PMNLs of rats with 1, 2 and 3 days of diabetes with and without treatment with amlodipine. Data from normal rats (□) are also shown. Each bar represents the mean (±1 SE) of 6–7 studies of diabetic rats and 21 studies of normal rats.

Similar effects of an elevation in basal levels of $[Ca^{2+}]_i$ may occur in diabetes mellitus. We have already mentioned that reduced ATP content and impairment in the activity of Ca^{2+}-ATPase, Na^+-K^+-ATPase, and the Na^+/Ca^{2+} exchanger have been observed in diabetes mellitus. Some data are also available incriminating the rise in $[Ca^{2+}]_i$ in alterations in the molecular machinery of cells. We

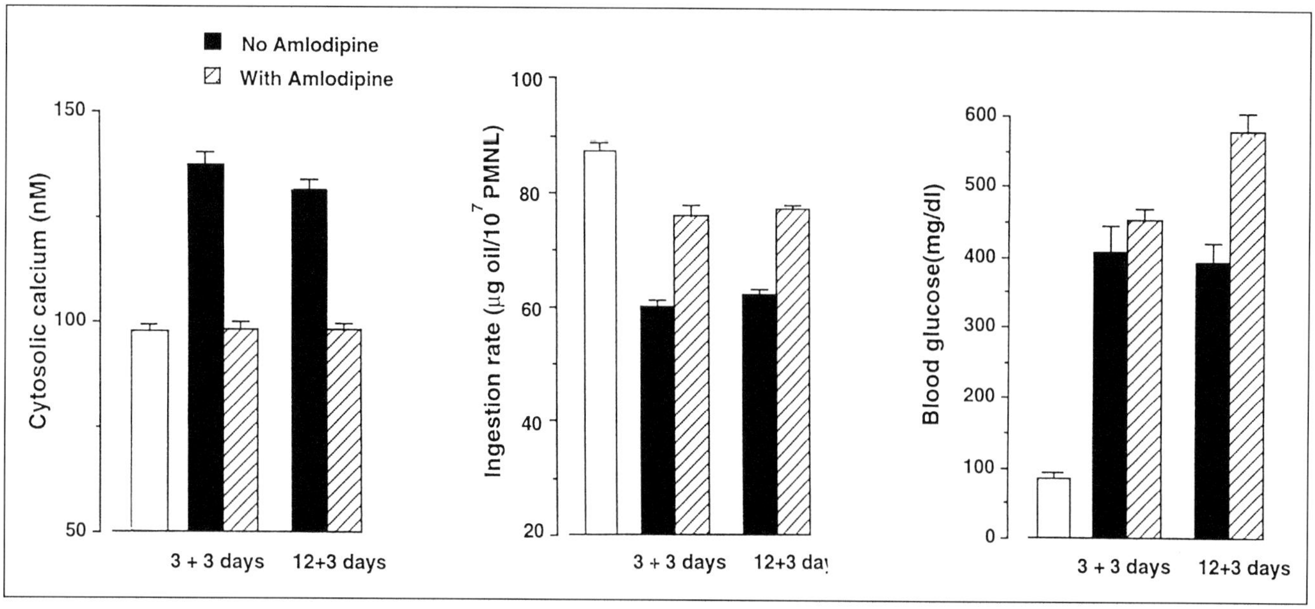

Fig. 7. Basal levels of $[Ca^{2+}]_i$ of PMNLs, phagocytosis, and blood glucose levels in 21 normal rats (□), and in 3–12-day diabetic rats which were then treated for 3 days with amlodipine. Each bar represents the mean (±1 SE) of 6 studies of diabetic rats.

examined the effect of streptozotocin-induced diabetes mellitus in rats on the $[Ca^{2+}]_i$ of the renal proximal tubular cells and on the expression of mRNA for the PTH-PTHrP, vasopressin, and angiotensin receptors in these cells. We found that the levels of $[Ca^{2+}]_i$ are elevated and the mRNAs for the hormone receptors are down-regulated [95]. Normalization of $[Ca^{2+}]_i$ levels in these cells by treatment of the rats with amlodipine normalized the expression of the mRNSs of these receptors [95]. Weiss et al. [96] also reported that diabetes mellitus in rats is associated with down-regulation of the insulin-like growth factor-I (IGF-I) gene and up-regulation of IGF-I receptor in rat kidneys. These observations again demmonstrate an effect of diabetes on the molecular machinery of cells. These authors, however, did not measure kidney cell $[Ca^{2+}]_i$, but concluded that these changes are related to the hyperglycemia. Studer and Ganas [97] found that the basal levels of $[Ca^{2+}]_i$ of hepatocytes from diabetic rats are elevated and the vasopressin-, angiotensin- and glucagon-induced rise in $[Ca^{2+}]_i$ is reduced. They did not measure the mRNA or the proteins of these receptors, and it is possible that the reduced calcium signal induced by these hormones is due to a reduction in their receptors. Mene et al. [98] found that the calcium signals induced by angiotensin II, prostaglandin $F_{2\alpha}$ and arginine-vasopressin in mesangial cells cultured for 5 days in media containing 22 m*M* glucose are significantly reduced. This observation is consistent with alterations in the receptors of these agonists in the mesangial cells. Tanaka et al. [99] reported that cell surface $[^3H^+]$-bunazosin binding by cardiac myocytes from rats with streptozotocin-induced diabetes is markedly reduced. This binding reflects the number and/or affinity of the α-adrenoreceptors. The generation of IP3 after stimulation of the cardiac myocytes with noradrenaline is also reduced. Treatment of these animals with verapamil corrected these abnormalities despite persistent hyperglycemia. These data are consistent with adverse effects of diabetes on the molecular machinery and on cell signalling. Since verapamil corrected these derangements, it is reasonable to suggest that elevated levels of $[Ca^{2+}]_i$ of cardiac myocytes are responsible for these derangements.

It is apparent from the above discussion that (1) derangements in cellular regulation of calcium are present in both IDDM and NIDDM, (2) these abnormalities could adversely affect cell function through many pathways, and (3) a potential therapeutic approach may be available to combat these disturbances and their effects on cell function. Thus, the comprehensive study of the abnormalities in cellular calcium homeostasis in diabetes, an understanding of their mechanisms, and the evaluation of potential approaches for their prevention and/or management are of extreme clinical importance, especially for those patients with uncontrolled hyperglycemia. The data which would be generated by such studies may yield information enabling the formulation of new therapeutic approaches that may be useful in the treatment of many of the complications in patients with uncontrolled hyperglycemia.

Acknowledgment

This work was supported by Grant No. DK29955 from the National Institute of Diabetes, Digestive and Kidney Diseases.

References

1 Pirart J: Diabetes mellitus and its degenerative complications: A prospective study of 4,400 patients observed between 1947 and 1973. Diabetes Care 1978;1:168–188.

2 Engerman R, Bloodworth JM Jr, Nelson S: Relationship of microvascular disease in diabetes to metabolic control. Diabetes 1977;26:760–769.

3 Cohen AJ, McGill PD, Rossetti RG, Guberski DL, Like AA: Glomerulopathy in spontaneously diabetic rat: Impact of glycemic control. Diabetes 1987; 36:944–951.

4 Diabetes Control and Complications Trial Research Group: The effect of intensive treatment of diabetes on the development and progression of long-term complications in insulin-dependent diabetes mellitus. N Engl J Med 1993;329:977–986.

5 Nathan DM, Singer DE, Hurxthal K, Goodson JD: The clinical information value of the glycosylated hemoglobin assay. N Engl J Med 1984;310:341–346.

6 Klein R, Klein BE, Moss SE, Davis MD, Demets DL: The Wisconsin epidemiologic study of diabetic retinopathy. III: Prevalence and risk of diabetic retinopathy when age at diagnosis is 30 or more years. Arch Ophthalmol 1984;102:527–532.

7 Nathan DM, Singer DE, Godine JE, Harrington CH, Perlmuter LC: Retinopathy in older type II diabetics: Association with glucose control. Diabetes 1985;35:797–801.

8 Klein R, Klein BE, Moss SE, Davis MD, DeMets DL: Glycosylated hemoglobin predicts the incidence and progression of diabetic retinopathy. JAMA 1988;260:2864–2871.

9 Singer DE, Nathan DM, Anderson KM, Wilson PW, Evans JC: Association of HbA_1C with prevalent cardiovascular disease in the original cohort of the Framingham Heart Study. Diabetes 1992;41: 202–208.

10 Gabby KH, Merola LO, Field RA: Sorbitol pathway: Presence in nerve and cord with substrate accumulation in diabetes. Science 1966;151:209–210.

11 Gabby KH: The sorbitol pathway and the complications of diabetes. N Engl J Med 1973;288:831–836.

12 Greene DA, Lattimer SA, Sima AA: Sorbitol, phosphoinositides, and sodium-potassiumATPase in the pathogenesis of diabetic complications. N Engl J Med 1987;316:599–606.

13 Greene DH, Lattimer SA, Sima AA: Are disturbance of sorbitol, phosphoinositide, and Na^+-K^+-ATPase regulation involved in pathogenesis of diabetic neuropathy? Diabetes 1988;37:688–693.

14 Beyer-Mears A, Cruz E: Reversal of diabetic cataract by sorbinil, an aldose reductase inhibitor. Diabetes 1985;34:15–21.

15 Engerman RL, Kern TS, Larson ME: Nerve conduction and aldose reductase inhibition during 5 years of diabetic or galactosaemia in dogs. Diabetologia 1994;37:141–144.

16 Fagius J, Brattberg A, Jameson S, Berne C: Limited benefit of treatment of diabetic polyneuropathy with an aldose reductase inhibitor: A 24-week controlled trial. Diabetologia 1985;28:323–329.

17 Krentz AJ, Monigsberger L, Ellis SH, Hardman M, Nattrass M: A 12 month randomized controlled study of the aldose reductase inhibitor ponalrestat in patients with chronic synaptosomes diabetic neuropathy. Diabetic Med 1992;9:483–488.

18 Sima AA, Prashar A, Nathaniel V, Bel V, Werb MR, Green DA: Overt diabetic neuropathy: Repair of axo-glial dysfunction and axonal atrophy by aldose reductase inhibition and its correlation to improvement in nerve conduction velocity. Diabetic Med 1993;10:115–121.

19 Brownlee M, Vlassara H, Cerami A: Nonenzymatic glycosylation and the pathogenesis of diabetic complications. Ann Intern Med 1984;101:527–537.

20 Maillard LC: Action des acides amines sur les sucres: Formation des melanoidines par voie methodique. CR Acad Sci 1912;154:66–68.

21 Patton AR, Hill EG: Inactivation of nutrients by heating with glucose. Science 1948;107:68–69.

22 Mohammad A, Fraenkel-Conrat H, Olcott HS: The 'browing' reaction of proteins with glucose. Arch Biochem 1949;24:157–158.

23 Bunn HF, Haney DN, Gabbay KM, Gallop PM: Further identification of the nature and linkage of the carbohydrate in hemoglobin A1c. Biochem Biophys Res Commun 1975;67:103–109.

24 Bucala R, Cerami A, Vlassara H: Advanced glycosylation end products in diabetic complications. Diabetes Rev 1995;3:258–268.

25 Bucala R, Cerami A: Advanced glycosylation: Chemistry, biology, and implications for diabetes and aging. Adv Pharmacol 1993;23:1–34.

26 Ahmed MV, Thrope SR, Baynes JW: Identification of carboxymethyllysine as a degradation product of fructosyllysine in glycated protein. J Biol Chem 1986;261:4889–4894.

27 Ahmed MV, Dunn JA, Walla MD, Thrope SR, Baynes JW: Oxidative degradation of glucose adducts to protein. J Biol Chem 1988;263:8816–8821.

28 Gugliucci A, Bendayan M: Renal fate of circulating advanced glycated end products (AGE): Evidence for reabsorption and catabolism of AGE-peptides by renal proximal tubular cells. Diabetologia 1996; 39:149–160.

29 Malita Z, Radoff S, Rayfield EJ, Yang Z, Skolnik E, Delaney V, Friedman EA, Cerami A, Vlassara H: Advanced glycosylation end products in patients with diabetic nephropathy. N Engl J Med 1991;325: 836–842.

30 Makita Z, Bucala R, Rayfield EJ, Friedman EA, Kaufman AM, Korbet SM, Barth RH, Winston JA, Fah H, Manogue K, Cerami A, Vlassara M: Diabetic-uremic serum advanced glycosylation end products are chemically reactive and resistant to dialysis therapy: Role in mortality of uremia. Lancet, in press.

31 Resnik LM, Gupta RK, Bhargava KK, Gruenspan H, Alderman MH, Laragh JH: Cellular ions in hypertension diabetes and obesity: A nuclear magnetic resonance spectroscopic study. Hypertension 1991;17:951–957.

32 Tschöpe D, Rösen P, Gried FA: Increase in the cytosolic concentration of calcium in platelets of diabetic type II. Thromb Res 1991;62:421–428.

33 Pellegatta F, Folli F, Ronchi P, Caspani L, Galli L, Vicari AM: Deranged platelet calcium homeostasis in poorly controlled IDDM patients. Diabetes Care 1993;16:178–183.

34 Segal S, Lloyd S, Sherman N, Sussman KE, Draznin B: Postprandial changes in cytosolic free calcium and glucose uptake in adipocytes in obesity and non-insulin-dependent diabetes mellitus. Horm Res 1990;34:39–44.

35 Draznin B, Sussman KE, Eckel RH, Kao M, Yost T, Sherman NA: Possible role of cytosolic free calcium concentrations in mediating insulin resistance of obesity and hyperinsulinemia. J Clin Invest 1988; 28:1848–1852.

36 Studer RK, Gansas L: Effect of diabetes on hormone-stimulated and basal hepatocyte calcium metabolism. Endocrinology 1989;125:2421–2433.

37 Allo SN, Lincoln TM, Wilson GL, Green FJ, Watanabe AM, Schaffer SW: Non-insulin-dependent diabetes induced defects in cardiac cellular calcium regulation. Am J Physiol 1991;260:C1165–C1171.

38 Schaffer SW: Cardiomyopathy associated with non-insulin-dependent diabetes. Mol Cell Biochem 1991;107:1–20.

39 Ohara T, Sussman KE, Draznin B: Effect of diabetes on cytosolic free Ca^{2+} and Na^+-K^+-ATPase in rat aorta. Diabetes 1991;40:1560–1563.

40 Levy J, Gavin JR III, Sowers JR: Diabetes mellitus: A disease of abnormal cellular calcium metabolism. Am J Med 1994;96:260–273.

41 Massry SG, Smogorzewski M: Mechanisms by which parathyroid hormone mediates its deleterious effects on organ function in uremia. Semin Nephrol 1994;14:219–231.

42 Massry SG, Fadda GZ, Perna AF, Kiersztejn M, Smogorzewski M: Mechanisms of organ dysfunction in phosphate depletion: A critical role for a rise in cytosolic calcium. Miner Electrolyte Metab 1992;18:133–140.

43 Alexiewicz JA, Kumar D, Smogorzewski M, Klin M, Massry SG: Polymorphonuclear leucocytes in non-insulin-dependent diabetes mellitus: Abnormalities in metabolism and function. Ann Intern Med 1995;123:919–924.

44 Alexiewicz JM, Kumar D, Smogorzewski M, Massry SG: Elevated cytosolic calcium and impaired proliferation of B lymphocytes in type II diabetes mellitus. Am J Kidney Dis, in press.

45 Hoenig M, Sharp GW: Glucose induces insulin release and a rise in cytosolic calcium concentration in a transplantable rat insulinoma. Endocrinology 1985;119:2502–2507.

46 Fadda GZ, Massry SG: Impaired glucose-induced calcium signal in pancreatic islets in chronic renal failure. Am J Nephrol 1991;11:475–478.

47 Barbagallo M, Shan J, Pang PKT, Resnick LM: Glucose Glucose-induced alterations of cytosolic free calcium in cultured rat tail artery vascular smooth muscle cells. J Clin Invest 1995;95:763–767.

48 Demerdash TM, Seyrek N, Smogorzewski M, Marchinowski W, Nasser-Moaddeli S, Massry SG: The pathways through which glucose induces a rise in $[Ca^{2+}]_i$ of PMNLs of rats. Kidney Int, in press.

49 Resnick LM: Ionic basis of hypertension, insulin resistance, and related disorder: Mechanisms of syndrome X. Am J Hypertens 1993;5:296–301.

50 Draznin B, Leitner JW, Sussman KE, Sherman NA: Insulin and glucose modulate protein kinase C activity in rats adipocytes. Biochem Biophys Res Commun 1988;156:570–575.

51 Lee TS, Saltsman KA, Ohashi H, King GL: Activation of protein kinase C by elevation of glucose concentration: Proposal for a mechanism in the development of diabetic vascular complications. Proc Natl Acad Sci USA 1989;86:5141–5145.

52 Sahai A, Fadda GZ, Massry SG: Parathyroid hormone activates protein kinase C of pancreatic islets. Endocrinology 1992;131:1889–1894.

53 Stojceva-Taneva O, Fadda GZ, Smogorzewski M, Massry SG: Parathyroid hormone increases cytosolic calcium of thymocytes. Nephron 1993;64:592–599.
54 Fadda GZ, Thanakitcharu M, Smogorzewski M, Massry SG: Prathyroid hormone raises cytosolic calcium in pancreatic islets: Study on mechanism. Kidney Int 1993;43:554–560.
55 Klin M, Smogorzewski M, Khilnani H, Michnowska M, Massry SG: Mechanisms of PTH-induced rise in cytosolic calcium in adult rat hepatocytes. Am J Physiol 1994;267:G754–G763.
56 Ni Z, Smogorzewski M, Massry SG: Effect of PTH on $[Ca^{2+}]_i$ of rat adipocytes. Endocrinology 1994; 135:1837–1844.
57 Tanaka H, Smogorzewski M, Koss M, Massry SG: Pathways involved in PTH-induced rise in cytosolic Ca^{2+} concentration of rat renal proximal tubule. Am J Physiol 1995;268:F330–F337.
58 Carafoli E: Calcium pumps of the plasma membrane. Physiol Rev 1991;71:129–153.
59 Makino N, Dhalla KS, Elimban V, Dhalla NS: Sarcolemmal Ca^{2+} transport in streptozotocin-induced diabetic cardiomyopathy in rats. Am J Physiol 1987; 253;E202–E207.
60 Zemel MB, Sowers JR, Shehin S, Walsh MF, Levy J: Impaired calcium metabolism associated with hypertension in Zuker obese rats. Metabolism 1990; 39:704–708.
61 Green DA, Lattimer SA: Impaired rat sciatic nerve sodium-potassium adenosine triphosphatase in acute streptozotocin diabetes and its correction by dietary myoinositol supplemen- tation. J Clin Invest 1983;72:1958–1963.
62 Mazzanti L, Rabini RA, Faloia E, Fumelli P, Bertoli E, De Pirro R: Altered cellular Ca^{2+} and Na^+ transport in diabetes mellitus. Diabetes 1990; 39:850–854.
63 Kuriyama S, Tokudome G, Tomonari H, Utsunomiya Y, Matsui K, Hashimoto T, Sakai O: Differential regulation of cation transport of vascular smooth muscle in a high glucose concentration milieu. Diabetes Res Clin Pract 1994;29:77–84.
64 Denton RM, McCormack JG: Ca^{2+} transport by mammalian mitochondria and its role in hormone action. Am J Physiol 1985;249:E543–E554.
65 Trump BE, Berezsky IK: The role of ion deregulation in toxic cell injury. Adv Mod Environ Toxicol 1987;14:27–50.
66 Seyrek N, Marcinkowski W, Smogorzewski M, Demerdash TM, Massry SG: Amlodypine prevents and reverses the elevation in $[Ca^{2+}]_i$ and the impaired phagocytosis of PMNL of diabetic rats. Nephrol Dial Transplant, in press.
67 Pierce GN, Afzal N, Kroeger EA, Lockwood MK, Kutryk MJB, Eckhert CD, Dhalla NS: Cataract formation is prevented by administration of verapamil to diabetic rats. Endocrinology 1989;125:730–735.
68 El-Belbessi S, Brautbar N, Anderson K, Campese VM, Massry SG: Effect of chronic renal failure on heart. Am J Nephrol 1986;6:369–375.
69 Smogorzewski M, Koureta P, Fadda GZ, Perna AF, Massry SG: Chronic parathyroid hormone excess in vivo increases resting levels of cytosolic calcium in brain synaptosomes: Studies in presence and absence of chronic renal failure. J Am Soc Nephrol 1991;1:1162–1168.
70 Hajjar SM, Smogorzewski M, Zayed A, Fadda GZ, Massry SG: Effect of chronic renal failure on Ca^{2+} ATPase of brain synaptosomes. J Am Soc Nephrol 1991;2:1115–1121.
71 Massry SG, Hajjar SM, Koureta P, Fadda GZ, Smogorzewski M: Phosphate depletion increases cytosolic calcium of brain synaptosomes. Am J Physiol 1991;260:F12–F18.
72 Zhou XJ, Fadda GZ, Perna AF, Massry SG: Phosphate depletion impairs insulin secretion by pancreatic islets. Kidney Int 1991;39:120–128.
73 Fadda GZ, Hajjar SM, Perna AF, Zhou XJ, Lipson LG, Massry SG: On the mechanism of impaired insulin secretion in chronic renal failure. J Clin Invest 1991;87:255–261.
74 Alexiewicz JM, Smogorzewski M, Fadda GZ, Massry SG: Impaired phagocytosis in dialysis patients: Studies on mechanisms. Am J Nephrol 1991; 11:102–111.
75 Chervu I, Kiersztejn M, Alexiewicz JM, Fadda GZ, Massry SG: Impaired phagocytosis in chronic renal failure is mediated by secondary hyperparathyroidism. Kidney Int 1992;41:1501–1505.
76 Kiersztejn M, Chervu I, Smogorzewski M, Fadda GZ, Alexiewicz JM, Massry SG: On the mechanisms of impaired phagocytosis in phosphate depletion. J Am Soc Nephrol 1992;2:1484–1489.
77 Zhang YB, Smogorzewski M, Ni Z, Massry SG: Altered cytosolic calcium homeostasis in rat cardiac myocytes in CRF. Kidney Int 1994;45:1113–1119.
78 Ni Z, Smogorzewski M, Massry SG: Elevated cytosolic calcium of adipocytes in chronic renal failure. Kidney Int 1995;47:1624–1629.
79 Klin M, Smogorzewski M, Massry SG: Chronic renal failure increases cytosolic Ca^{2+} of hepatocytes. Am J Physiol 1995;269:G103–G109.
80 Smogorzewski M, Campese VM, Massry SG: Abnormal norepinephrine uptake and release in brain synaptosomes in chronic renal failure. Kidney Int 1989;36:458–465.
81 Hajjar SM, Fadda GZ, Thanakitcharu P, Smogorzewski M, Massry SG: Reduced activity of Na^+-K^+ ATPase of pancreatic islets in chronic renal failure: Role of secondary hyperparathyroidism. J Am Soc Nephrol 1992;2:1355–1359.
82 Islam A, Smogorzewski M, Zayed A, Massry SG: Effect of chronic renal failure with and without secondary hyperparathyroidism on the activities of synaptosomal tyrosine hydroxylase and monoamine oxidase. Nephron 1992;61:32–36.
83 Ni Z, Smogorzewski M, Massry SG: Derangements in acetylcholine metabolism in brain synaptosomes in chronic renal failure. Kidney Int 1993;44:630–637.
84 Klin M, Smogorzewski M, Ni Z, Zhang G, Massry SG: Abnormalities in hepatic lipase in chronic renal failure: Role of excess parathyroid hormone. J Clin Invest 1996;97:2167–2173.
85 Akmal M, Perkins S, Kasim SE, Oh HY, Smogorzewski M, Massry SG: Verapamil prevents chronic renal failure-induced abnormalities in lipid metabolism. Am J Kidney Dis 1993;22:1158–1163.
86 Akmal M, Kasim SE, Soliman AR, Massry SG: Excess parathyroid hormone adversely affects lipid metabolism in chronic renal failure. Kidney Int 1990;37:854–858.
87 Perna AF, Fadda GZ, Massry SG: Inhibition of phosphofructokinase activity in pancreatic islets in phosphate depletion. Miner Electrolyte Metab 1991;17:8–11.
88 Michnowska M, Smogorzewski M, Massry SG: Impaired Na^+-H^+ exchanger activity of hepatocytes in CRF. J Am Soc Nephrol, in press.
89 Islam A, Smogorzewski M, Massry SG: Effect of chronic renal failure and parathyroid hormone on phospholipid content of brain synaptosome. Am J Physiol 1989;256:F705–F710.
90 Smogorzewski M, Islam A, Koureta P, Fadda GZ, Massry SG: Reduced phospholipid content of brain synaptosomes in phosphate depletion. Am J Physiol 1991;261:E742–E747.
91 Brautbar N, Tabernero-Romo J, Coats J, Massry SG: Impaired myocardial lipid metabolism in phosphate depletion. Kidney Int 1984;26:18–23.
92 Tian J, Smogorzewski M, Kedes L, Massry SG: PTH-PTHrP receptor mRNA is down-regulated in chronic renal failure. Am J Nephrol 1994;14:41–46.
93 Marcinkowska W, Smogorzewski M, Zhang G, Ni Z, Kedes L, Massry SG: Renal mRNA of PTH-PTHrP receptor, $[Ca^{2+}]_i$ and phosphaturic response to PTH in phosphate depletion. J Am Soc Nephrol, in press.
94 Smogorzewski M, Tian J, Massry SG: Down-regulation of PTH-PTHrP receptor of heart in CRF: Role of $[Ca^{2+}]_i$. Kidney Int 1995;47:1182–1186.
95 Massry SG, Klin M, Ni Z, Tian J, Kedes L, Smogorzewski M: Impaired agonist-induced calcium signaling in hepatocytes from chronic renal failure rats. Kidney Int 1995;48:1324–1331.
96 Weiss O, Anner H, Nephesh I, Alayoff A, Bursztyn M, Raz I: Insulin-like growth factor-I (IGF-I) and IGF-I receptor gene expression in the kidney of the chronically hypoinsulinemic rat and hyperinsulinemic rat. Metabolism 1995;8:982–986.
97 Studer RK, Ganas L: Effect of diabetes on hormone stimulated and basal hepatocytes calcium metabolism. Endocrinology 1989;125:2421–2433.
98 Mene P, Pugliese G, Pricci F, Di Mario U, Cinotti GA, Pugliese F: High glucose inhibits cytosolic calcium signaling in cultured rat mesangial cells. Kidney Int 1993;43:585–591.
99 Tanaka Y, Kashiwage A, Saeki Y, Takagi Y, Asahina T, Kikkawa R, Shigests Y: Effects of verapamil on the cardiac alpha 1-adrenoreceptor signalling system in diabetic rats. Eur J Pharmacol 1993;244: 105–109.

Mineral and Electrolyte Metabolism

Miner Electrolyte Metab 1997;23:261–264

Gary C. Curhan

Channing Laboratory,
Department of Medicine,
Brigham and Women's Hospital,
the Department of Nutrition,
Harvard School of Public Health,
and the Renal Unit,
Massachusetts General Hospital,
Harvard Medical School, Boston, Mass., USA

Dietary Calcium, Dietary Protein, and Kidney Stone Formation

Key Words
Animal protein
Calcium
Diet
Kidney stones
Prospective studies

Abstract
Kidney stone disease is common and is a major cause of morbidity involving the urinary tract. Rising incidence rates of calcium oxalate stone disease, the most common type of kidney stone, have focused attention on dietary habits and their potential role in the development of nephrolithiasis. Traditionally, calcium restriction had been recommended to reduce the likelihood of calcium stone formation, but recent evidence suggests that dietary calcium restriction may actually increase the risk. Observational and experimental data suggest that restriction of animal protein may lower the risk of stone formation, but a randomized trial did not confirm this finding. Dietary modification may play an important role in reducing the likelihood of stone recurrence. Notably, dietary calcium restriction should be avoided in patients who have had a calcium oxalate kidney stone.

Introduction

Nephrolithiasis, or kidney stone disease, is a major cause of morbidity and a rare cause of mortality involving the urinary tract. Kidney stones often result in intense pain and suffering in patients of all ages and may have significant economic consequences as well. Approximately 12% of the US population will form a kidney stone at some time [1, 2]. While bladder stones historically were the predominant type of urinary tract stone, stone formation in the upper urinary tract has become more common following the industrialization and associated dietary changes that have occurred in the 20th century. Rising incidence rates of calcium stone disease in Western countries and Japan have focused attention on dietary habits and their potential role in the development of nephrolithiasis.

Kidney stones may form when the concentration or urinary constituents exceeds their solubility. Hypercalciuria, hyperoxaluria, and hyperuricosuria are all common conditions which promote crystal formation. Under normal conditions, natural urinary inhibitors such as citrate help maintain the stability of the saturated solution to prevent crystal formation. While crystalluria itself does not always lead to stone formation, crystals that attach to the renal epithelium may grow into stones.

The incidence of kidney stone disease begins to rise in the early teenage years, peaks between ages 30 and 60 and declines in later life [2, 3]. The incidence rate for males during the peak years is 3 cases/1,000 men per year, while the incidence rate for similarly aged women is 1 case/1,000 women per year. Recurrence rates are high with up to 50% of patients experiencing a recurrent stone within 10 years of their initial episode.

Less than 10% of stones in men and 25% in women may be attributed to urinary tract infection or well-defined metabolic disorders such as hyperparathyroidism or cystinosis. Over 85% of stones in men and 70% in women contain calcium [2, 4], most commonly as calcium oxalate. Moreover, evidence suggests that dietary factors play a particularly significant role in the process of calcium stone formation.

Dietary Factors

Based on per capita consumption data and principles of renal pathophysiology, several dietary factors have been implicated in the process of kidney stone formation. It has been suggested that dietary calcium, animal protein [5], oxalate [6], sodium [7], and sucrose [8] increase the likelihood of stone formation, while fiber [9], magnesium [10], and potassium [11] intake may reduce the risk (table 1). Furthermore, while a high fluid intake has long been an accepted method of kidney stone prevention, information on the relative effect of individual beverages has only recently been studied. However, recent studies have not always supported the purported positive and negative effects of the nutrients listed in table 1.

To most effectively assess the relative impact of different nutrients on the process of kidney stone formation, prospective or experimental studies are required. Unfortunately, the interpretation of

0378–0392/97/0236–0261 $12.00/0

Gary C. Curhan, MD, ScD
Renal Unit, Massachusetts General Hospital
Fruit Street, Boston, MA 02114 (USA)
Tel. +1 617 726 4900, fax +1 617 726 7491

Table 1. Dietary factors reported to be associated with risk of formation of calcium-containing kidney stones

Increased risk	Decreased risk
Calcium	Calcium
Animal protein	Potassium
Sodium	Magnesium
Sucrose	Fiber
Oxalate	Fluid intake
Ascorbic acid (vitamin C)	Pyridoxine (vitamin B_6)

much of the previous research is limited by study design. For example, retrospective dietary studies that have evaluated the association between diet and recurrent stone formation are limited by the potential bias of dietary recall or changes in dietary habits that result from the disease. Moreover, the majority of prospective studies have been limited to investigating changes in the lithogenic composition of urine as opposed to changes in stone formation rates. Fortunately, information from well-designed studies is now becoming available for calcium oxalate stones which will be the focus of this article.

Dietary Calcium

Dietary calcium has long been suspected of raising the risk of kidney stone formation. This hypothesis derived largely from the finding of hypercalciuria in 20–40% of individuals with recurrent calcium kidney stones. Consequently, patients traditionally have been advised to restrict their intake of dietary calcium to prevent stone recurrence. Although a reduction in urinary calcium may be seen as a result of dietary calcium restriction in individuals with and without idiopathic hypercalciuria, there are no data demonstrating that these dietary changes reduce the rate of stone formation.

Surprisingly, recent evidence suggests that calcium restriction actually increases the risk of stone formation and that a diet high in calcium may even be protective [12]. Although case-control studies of diet and nephrolithiasis have not demonstrated that patients with calcium kidney stones have a different mean intake of calcium [13–16], the interpretation of these studies is complicated by the lack of control for other dietary risk factors associated with stone formation. Notably, data from a prospective study of dietary factors and the risk of incident stone disease in a cohort of 45,619 male health professionals (the Health Professionals Follow-Up Study) demonstrated that men with a higher intake of dietary calcium had a 34% *lower* risk of incident stone formation [12] (table 2). This inverse association between dietary calcium intake and risk of stone formation does not appear to be limited to men. Preliminary results from a cohort of over 100,000 female nurses (the Nurses' Health Study) demonstrated that women with the highest intake of dietary calcium compared with those with the lowest intake had a 28% lower risk [17].

While the mechanism of this effect remains to be elucidated, it may be related to the effect of dietary calcium on oxalate absorption in the gastrointestinal tract. Experimental evidence suggests that urinary oxalate plays an even more significant role in stone formation than calcium, as calcium oxalate saturation of urine increases rapidly with small increases in oxalate concentration [18]. Moreover, it has been shown that oxalate absorption in the gastrointestinal tract is greatest at low levels of calcium intake [19]. Dietary calcium presumably prevents absorption of oxalate by binding with it in the intestinal lumen leading to its excretion in the stool. Consequently, calcium restriction actually may be harmful in that it results in increased urinary oxalate excretion.

Supplemental Calcium

Calcium supplements are commonly used by adults in the US and other countries for the prevention of a variety of disorders such as osteoporosis and preeclampsia [20]. However, there has been concern that individuals consuming large doses of supplemental calcium may be at increased risk of kidney stone formation. The use of supplemental calcium will increase urinary calcium excretion, and if not taken with meals, would not lower urinary oxalate, and thus would increase the risk of stone formation.

In the prospective study of diet and stones in men [12], the relative risk of incident kidney stone formation in men consuming more than 500 mg of supplemental calcium daily compared with men not taking supplements was 1.23 (95% confidence interval 0.84–1.79). Although this result was not statistically significant, indicated by the confidence interval including a risk of 1.0, there was no evidence that calcium supplement use had the same protective effect as dietary calcium. The lack of a significant association between supplemental calcium use and risk of stone formation has been seen after 8 years of follow-up of this group of men (unpublished data).

The preliminary results in the prospective study in women demonstrated a 20% increased risk of stone formation in women using a calcium supplement daily compared with nonusers [17]. Although the increase in risk in the female cohort was statistically significant, the overall importance in the general population needs clarification. The baseline risk of incident stone formation in adult women is 1 case/1,000 women per year. For a woman with no history of kidney stones using a calcium supplement, the incidence rate would increase to only 1.2 cases/1,000 women per year.

The apparent contradiction between the protective effect of dietary calcium and the detrimental effect of supplement calcium may be due to differences in the timing of ingestion. In these two cohorts, more than 65% of the men and women did not take the supplement with meals, or took it only with breakfast when the oxalate content of the meal is likely to be low. If the calcium supplement is not taken with meals containing oxalate, a protective effect would not be expected. Rather the observed increase in risk might be due to increased urinary calcium excretion.

Animal Protein

Foods containing animal protein include meat, poultry, fish, and dairy products. Dietary intake of animal protein leads to increased urinary excretion of uric acid, a risk factor for both uric acid and calcium stone formation [21]. In addition, the increased acid load from the metabolism of sulfur-containing amino acids results in

Table 2. Multivariate relative risk of incident symptomatic kidney stones among 45,510 male health professionals, according to dietary calcium intake

Variable	Dietary calcium					p (for trend)
	quintile 1 (n=8,861)	quintile 2 (n=9,029)	quintile 3 (n=9,106)	quintile 4 (n=9,184)	quintile 5 (n=9,330)	
Dietary calcium intake, mg/day	<605	605–722	723–848	849–1,049	≥1,050	
Relative risk	1.0	0.74	0.68	0.68	0.66	0.02
95% confidence interval	referent	0.57–0.97	0.52–0.90	0.51–0.90	0.49–0.90	

The multivariate model included age, profession, use of thiazide diuretics, alcohol, and dietary intake of animal protein, potassium, and fluid. [Adapted from ref. 12.]

Table 3. Multivariate relative risk of incident symptomatic kidney stones among 45,510 male health professionals, according to animal protein intake

Variable	Animal protein					p (for trend)
	quintile 1	quintile 2	quintile 3	quintile 4	quintile 5	
Animal protein intake, g/day	<51	51–58	59–66	67–76	≥77	
Relative risk	1.0	0.97	1.41	1.07	1.33	0.05
95% confidence interval	referent	0.73–1.31	1.08–1.85	0.79–1.44	1.00–1.77	

The multivariate model included age, profession, use of thiazide diuretics, alcohol, and dietary intake of dietary calcium, animal protein, potassium, and fluid. [Adapted from ref. 12.]

lower levels of urinary citrate, an important inhibitor of calcium oxalate crystallization, and may increase urinary calcium excretion as well [19]. Accordingly, per capita consumption data from England revealed a direct relation between dietary animal protein and stone disease [5]. Similarly, prospective data from the Health Professionals Follow-Up Study demonstrated that men with the highest intake of animal protein had a 33% greater risk of stone formation than those with the lowest intake [12] (table 3). Thus, lowering the intake of animal protein appears to decrease the risk of stone formation.

There has been only one published randomized controlled trial of dietary modification and the risk of stone formation. Experienced investigators randomly assigned 99 individuals who had a calcium oxalate stone for the first time to a low-animal-protein, high-fiber diet and instruction on high fluid and adequate calcium intake, or to instruction on high fluid and adequate calcium intake alone [22]. After following individuals in both groups for up to 4.5 years, the risk of a recurrent stone in the low-animal-protein group was 5.6-fold *higher* than in the control group. The reason for the apparent discrepancy between the observational data and this trial is unknown, but there are several possible explanations. First, dietary calcium intake was not measured directly, so possibly the intake in the low-protein group was lower than in the control group. Second, reported fluid intake was higher in the controls in the first 2 years of the study when most of the stone events occurred. Finally, the recurrence rates were much lower than expected in both the case and control groups, suggesting that some other unknown factors influenced the subjects.

Conclusion

Kidney stone disease is a substantial health problem associated with significant pain and suffering as well as economic costs. Over US $2 billion were spent in 1986 on stone disease, most of this on treatment of existing stones and only a small percentage on prevention. Clearly, efforts to prevent or at least reduce the likelihood of developing a kidney stone would be an important component in the care of patients at risk. In particular, modifiable dietary factors appear to play an important role in the formation of calcium oxalate stones, the type of stone formed in the vast majority of cases. Once secondary causes of stone formation, such as hyperparathyroidism and renal tubular acidosis, are excluded, dietary counselling is a prudent and effective element of the therapeutic regimen and kidney stone prevention program. Specifically, for individuals who have a history of a calcium-containing kidney stone, calcium restriction should be avoided (except in the rare instances of excessive intake of greater than several grams per day). We recommend a dietary intake of elemental calcium of at least 800 mg/day (the current USRDA for adults) to prevent a negative calcium balance, bone mineral loss, and increased intestinal absorption of oxalate. At present, there is no evidence to support the belief that calcium restriction is beneficial, and current data suggest that it may in fact be harmful. Even in patients with hypercalciuria, dietary calcium restriction may not be appropriate. There are no studies documenting a protective effect of calcium restriction on stone formation rates and restricting these individuals may be more likely to lead to a negative calcium balance. Based on observational and experimental data, restriction of animal protein appears beneficial; however, the results from the outpatient intervention trial did not support dietary restriction and, thus, further study is required. All patients should be encouraged to achieve adequate fluid intake to produce at least 2 liters of urine per day. Other dietary recommendations that may be protective include a low-sodium diet and a high-potassium diet. At present, dietary modification may play an important role in reducing the likelihood of stone recurrence.

References

1 Sierakowski R, Finlayson B, Landes RR, Finlayson CD, Sierakowski N: The frequency of urolithiasis in hospital discharge diagnoses in the United States. Invest Urol 1978;15:438–441.
2 Johnson CM, Wilson DM, O'Fallon WM, Malek RS, Kurland LT: Renal stone epidemiology: A 25-year study in Rochester, Minnesota. Kidney Int 1979;16:624–631.
3 Hiatt RA, Dales LG, Friedman GD, Hunkeler EM: Frequency of urolithiasis in a prepaid medical care program. Am J Epidemiol 1982;115:255–265.
4 Coe FL, Parks JH: Nephrolithiasis: Pathogenesis and Treatment. Chicago, Year Book Medical Publishers, 1988.
5 Robertson WG, Peacock M, Hodgkinson A: Dietary changes and the incidence of urinary calculi in the U.K. between 1958 and 1976. J Chronic Dis 1979;32:469–476.
6 Larsson L, Tiselius HG: Hyperoxaluria. Miner Electrolyte Metab 1987;13:242–250.
7 Muldowney FP, Freaney R, Moloney MF: Importance of dietary sodium in the hypercalciuria syndrome. Kidney Int 1982;22:292–296.
8 Lemann J Jr, Piering WF, Lennon EJ: Possible role of carbohydrate-induced calciuria in calcium oxalate kidney stone formation. N Engl J Med 1969; 280:232–237.
9 Gleeson MJ, Thompson AS, Mehta S, Griffith DP: Effect of unprocessed wheat bran on calciuria and oxaluria in patients with urolithiasis. Urology 1990; 35:231–234.
10 Johansson G, Backman U, Danielson BG, Fellstrom B, Ljungall S, Wikstrom B: Biochemical and clinical effects of the prophylactic treatment of renal calcium stones with magnesium hydroxide. J Urol 1980;124:770–774.
11 Lemann J Jr, Pleuss JA, Gray RW, Hoffmann RG: Potassium administration increases and potassium deprivation reduces urinary calcium excretion in healthy adults. Kidney Int 1991;39:973–983.
12 Curhan GC, Willett WC, Rimm EB, Stampfer MJ: A prospective study of dietary calcium and other nutrients and the risk of symptomatic kidney stones. N Engl J Med 1993;328:833–838.
13 Griffith HM, O'Shea B, Kevany JP, McCormick JS: A control study of dietary factors in renal stone formation. Br J Urol 1981;53:416–420.
14 Power C, Barker DJP, Nelson M, Winter PD: Diet and renal stones: A case-control study. Br J Urol 1984;56:456–459.
15 Fellstrom B, Danielson BG, Karlstrom B, Lithell H, Ljunghall S, Vessby B: Dietary habits in renal stone patients compared with healthy subjects. Br J Urol 1989;63:575–580.
16 Trinchieri A, Mandressi A, Luongo P, Longo G, Pisani E: The influence of diet on urinary risk factors for stones in healthy subjects and idiopathic renal calcium stone formers. Br J Urol 1991;67: 230–236.
17 Curhan GC, Willett WC, Speizer FE, Stampfer MJ: A prospective study of dietary and supplemental calcium and vitamin C and the risk of kidney stones in women (abstract). J Urol 1996;155:641A.
18 Borsatti A: Calcium oxalate nephrolithiasis: Defective oxalate transport. Kidney Int 1991;39:1283–1298.
19 Zarembski PM, Hodgkinson A: Some factors influencing the urinary excretion of oxalic acid in man. Clin Chim Acta 1969;25:1–10.
20 NIH Consensus Development Panel on Optimal Calcium Intake: Optimal calcium intake. JAMA 1994;272:1942–1948.
21 Breslau NA, Brinkley L, Hill KD, Pak CYC: Relationship of animal protein-rich diet to kidney stone formation and calcium metabolism. J Clin Endocrinol Metab 1988;66:140–146.
22 Hiatt RA, Ettinger B, Caan B, Quesenberry CP Jr, Duncan D, Citron JT: Randomized controlled trial of a low animal protein, high fiber diet in the prevention of recurrent calcium oxalate kidney stones. Am J Epidemiol 1996;144:25–33.

Mineral and Electrolyte Metabolism

Miner Electrolyte Metab 1997;23:265–268

Salt Intake, Urinary Sodium, and Hypercalciuria

Massimo Cirillo[a, b]
Carolina Ciacci[c]
Martino Laurenzi[a, e]
Mariacristina Mellone[d]
Gabriele Mazzacca[c]
Natale G. De Santo[b]

[a] Department of Preventive Medicine, Northwestern University Medical School, Chicago, Ill., USA;
[b] Chair of Nephrology, Second University of Naples,
[c] Chair of Gastroenterology, and
[d] Department of Biochemistry, 'Federico II' University, Naples, and
[e] Merck Sharp & Dohme, Center for Epidemiologic Research, Rome, Italy

Key Words
Hypercalciuria
Salt
Urinary calcium
Urinary sodium

Abstract

Several studies have reported that high sodium (Na) intake increases not only urinary Na but also urinary calcium (Ca), suggesting that high Na intake could be involved in the pathogenesis of hypercalciuria. No research data are available on the relationship of Na intake to the prevalence of hypercalciuria within the general population. Moreover, it is not clear if Na intake relates only to urinary Ca or also to other indices of Ca homeostasis, including intestinal Ca absorption. In the present paper, two distinct studies addressed these points using 24-hour urinary Na as an index of salt intake in individuals on their habitual unrestricted free diet. Study 1 analyzed the relationship between 24-hour urinary Na and hypercalciuria (24-hour urinary Ca ≥ 7.5 mmol in men, ≥ 6.25 mmol in women) in a population sample of 203 men and women, aged 20–59 years. Study 2 analyzed the relationship between 24-hour urinary Na and intestinal strontium (Sr) absorption, used as an index of intestinal Ca absorption, urinary (24-hour and fasting) and plasma Ca, and plasma parathyroid hormone in 36 healthy men and women, aged 18–65 years. Within the population sample (study 1), 24-hour urinary Na was directly and significantly correlated with prevalence of hypercalciuria when controlling for gender, age, weight, and urinary creatinine: the relationship was continuous and linear for urinary Na ranging between 40 and 200 mmol/24 h. In the 36 volunteers (study 2), 24-hour urinary Na was related to 24-hour and fasting urinary Ca ($p < 0.001$) but not to intestinal Sr absorption: the relationship between 24-hour urinary Na and urinary Ca (both 24 h and fasting) was also significant, controlling for other variables. The results indicate that in adults on their habitual diet, urinary Na, which reflects dietary salt intake, correlates with the prevalence of hypercalciuria independently of intestinal Ca absorption and mainly via renal mechanisms.

Introduction

Hypercalciuria is considered a major risk factor for urinary stone formation [1]. Many factors affect urinary calcium (Ca) excretion and could be involved in the pathogenesis of hypercalciuria [1, 2]. The effect of dietary intake of sodium (Na) chloride as common salt on urinary Ca excretion was first described more than 30 years ago by Kleeman et al. [3]. After that landmark paper, other authors confirmed that urinary Ca parallels Na intake when Na intake is increased or decreased for several days. The mechanisms underlying the effect of dietary Na on urinary Ca involve the relationships linking urinary Na excretion to Na intake, and urinary Ca to urinary Na excretion [7].

These classic studies based on the stable modification of dietary Na intake abolished the day-to-day intraindividual variability in urinary Na excretion which is observed on a habitual free diet [8–11], and which reflects variability in the amount of Na present in the food, used for cooking, and added at the table [12]. However, the idea that dietary Na intake contributes to the regulation of urinary Ca excretion also in the absence of stable Na restriction or supplementation is supported by the consistent finding of a positive correlation between 24-hour urinary Na and Ca in individuals on their habitual free diet [13–15].

Despite the number of studies on the relationship between urinary Na and Ca, no research data are available on the relationship between urinary Na and the frequency of hypercalciuria within

KARGER
E-Mail karger@karger.ch
Fax +41 61 306 12 34
http://www.karger.ch

0378–0392/97/0236–0265 $12.00/0

Dr. Massimo Cirillo, MD
Cattedra di Nefrologia, Seconda Università di Napoli
Nuovo Policlinico (Pad. 17), via Sergio Pansini, 5
I–80131 Naples (Italy)
Tel. +39 81 566 6650, fax +39 81 566 6655

Table 1. Descriptive statistics: hypercalciuria and other variables in a population sample

	Men	Women
Number	101	102
Hypercalciuria	16 (15.8%)	10 (9.8%)
Age, years	39.3±11.6	39.9±11.5
Weight, kg	74.3±10.6	63.6±11.3
Height, m	1.70±0.07	1.59±0.05
24-Hour urinary Ca, mmol	5.18±2.08	4.23±1.72
24-Hour urinary Na, mmol	188±69	161±55
24-Hour urinary creatinine, mmol	13.7±2.3	9.5±1.7

a population. Moreover, it is still debated whether or not the relationship of Na homeostasis to urinary Ca excretion also implies an effect on intestinal Ca handling [16–18]. The present paper deals with these issues and analyzes data collected in individuals on their habitual diet to investigate the relationship between 24-hour urinary Na and the frequency (prevalence) of hypercalciuria, and to indices of overall Ca homeostasis, including intestinal Ca absorption.

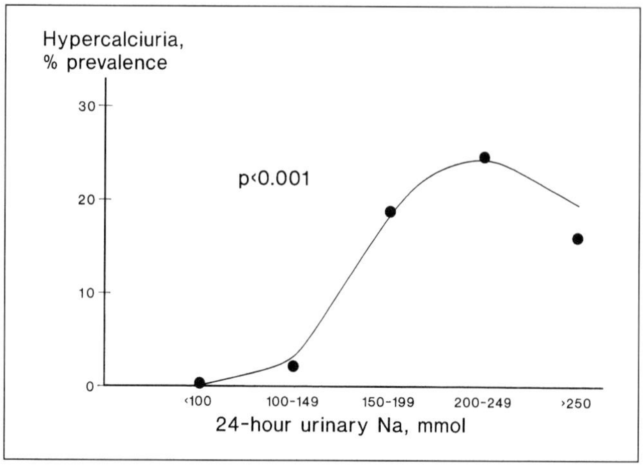

Fig. 1. Percentage prevalence of hypercalciuria (24-hour urinary Ca excretion ≥7.5 mmol in men, ≥6.25 mmol in women) in five strata of 24-hour urinary Na excretion (p value given by χ^2 analysis among all strata using the Mantel-Haenszel test for linearity). Data for a population sample of 203 men and women; numbers in the five strata are 21, 57, 64, 41, and 20 from the lowest to the highest. The range of urinary Na (min–max) is 40–403 mmol/24 h.

Material and Methods

Two distinct sets of data were analyzed. In study 1, the relationship between urinary Na and hypercalciuria was analyzed using data collected for a population sample comprising 203 men and women, aged 20–59 years, who participated in the INTERSALT study for the city of Gubbio (northern central Italy) [19, 20]; the analysis included anthropometric data and 24-hour urinary excretion of Na, Ca, and creatinine as described elsewhere [19, 20].

In study 2, the relationship of urinary Na to indices of Ca homeostasis including intestinal Ca absorption was studied in a group of 36 healthy, nonobese individuals (body mass index <30, weight/height, kg/m^2), aged 18–65 years, with plasma creatinine <106 μmol/l (<1.20 mg/100 ml), normal urinalysis, and no history of renal disease or disorders of Ca homeostasis. The 36 volunteers were selected from the staff of the hospital or among the friends of patients admitted to our division. The study protocol included anthropometric measures, the strontium (Sr) test, used as an index of intestinal Ca absorption [21–26], collection of 24-hour urine under fed conditions and blood and urine samples after an overnight fast. The protocol had been approved by the Ethical Committee of the University of Naples, and informed consent was obtained from all individuals. Briefly, after the completion of the 24-hour urine collection, participants were invited to come to the clinic early in the morning to undergo a 2-hour urine collection after an overnight fast (fasting condition), blood sampling, and the Sr test. The Sr test, based on the oral administration of Sr chloride, was performed according to the procedure described by Milsom et al. [21] modified as previously reported [27]. Blood venous samples were taken 15, 30, and 45 min after Sr administration to measure the plasma Sr concentration; the rate of intestinal Sr absorption was estimated by linear regression analysis of plasma Sr concentration over time. The 24-hour urine collection was used to measure the excretion under the fed condition of Na, Ca, creatinine, and urea, used to estimate the dietary protein intake [28]. Urinary and plasma total Ca, and plasma Sr were measured by atomic absorption spectrophotometry. Serum ionized Ca was measured by a Ca-specific electrode in blood samples withdrawn without stasis. Urinary urea was measured by automated biochemistry. Plasma and urinary creatinine were measured by the picric acid colorimetric method. The plasma concentration of intact parathyroid hormone (PTH) was measured by radioimmunoassay.

Data were analyzed using ANOVA, correlation analysis, logistic regression, χ^2 analysis, and multiple linear regression, as appropriate.

Results

Study 1 – Relationship between 24-Hour Urinary Na and Hypercalciuria within a Population Sample

Table 1 shows descriptive statistics by gender for hypercalciuria, age, anthropometric data, and 24-hour urinary excretion of Na, Ca, and creatinine in the population sample. When prevalence of hypercalciuria (dependent variable) was regressed over 24-hour urinary Na by logistic analysis, the relationship was significant for men (regression coefficient ±SE=0.0053±0.0025, p=0.049), women (0.0153±0.0062, p=0.013), and men and women combined controlling for gender (0.0082±0.0032, p=0.011). Exponentiation of the coefficient found in the sex-controlled analysis indicated that an increase of 60 mmol in 24-hour urinary Na (about 1 SD for men and women of this sample, see table 1) related to a 1.63-fold increase in the relative risk for hypercalciuria (95% confidence interval= 1.12–2.38). When the population sample was divided in two groups based on the median 24-hour urinary Na (169 mmol/24 h), individuals above the median had a higher prevalence of hypercalciuria than those below the median (21.8 and 3.9%, respectively, p<0.001; odds ratio =6.8, 95% confidence interval=2.3–20.6). The difference between individuals above and below the median was also significant in an analysis controlled for gender, age, weight, and 24-hour urinary creatinine/kg body weight, used as an index of completeness in urine collection (odds ratio=4.3, 95% confidence interval=1.3–14.3).

Figure 1 shows the percentage prevalence of hypercalciuria across five strata of 24-hour urinary Na. The prevalence of hypercal-

Table 2. Descriptive statistics, indices of Ca homeostasis, and other variables in the group of healthy volunteers

Men/women	18/18
24-Hour urinary Ca, mmol	5.38±2.19
Fasting urinary Ca, μmol/min	1.79±1.06
Serum total Ca, mmol/l	2.26±0.12
Serum ionized Ca, mmol/l	1.21±0.06
Plasma PTH, pg/ml	30.8±8.9
Intestinal Sr absorption, μmol/l per minute	0.232±0.081
Age, years	40.4±12.4
Weight, kg	62.4±13.0
Height, m	1.66±0.12
24-Hour urinary Na, mmol	183±83
Dietary protein intake, g/24 h	59.3±7.1
Cretinine clearance, ml/min per 1.73 m^2 of body surface area	114±18

ciuria increased linearly with rising levels of urinary Na from 40 to 200 mmol/24 h.

Study 2 – Relationship between Urinary Na and Indices of Ca Homeostasis in Healthy Volunteers

Descriptive statistics for indices of Ca homeostasis and other variables are shown for the group of healthy volunteers in table 2: gender distribution, age, anthropometric data, and other characteristics of this group were similar to those of the population sample analyzed in study 1. Table 3 shows correlation coefficients between 24-hour urinary Na and indices of Ca homeostasis. Correlation coefficients were significant for 24-hour and fasting urinary Ca, but not for serum Ca, plasma PTH, and intestinal Sr absorption.

Multiple linear regression analyses were done to assess if the relationships of 24-hour urinary Na to 24-hour and fasting urinary Ca were dependent on other variables. Table 4 shows the results of multiple liner regression coefficients in two models (1 and 2) where 24-hour urinary Ca and fasting urinary Ca, respectively, were used alternatively as the dependent variable, using 24-hour urinary Na and other parameters as independent variables. In both models 1 and 2, the regression coefficient of 24-hour urinary Na to urinary Ca was significant, also when controlling for other variables. Moreover, in model 1 (dependent variable: 24-hour urinary Ca), regression coefficients were significant and positive for plasma ionized Ca and intestinal Sr absorption, significant and negative for plasma PTH. In model 2 (dependent variable: fasting urinary Ca), findings were similar to those of model 1 except for intestinal Sr absorption, which was not related to fasting urinary Ca.

Discussion

The present study shows the novel finding that an index of dietary intake of Na positively and independently relates to the prevalence of hypercalciuria in a population sample of adult men and women studied on their habitual free diet. The relationship is independent of age, weight, and completeness of urine collection and almost linear when 24-hour urinary Na ranged between 40–200 mmol. In this range of 24-hour urinary Na, a difference of 100 mmol relates to a 4-fold increase in the prevalence of hypercalciuria; a further increase in 24-hour urinary Na did not correlate with increases in the prevalence of hypercalciuria. The observation that the relationship between urinary Na and hypercalciuria is not linear over the entire range of urinary Na does not support the use of a linear nomogram to plot hypercalciuria and urinary Na excretion, as proposed by Sabto et al. [14]. Dietary Ca intake was not analyzed in the present study, but it cannot explain the relationship between urinary Na and Ca since a significant relationship between urinary Na and Ca was also found after prolonged fasting [7]. The correlation between urinary Na and urinary Ca excretion measured in two different urine collections – 24-hour urine and fasting urine – indicates that the relationship between urinary Na and Ca does not reflect only the 'correlated error' due to the measurement of the concentration of the two ions in the same urine sample. In the group of healthy volunteers, protein intake was not related to urinary Ca excretion, a finding apparently in contrast with other studies [29, 30]; in fact, the protein intake in this group of healthy individuals was lower than that reported to affect urinary Ca [29, 30].

Table 3. Correlation coefficients of 24-hour urinary Na with indices of Ca homeostasis in 36 healthy volunteers

24-Hour urinary Ca	+0.564*
Fasting urinary Ca	+0.522*
Serum total Ca	+0.151
Serum ionized Ca	−0.260
Plasma PTH	−0.232
Intestinal Sr absorption	+0.178

*$p < 0.001$.

The finding that urinary Na related to fasting renal Ca excretion and not to the index of intestinal Ca absorption suggests that a high salt intake could also accelerate the release of Ca ions from the bone. This finding is in keeping with data reported by McParland et al. [16] and is in contrast with those on intestinal ^{47}Ca absorption reported by Breslau et al. [17]. Several factors can explain the discrepancies among these reports: the presence and the duration of a stable intervention in dietary salt intake, the use of Sr instead of radioactive Ca to measure intestinal absorption, the age of the individuals under study [18] and, finally, the level of salt intake explored.

In summary, the data of the present study support the idea that a high salt intake could be included in the list of factors causing and/or maintaining hypercalciuria. Salt intake should be lowered in patients with or at risk for hypercalciuria, given, in addition, the evidence for a positive association of urinary Na excretion with urinary stone disease in population-based studies [31]. The possibility that high salt intake could also favor a Ca-deficient status needs to be investigated further.

Acknowledgments

Thanks are expressed to the INTERSALT Cooperative Research Group and particularly to Dr. Paul Elliot, Department of Public Health and Policy,

Table 4. Multivariate linear regression coefficients of 24-hour urinary Na and other variables to (dependent variable) 24-urinary Ca (model 1) and fasting urinary Ca (model 2) in 36 healthy volunteers

Independent variable	Dependent variable	
	model 1 24-hour urinary Ca (μmol/min)	model 2 fasting urinary Ca (μmol/min)
Gender (men = 1, women = 0)	+0.592	+0.579
Age, years	+0.00292	−0.00393
Weight, kg	−0.0299	−0.0553
24-hour urinary Na, mmol	+0.0133**	+0.0144***
24-hour urinary urea, mmol	+0.108	+0.116
Plasma ionized Ca, mmol/l	+8.977*	+6.339*
Plasma PTH, pg/ml	−0.1179***	−0.0708***
Intestinal Sr absorption, μmol/l per minute	+6.750**	+2.051
Creatinine clearance, ml/min per 1.73 m^2	−0.0116	−0.0177

*$p<0.05$, **$p<0.01$, ***$p<0.001$.

London School of Hygiene and Tropical Medicine, UK, for making available data from 24-hour urine collections for the 203 men and women participating in the INTERSALT Study for the city of Gubbio, Italy. The present study was supported by grants from the Consiglio Nazionale delle Ricerche (CNR) and by the Ministero della Università e della Ricerca Scientifica e Tecnologica (MURST, 40% and 60%).

References

1 Goldfarb S: Dietary factors in the pathogenesis and prophylaxis of calcium nephrolithiasis. Kidney Int 1988;34:544–555.

2 Lemann J, Adams ND, Gray RW: Urinary calcium excretion in human beings. N Engl J Med 1979; 301:535–541.

3 Kleeman CR, Bohannan J, Bernstein D, Maxwell MH: Effect of variation in sodium intake on calcium excretion in normal humans. Proc Soc Exp Biol Med 1964;115:29–32.

4 McCarron DA, Rankin LI, Bennet WM, Krutzik S, McClung MR, Luft FC: Urinary calcium excretion at extremes of sodium intake in normal man. Am J Nephrol 1981;1:84–90.

5 Muldowney FP, Freaney R, Moloney MF: Importance of dietary sodium in the hypercalciuria syndrome. Kidney Int 1982;22:292–296.

6 Silver J, Friedlander MM, Rubinger D, Popovtzer MM: Sodium-dependent idiopathic hypercalciuria in renal stone-formers. Lancet 1983;ii:484–485.

7 Sutton RAL, Dirk JH: Calcium and magnesium: Renal handling and disorders of metabolism; in Brenner DM, Rector FC (ed): The Kidney, ed 3. Philadelphia, Saunders, 1986, pp 592–638.

8 Liu K, Cooper R, McKeever J, McKeever P, Byington R, Soltero I, Stamler R, Gosh F, Stevens E, Stamler J: Assessment of the association between salt intake and high blood pressure: Methodological problems. Am J Epidemiol 1979;110:219–226.

9 Liu K, Cooper R, Soltero I, Stamler J: Variability in 24-hour urine sodium excretion in children. Hypertension 1979;1:631–636.

10 Lisheng L, Deyu Z, Lue J, Liao Y, Liu K, Stamler J: Variability of urinary sodium and potassium excretion in north Chinese men. J Hypertens 1987;5: 331–335.

11 Siani A, Iacoviello L, Giorgione N, Iacone R, Strazzullo P: Comparison of variability of urinary sodium, potassium, and calcium in free-living men. Hypertension 1989;13:38–42.

12 Stamler J: Dietary salt and blood pressure. Ann NY Acad Sci 1993;676:122–156.

13 Strazzullo P, Nunziata V, Cirillo M, Giannattasio R, Ferrara LA, Mattioli PL, Mancini M: Abnormalities of calcium metabolism in essential hypertension. Clin Sci 1983;65:137–141.

14 Sabto J, Powell MJ, Breidhal MJ, Gurr W: Influence of urinary sodium on calcium excretion in normal individuals – A redefinition of hypercalciuria. Med J Aust 1984;140:354–356.

15 Wasserstein AG, Stolley PD, Soper KA, Goldfarb S, Maislin G, Agus Z: Case-control study of risk factors for idiopathic calcium nephrolithiasis. Miner Electrolyte Metab 1987;13:85–95.

16 McParland BE, Goulding A, Campbell AJ: Dietary salt affects biochemical markers of resorption and formation of bone in ederly women. BMJ 1989;299: 834–835.

17 Breslau NA, McGuire NA, Zerwekh JE, Pak CYC: The role of dietary sodium on renal excretion and intestinal absorption of calcium and vitamin D metabolism. J Clin Endocrinol Metab 1982;55:369–373.

18 Breslau NA, Sakhaee K, Pak CYC: Impaired adaptation to salt-induced urinary calcium losses in postmenopausal osteoporosis. Trans Assoc Am Phys 1985;98:107–115.

19 INTERSALT Cooperative Research Group: INTERSALT: An international study of electrolyte excretion and blood pressure: Results for 24-hour urinary sodium and potassium excretion. BMJ 1988;297:319–328.

20 INTERSALT. J Hum Hypertens 1989;3(special issue 5):279–331.

21 Milsom S, Ibbertson K, Hannan S, Show D, Pybus J: Simple test of intestinal calcium absorption measured by stable strontium. BMJ 1987;295:231–234.

22 Reid IR, Pybus J, Lim TMT, Hannon S, Ibbertson HK: The assessment of intestinal calcium absorption using stable strontium. Calcif Tissue Int 1986; 38:303–305.

23 Porzig H: Transmembrane strontium movements and strontium-induced changes in membrane cation permeability in nonexcitable cell membranes; in Skoryna SC (ed): Handbook of Stable Strontium. London, Plenum, 1981, pp 183–200.

24 Papworth DG, Patrick G: The kinetics of influx of calcium and strontium into rat intestine in vitro. J Physiol (Lond) 1970;210:999–1020.

25 Hendrix JZ, Alcock NW, Archibald RM: Competition between calcium, strontium and magnesium for absorption in the isolated rat intestine. Clin Chem 1963;9:734–744.

26 Gutteridge DH, Robinson CJ, Joplin GF: Delayed strontium absorption in post-menopausal osteoporosis and osteomalacia. Clin Sci 1968;34:351–363.

27 Ciacci C, Cirillo M, Mellone M, Basile F, Mazzacca G, De Santo NG: Hypocalciuria in overt and subclinical celiac disease. Am J Gastroenterol 1995;90: 1480–1484.

28 Isakson B: Urinary nitrogen output is a valid test in dietary surveys. Am J Clin Nutr 1980;33:4–12.

29 Licata AA, Bou E, Bartter FC, Cox J: Effects of dietary protein on urinary calcium in normal subjects and in patients with nephrolithiasis. Metabolism 1979;28:895–901.

30 Linkswiler HM, Zemel MB, Hegsted M, Schuette S: Protein-induced hypercalciuria. Fed Proc 1981; 40:2429–2433.

31 Cirillo M, Laurenzi M, Panarelli W, Stamler J: Urinary sodium to potassium ratio and urinary stone disease. Kidney Int 1994;46:1133–1139.

Mineral and Electrolyte Metabolism

Miner Electrolyte Metab 1997;23:269–272

Calciotropic Hormones and Nephrolithiasis

A. D'Angelo
L. Calò
S. Cantaro
S. Giannini

Institute of Internal Medicine,
Division of Nephrology,
University of Padua, Italy

Key Words
Calcitonin
Calcium
Hypercalciuria
Hyperoxaluria
Nephrolithiasis
Parathyroid Hormone
Vitamin D

Abstract
In recurrent calcium stone formers interfering factors or changes in receptor sensitivity may alter the interrelationships among calcium-regulating hormones, and hormonal behavior often does not fit with the theoretical assumptions. The vitamin D system appears to have the most important metabolic and clinical effects. Abnormal up-regulation of the synthesis of calcitriol and the consequent parathyroid hormone (PTH) suppression can induce hypercalciuria. Consequently, the hypocalciuric effect of thiazide would be caused by an enhanced response to PTH and by a reduction in $1,25(OH)_2$-vit D. A negative role of vitamin D on the skeleton has been observed in the presence of a negative calcium balance. Moreover, vitamin D also plays a role in urine oxalate excretion. PTH seems not to be directly stimulated in hypercalciuria and recurrent calcium nephrolithiasis, and patients with hyperparathyroidism and recurrent calcium nephrolithiasis show a similar degree of bone demineralization, irrespective of the presence or absence of the so-called 'primary hyperparathyroidism.' Calcitonin plays a contributory role in the pathogenesis of recurrent calcium nephrolithiasis that seems to be strictly related to dietary calcium intake. A higher sensitivity of thyroid C cells, particularly in absorptive hypercalciuric patients, could be related to the pathogenesis of hypercalciuria and contribute to its persistence.

Calcium nephrolithiasis is a multifactorial disorder that involves a number of contributing factors such as excessive urinary solute concentration, reduced urinary excretion of inhibitors of crystal formation, and changes in urine pH. All these factors represent the so-called lithogen risk which is determined by genetic, metabolic, nutritional and/or environmental components.

Among these factors, hypercalciuria (HPC) is a common feature in stone formers and is therefore one of the most important determinants of nephrolithiasis. The importance of HPC in the pathophysiology of nephrolithiasis is documented by its presence in about 50% of renal stone patients and in more than 70% of those with recurrent calcium nephrolithiasis (RCN). Idiopathic HPC is actually classified as 'diet dependent' or 'absorptive' (AHPC) and 'diet independent' or 'fasting' (FHPC) and it is considered to be a primary metabolic abnormality that results from an abnormal calcium metabolism by calcium-regulating hormones [1–3]. However, the mechanisms involved in the determination of this common metabolic disorder are not yet fully understood and the hormonal behavior often does not fit with the theoretical assumptions. In fact, the interrelationships among the various calcium-regulating hormones to maintain extracellular ionized calcium are only theoretically clarified, and a role for interfering factors or changes in receptor sensitivity cannot be ruled out.

In renal idiopathic HPC stone formers, an oral calcium load tends to increase the calcitriol level and the increased calcium excretion correlates with the circulating vitamin D level; on the other hand parathyroid hormone (PTH) is suppressed. These effects on calcitriol and PTH indicate that HPC could be induced by both an abnormal up-regulation of calcitriol synthesis and by consequent PTH suppression [4]. In contrast, a low-calcium diet effectively reduces urinary calcium even if fasting calcium excretion is apparently not affected by changes in dietary calcium content. On the other hand, a low-calcium diet may induce a substantial increase in bone resorption, that can be reverted by an oral calcium load.

The marked increase in bone resorption induced by a low-calcium diet is not accompanied by apparent signs of increased parathyroid activity or by a significant difference in calcitriol concentration [5]. There are no convincing explanations for this finding. It is possible, however, that a few days (15) of calcium deprivation were not sufficient to stimulate the PTH-vitamin D system, and therefore the increased bone turnover was certainly not dependent on PTH stimulation. On the other hand, a very impressive reduction in calcitonin levels was found when patients were switched from a free-calcium to a low-calcium diet [5], suggesting a relationship between calcitonin and urinary calcium excretion and confirming the possibility that an increased dietary calcium content can stimu-

KARGER
E-Mail karger@karger.ch
Fax +41 61 306 12 34
http://www.karger.ch

0378–0392/97/0236–0269 $12.00/0

Angela D'Angelo, MD
Institute of Internal Medicine, Division of Nephrology
University of Padova
Via Giustiniani, 2, I–35128 Padova (Italy)
Tel. +39 49 8212150, fax +39 49 8212151

late, and its reduction can reduce, calcitonin levels. It is difficult, however, to suggest a primary role for calcitonin in modulating the bone resorption rate in relation to variable dietary calcium content, since an acute oral calcium load does not induce any change in calcitonin levels, and this 'primary' increased bone turnover rate certainly needs a thorough investigation of other aspects such as bone receptor sensitivity to calciotropic hormones and/or local calcium-regulating factors. On the other hand, in patients with idiopathic HPC, we recently observed [6] higher calcitonin levels, with PTH and $1,25(OH)_2$ vit D levels still in the normal range; however, under a low-calcium diet, calcitonin decreased (more in AHPC than in FHPC patients). These observations could indicate a higher sensitivity of thyroid C cells in HPC, and particularly in AHPC; this higher calcitonin production could thus be related to the pathogenesis of hypercalciuria and contribute to its persistence. Increased mobilization of bone calcium has been suggested in idiopathic HPC and a reduced mineral content has been shown in this condition [7–10].

Hypercalciuric stone formers show lower PTH and slightly elevated $1,25(OH)_2$ vit D levels than normocalciuric patients and control subjects with a positive correlation between circulating calcitriol, renal mass, creatinine clearance, protein, and NaCl intakes. This relative hypoparathyroidism in hypercalciuric patients suggests an influence of high protein and sodium intake in up-regulating calcitriol synthesis and in the subsequent relative low levels of PTH [11, 12].

The increased bone resorption and the reduction in bone density observed in patients with FHPC has been ascribed to an increased production of interleukin-1 [13], a potent stimulator of bone resorption that may be involved in the pathogenesis of idiopathic HPC [17]; however, a reduced bone mineral content was also shown in patients with AHPC, suggesting a causative role for high vitamin D levels in the increased bone resorption [14]. Our data on bone mineral content [6] show that the vertebral mineral content of patients with recurrent calcium nephrolithiasis (RCN) was significantly reduced, with FHPC patients more compromised. In these patients, bone demineralization correlated both with plasma pH and increased urinary hydroxyproline excretion, in the absence of PTH and calcitriol changes. Our observations, however, do not rule out a causative role for vitamin D, and could favor a negative role of vitamin D production on the skeleton only in the presence of a negative calcium balance. In contrast, other studies have shown a negative correlation between fasting vitamin D level and urinary calcium excretion, while a positive correlation was instead found between vitamin D and vertebral bone mineral density [15], therefore suggesting possible bone protection exerted by calcitriol.

Additional informations on the hormonal pattern in idiopathic HPC could be obtained by studies on the effect of drugs known for their hypocalciuric action. Thiazide diuretics have been widely used to prevent renal stone recurrences owing to their hypocalciuric action [16] and their positive effect on the calcium balance [17]. Either an enhanced response to PTH or a direct action of hydrochlorothiazide (HCTZ) itself have been suggested as possible mechanisms [18]. However, this drug has also been demonstrated to reduce plasma levels of $1,25(OH)_2$ vit D in idiopathic HPC suggesting more complex effects [19].

Renal prostaglandin E_2 (PGE_2) is one of the factors known to influence urinary calcium excretion; urinary PGE_2, in fact, increased in idiopathic HPC and prostaglandin synthetase inhibitors have been shown to normalize hypercalciuria [20, 21]. The results of acute experiments, however, are difficult to interpret, since urinary PGE_2 is reported to be either increased or unchanged but chronic administration of HCTZ lowered both urinary calcium and PGE_2 excretion [22].

The decrease in plasma $1,25(OH)_2$ vit D we observed after HCTZ [22] is in keeping with the observations reported by others [19], and $1,25(OH)_2$ vit D and PGE_2 metabolites were also correlated before and after HCTZ [22]. The HCTZ-mediated PGE_2 reduction could, in fact, effectively remove a stimulus to 1α hydroxylase activity for conversion of 25(OH) D to $1,25(OH)_2$ vit D, as also suggested by the demonstration that the modifications induced by HCTZ in 25(OH) D and $1,25(OH)_2$ vit D levels were inversely correlated [22]. Finally, in this study we also found an increased HCTZ-mediated PTH level [22]. The following sequence of events could therefore be postulated: inhibition of PGE_2 synthesis by HCTZ results in a reduction in 1α hydroxylase activity causing a fall in plasma $1,25(OH)_2$ vit D. This fall would result in a decrease in gut intestinal Ca^{2+} absorption and a consequent fall in plasma Ca^{2+}. The fall in $1,25(OH)_2$ vit D, together with low plasma Ca^{2+}, stimulates PTH secretion, which results in increased renal calcium reabsorption and lowered urinary calcium excretion [22].

In another study we also confirmed the existence of two different functional responses of RCN patients long term treated with HCTZ [23]. In 'responders,' HCTZ lowers calcium excretion with a small reduction in vitamin D and normal bone metabolism. In 'nonresponders,' the lack of effect on calcium excretion was accompanied by higher values for $1,25(OH)_2$ vit D and no change in PTH, followed by increased bone turnover and lower bone mineral content [23].

Nifedipine, a calcium channel blocker, that also has a hypocalciuric effect, decreased $1,25(OH)_2$ vit D and PGE_2, with no effect on PTH [24]. It was postulated that nifedipine altered the characteristics of mitochondrial calcium transport and, consequently, 1α hydroxylase activity [24].

In idiopathic HPC with RCN, the bone mineral content may be reduced. In HPC we observed a reduced bone density in lumbar spine, with no difference between males and females after thiazide administration [25]. HPC was more evident in males than in females and in the overall hypercalciuric patients it was correlated only with bone turnover indexes, with a lower bone mineral content only in more severe HPC [25].

The changes in biological markers of bone remodeling apparent with thiazides, at variance with the lower skeletal mass observed in females in the normal population [23], confirm that in idiopathic HPC there is a high rate of bone turnover, as assessed by morphological, biochemical, and densitometric evaluations, and the increment in bone mass in both trabecular and cortical sites may be due to a reduction in bone remodeling produced by thiazide. When RCN patients were compared with RCN after long-term HCTZ treatment, the former had significantly higher serum levels of $1,25(OH)_2$ vit D and higher urinary hydroxyproline and calcium excretion [23]. These data seem to suggest that during HCTZ treatment, the rate of recurrence of nephrolithiasis could be influenced by the status of the bone and by even slightly increased levels of $1,25(OH)_2$ vit D. On the other hand, it is also possible that some patients with RCN under long-term thiazide treatment do not show any inhibition of bone resorption which, in turn, may explain the recurrence of nephrolithiasis. Therefore, in RCN with HPC, bone mineral content is significantly reduced and this reduction is even more significant when only FHPC patients are considered [6]. Furthermore, patients with FHPC showed an increased bone remodeling rate and a lower blood pH than AHPC patients. No significant modifications of PTH and calcitriol were observed [6]. Based on these data, a primary increment of bone tunover with prevalent osteoclastic activity can be suggested, probably substained by subclinical metabolic acidosis [6].

To complicate the possible role of PTH in idiopathic hypercalciuria even more, patients with HPC and RCN showed a similar degree of bone demineralization, irrespective of the presence or absence of primary hyperparathyroidism and/or of the correlation with either PTH or $1,25(OH)_2$ vit D [9].

An increasing number of patients with renal stones present primary hyperparathyroidism. The higher frequency of females in both stone formers and non-stone formers hyperparathyroid patients suggests a prevalence of endocrine disorder over kidney stone disease, the latter being more frequent in males. Nevertheless, only in non-stone formers patients with a long duration of disease, the x-ray skeletal survey showed severe and typical bone lesions, suggesting an important difference between the two groups [3]. Stone former patients had serum and urinary calcium levels and serum alkaline phosphatase lower than stone-free patients [3]. A different hormonal pattern could be responsible for the clinical and metabolic differences between these two groups. PTH and $1,25(OH)_2$ vit D levels, however, were higher in both groups than in controls and the 25(OH) vit D concentration apparently had almost the same statistical weight as PTH in determining the circulating $1,25(OH)_2$ vit D levels [3]. Therefore, in hyperparathyroid patients, $1,25(OH)_2$ vit D appears to be stimulated by the same parameters that regulate its serum concentration in normal subjects. Consequently, this hormone does not seem to possess any autonomous role in primary hyperparathyroidism [3]. Our studies do not exclude the possibility that patients with primary hyperparathyroidism and kidney stones may represent well-separated metabolic and clinical entities even if a predominantly cortical osteopenia seems to occur to the same extent and frequency in patients with and without nephrolithiasis [26]. In fact, patients with renal stones have been diagnosed as affected by primary hyperparathyroidism many years after the onset of nephrolithiasis, while non-stone formers are classified as hyperparathyroid patients within a shorter period of time. In addition, the findings of predominant parathyroid hyperplasia and the clinical evidence of recurrence of primary hyperparathyroidism after successful surgery only in stone formers suggest the possibility that the endocrine disorder might be the consequence over time, rather than the cause of nephrolithiasis [3, 27].

As to the origin of recurrent hyperparathyroidism, an attractive hypothesis suggests abnormalities in the PTH gene as the possible explanation for 'de novo' appearance of hyperparathyroidism after successful surgery [28]. It is likely that only slightly elevated levels of PTH in genetically predisposed subjects with a loose cellular calcium control may worsen the disease [29]. Therefore, even if there is no agreement about the existence of secondary hyperparathyroidism in patients with idiopathic HPC, PTH seems not to be primarily involved in the pathogenesis of idiopathic HPC [30].

Besides HPC, idiopathic hyperoxaluria has also demonstrated a relevant role in the pathogenesis of kidney stone formation [31], and it seems that intestinal hyperabsorption of oxalate is more evident in hypercalciuric patients [32]. As the increased intestinal calcium absorption might reduce the binding of calcium to oxalate in the gut, it might enhance the intestinal uptake of free oxalate, thus making more oxalate available for absorption.

In an attempt to evaluate the possible role of the vitamin D system in the pathogenesis of hyperoxaluria, we observed that calcium excretion was not significantly different in patients with associated hyperoxaluria and HPC and those with isolated HPC [1]. In contrast, patients with the combined defect showed significantly higher levels of urine oxalate excretion than those with isolated hyperoxaluria. Further, we found higher vitamin D levels in patients with the combined defect compared with isolated hyperoxaluric patients. 25(OH) D and calcitriol correlated both with urinary calcium and oxalate excretion, supporting a role for vitamin D in the pathogenesis of hyperoxaluria and idiopathic HPC in RCN.

Conclusions

The following major points may summarize the relationships between calcium-regulating hormones in nephrolithiasis. (1) The vitamin D system has the most important metabolic and clinical effects in patients with idiopathic HPC and RCN; in addition, vitamin D has been shown to possess a previously unrecognized role in urinary oxalate excretion. (2) PTH seems to be secondarily stimulated in RCN patients even when renal function is normal; hyperparathyroidism could develop as an adaptation to primary metabolic alterations. (3) Calcitonin plays a contributory role in the pathogenesis of RCN that seems to be strictly correlated with dietary calcium intake; thus a higher production of and sensitivity to calcitonin could be related to the persistence of HPC.

References

1 Giannini S, Nobile M, Castrignano R, Pati T, Tasca A, Villi G, Pellegrini F, D'Angelo A: Possible link between vitamin D and hyperoxaluria in patients with renal stone disease. Clin Sci 1993;84:51–54.

2 Pak CYC, Britton F, Peterson R, Ward D, Northcutt C, Breslan NA, McGuire J, Sakhall K, Bush S, Nicar M, Norman DA, Peter P: Ambulatory evaluation of nephrolithiasis: Classification, clinical presentation and diagnostic criteria. Am J Med 1980;69:19–30.

3 D'Angelo A, Lodetti MG, Giannini S, Castrignano R, Al Awady M, Malvasi L, Fabris A, Maschio G: Hyperparathyroidism: Cause or consequence of recurrent calcium nephrolithiasis? Miner Electrolyte Metab 1992;18:359–364.

4 Hess B, Lippuner K, Takkinen R, Villiger L, Ackermann D, Jaeger P: Effect of oral calcium loading on intact PTH and calcitriol in idiopathic renal calcium stone formers and healthy controls. Nephrol Dial Transplant 1995;10:1614–1619.

5 Messa P, Mioni G, Franzon R, Messa M, Cruciatti A, Giannini S, D'Angelo A: Factors affecting fasting urine calcium excretion in stone former patients on different dietary calcium intake. Scan Microsc 1992;6:239–246.

6 Nobile M, Calò L, Giannini S, Cantaro S, Carraro G, Vianello D, Tasca A, D'Angelo A: Bone mineral density and metabolism in patients with idiopathic hypercalciuria and recurrent calcium nephrolithiasis; in Pak CYC, Resnick MI, Preminger GM (eds): Urolithiasis 96. Dallas, Millet, 1996, pp 488–489.

7 Alhava EM, Junti M, Karjalainen P: Bone mineral density in patients with urolithiasis. Scand J Urol Nephrol 1976;10:154–156.

8 Barkin J, Wilson DR, Manuel MA, Bayley A, Murray T, Harrison J: Bone mineral content in idiopathic calcium nephrolithiasis. Miner Electrolyte Metab 1985;11:19–24.

9 Malvasi L, Sartori L, Giannini S, Al Awady M, Musaio F, Varotto S, D'Angelo A: Mineral metabolism and bone mineral content in calcium nephrolithiasis with and without hyperparathyroidism (abstract). Urol Res 1988;16:190.

10 Borghi L, Meschi T, Guezzo A, Maninetti L, Pedrazzoni M, Marcato A, Vescovi P, Novarini A: Vertebral mineral content in diet-dependent and diet-independent hypercalciuria. J Urol 1991;146:1334–1338.

11 Jaeger P: Relative hupoparathyroidism and calcitriol up-regulation in hypercalciuric calcium renal stone formers: Impact of nutrition. Am J Nephrol 1993;13:18–26.

12 Hess B, Ackermann D, Essig M, Takkinen R, Jaeger P: Renal mass and serum calcitriol in male idiopathic calcium renal stone formers: Role of protein intake. J Clin Endocrinol Metab 1995;80:1916–1921.

13 Pacifici R, Rothstein M, Rifas L, Lau KW, Baylenk DJ, Avioli LV, Hruska K: Increased monocyte interleukin 1 activity and decreased vertebral bone density in patients with fasting idiopathic hypercalciuria. J Clin Endocrinol Metabol 1990;71:138–145.

14 Pietschmann F, Breslan NA, Pak CYC: Reduced vertebral bone density in hypercalciuric nephrolithiasis. J Bone Miner Res 1992;12:1383–1388.

15 Bataille P, Achard JM, Fournier A, Boudailliez B, Westeel PF, El Esper N, Bergot C, Jans I, Lacou JD, Petit J, Henon G, Laral Jeantet MA, Bouillon R, Schert JL: Diet, vitamin D and vertebral mineral density in hypercalciuric calcium stone formers. Kidney Int 1991;39:1193–1205.

16 Yendt ER, Cohanim M: Prevention of calcium stones with thiazides. Kidney Int 1978;13:397–409.

17 Transbol I, Christensen GF, Jensen GF, Christensen C, McNair P: Thiazide for the postponement of postmenopausal bone loss. Metabolism 1982;31:383–386.

18 Brickman AS, Massry SG, Coburn JW: Changes in serum and urinary calcium during treatment with hydrochlorothiazide: Studies on mechanisms. J Clin Invest 1972;51:945–954.

19 Zerwekh JE, Pak CYC: Selective effects of thiazide therapy on serum 1α, 25-dihydroxyvitamin D and intestinal calcium absorption in renal and absorptive hypercalciuria. Metabolism 1980;29:13–17.

20 Betend L, David M, Vincent M, Hermier M, Francois F: Successful indomethacin treatment of 2 pediatric patients with severe tubulopathies. Helv Pediatr Acta 1979;34:339–344.

21 Donker AJ, De Jong PE, Statius van Eps LW, Bretjens JR, Bakker K, Doorebos H: Indomethacin in Bartter's syndrome. Nephron 1977;19:200–213.

22 Calò L, Cantaro S, Marchini F, Giannini S, Castrignano R, Gambaro G, Antonello A, Baggio B, D'Angelo A, Williams HE, Borsatti A: Is hydrochlorothiazide-induced hypercalciuria due to inhibition of PGE_2 synthesis? Clin Sci 1990;78:321–325.

23 Ortalda V, Fabris A, D'Angelo A, Giannini S, Perobelli L, Maschio G: Correlation between recurrence of nephrolithiasis and persistence of bone resorption in patients with recurrent calcium nephrolithiasis on long term thiazide therapy; in Walker VR, Sutton RAL, Bert Cameron EC, Pak CYC, Robertson WG (eds): Urolithiasis. Dallas, Millet, 1989, pp 793–795.

24 Calò L, Cantaro S, Piccoli A, D'Angelo A, Giannini S, Williams HE, Borsatti A: Effect of nifedipine on urinary excretion of calcium and calcium-controlling hormones in essential hypercalciuria. Cardiovasc Drugs Ther 1990;4:983–986.

25 Castrignano R, Pati T, Tronca R, Nobile M, Lodetti MG, Benetollo P, Al Awady M, D'Angelo A: Nephrolithiasis and calcium excretion (abstract). 11th Int Congr Nephrol, Tokyo 1990.

26 Silverberg ST, Shane E, Jacobi TP, Siris ES, Gartemberg F, Seldin D, Clemens TL, Bilezikian JP: Nephrolithiasis and bone involvement in primary hyperparathyroidism. Am J Med 1990;89:327–334.

27 D'Angelo A, Fabris A, Panebianco R, Ortalda V, Sartori L, Giannini S, Maschio G, Crepaldi G: What happens to patients with primary hyperparathyroidism and calcium nephrolithiasis after successful parathyroidectomy? Miner Metab Res Italy 1984;5:23–26.

28 Arnold A, Staunton CE, Goo Kim H, Gaz RD, Kronenberg HM: Monoclonality and abnormal parathyroid hormone genes in parathyroid adenomas. N Engl J Med 1988;318:658–662.

29 Palmieri GM: Calcium, why and how much? Miner Electrolyte Metab 1995;21:236–241.

30 Weisinger JR: New insights into the pathogenesis of idiopathic hypercalciuria: The role of bone. Kidney Int 1996;49:1507–1518.

31 Borsatti A: Calcium oxalate nephrolithiasis: Defective oxalate transport. Kidney Int 1991;39:1283–1298.

32 Marangella M, Fruttero B, Bruno M, Linari F: Hyperoxaluria in idiopathic calcium stone disease: Further evidence of calcium hyperabsorption of oxalate. Clin Sci 1982;63:381–385.

Mineral and Electrolyte Metabolism

Miner Electrolyte Metab 1997;23:273–276

Serum Ferritin and Tissue Iron in Anemic Dialysis Patients

Peter Bárány[a]
Lennart C. Eriksson[b]
Rolf Hultcrantz[c]
Erna Pettersson[a]
Jonas Bergström[a–d]

[a] Division of Renal Medicine, Department of Clinical Science,
[b] Department of Clinical Pathology,
[c] Section for Gastroenterology, Department of Internal Medicine, and
[d] Baxter Novum, Department of Clinical Science, Huddinge and Karolinska University Hospitals, Karolinska Institute, Stockholm, Sweden

Key Words
Anemia
Chronic renal failure
Ferritin
Hemodialysis
Hemosiderosis
Iron
Uremia

Abstract
The relationship between serum ferritin and tissue iron was investigated in 26 dialysis patients (17 hemodialysis patients, 9 chronic peritoneal dialysis patients) with anemia (median hemoglobin 74 g/l, range 56–92 g/l). Serum ferritin ranged from 18 to 9,435 μg/l (median 450 μg/l). Tissue iron was assessed in the liver biopsies of 4 hemodialysis patients with iron overload (serum ferritin 1,150–9,435 μg/l), in the muscle biopsies of 5 patients with serum ferritin 170–9,435 μg/l, and in bone marrow aspirations (semiquantitative assessment). The mean liver iron concentration was 15.4 ± 8.0 μg Fe/mg protein (mean $\pm$ SD), which is similar to that previously found in patients with untreated idiopathic hemochromatosis. Four patients with serum ferritin 170–620 μg/l had muscle iron concentrations (0.33 ± 0.10 μg Fe/mg protein) similar to those found in controls (0.23 ± 0.10, means $\pm$ SD). One patient with serum ferritin 9,435 μg/l had a markedly increased muscle iron concentration (1.3 μg Fe/mg protein). The bone marrow iron was assessed as negative in 3 patients (serum ferritin 44–85 μg/l), positive in 8 (serum ferritin 18–379 μg/l), increased in 11 patients (serum ferritin 222–4,210 μg/l), and was markedly increased in 2 patients (serum ferritin 4,550 and 9,435 μg/l). Bone marrow iron correlated significantly with serum ferritin concentrations (spearman rank correlation coefficient rho – 0.89, $p < 0.001$). These results show that in dialysis patients with a stable iron balance and unstimulated erythropoiesis, i.e., patients without erythropoietin treatment and parenteral iron, serum ferritin is a useful indicator of iron stores. Our findings also suggest that the relationship between tissue iron and serum ferritin levels in end-stage renal disease is altered, i.e., a relative increase in serum ferritin levels unrelated to iron stores is observed in dialysis patients.

Introduction

The assessment of iron status is often difficult in chronic renal failure and there is no reliable single marker of iron status [1]. The most frequently used markers for monitoring iron metabolism are serum ferritin, serum iron, and transferrin saturation. Serum ferritin is secreted mainly by the reticuloendothelial tissue, and, in equilibrium, it reflects iron stores in both healthy subjects [2] and in patients with chronic renal failure [3–7]. In certain diseases such as infection, inflammatory disease, and liver disease there is an increase in serum ferritin levels which is unrelated to iron stores [2]. Similarly, the relationship between serum ferritin and iron stores in the tissues may be altered in chronic renal failure.

The aim of the present study was to measure iron stores in muscle, liver, and bone marrow in anemic dialysis patients (without erythropoietin treatment or iron supplements) and to compare these with serum ferritin measurements.

Material and Methods

Patients

Twenty-six dialysis patients (17 females, 9 males, median age 55 years, range 24–73 years) with anemia (median hemoglobin 73 g/l, range 56–92) participated in the study. Seventeen patients were on maintenance hemodialysis treatment and nine on continuous peritoneal dialysis (7 continuous

KARGER
E-Mail karger@karger.ch
Fax +41 61 306 12 34
http://www.karger.ch

0378–0392/97/0236–0273 $12.00/0

Dr. Peter Bárány
Division of Renal Medicine
Huddinge University Hospital
S-141 86 Huddinge (Sweden)
Tel. +46 8 746 10 00, fax +46 8 711 47 42

ambulatory, 2 continuous cyclic). The peritoneal dialysis patients were significantly older [median age 66 (50–73) years] than the hemodialysis patients [median age 48 (24–73) years, $p<0.01$]. The primary renal diagnoses were chronic glomerulonephritis (11 patients), chronic pyelonephritis (4 patients), diabetic nephropathy (1 patient), and other diagnoses in 10 patients. Three patients were bilaterally nephrectomized. The median time on dialysis treatment was 27 (range 5–120) months. The patients were studied before starting erythropoietin treatment and none of the patients were on iron therapy. The median number of units of erythrocyte concentrate transfused was 18 (range 0–90). During the 6 months prior to the study, a median of 4 (range 0–17) units had been transfused. The study was approved by the Ethics Committee of the Karolinska Institute. The patients participated voluntarily and were all fully informed about the purpose of the study and the risks and discomfort associated with the experiments.

Bone Marrow Examinations

Bone marrow aspiration was performed in 24 patients. Iron content was determined after Perl's blue iron staining by light microscopic examination of 5-μm sections from bone marrow aspirates, as well as in bone marrow fragments in smear preparations. Bone marrow iron content was semiquantitatively assessed as negative, positive (+), markedly increased (++), or markedly increased as well as iron positive in endothelial cells (+++).

Liver Biopsy Specimens

Four hemodialysis patients with iron overload (serum ferritin concentrations 1,150–9,435 μg/l) underwent liver investigations. Percutaneous liver biopsy, using a Menghini needle (1.6 mm in diameter), and light microscopy examination were performed in 2 patients (No. 6 and 8). A percutaneous fine-needle aspiration was performed in 2 other patients (No. 5 and 7). Iron content was determined after homogenization with the aid of a Pye Unicam atomic absorption spectrophotometer (SP 2900). Protein content was determined using the Lowry method [8]. The results were compared with a previous analysis [9, 10] of liver iron content in controls and in patients with idiopathic hemochromatosis.

Muscle Biopsies

In 5 hemodialysis patients whose serum ferritin concentrations ranged between 172 and 9,435 μg/l, biopsy samples of the quadriceps femoris muscle were obtained under local anesthesia by percutaneous needle biopsy [11]. Iron and protein content in the muscle specimens were determined using methods similar to those used for the liver biopsy specimens. Control values were obtained from an analysis of muscle tissue from autopsy material ($n=10$).

Blood Chemistry

Serum values of hemoglobin, iron, transferrin, and ferritin were determined using routine methods.

Statistics

The spearman rank correlation coefficient (rho) was used for analysis of correlations.

Results

Bone Marrow (fig. 1)

Bone marrow aspirations were obtained from 24 patients. The bone marrow iron was assessed as negative in 3 patients (serum ferritin 44–85 μg/l), positive (+) in 8 patients (serum ferritin 18–379 μg/l), increased (++) in 11 patients (serum ferritin 222–4,210 μg/l), and was markedly increased (+++) in 2 patients (serum ferritin 4,550 and 9,435 μg/l). The iron content in the bone marrow was significantly correlated with the serum ferritin level (spearman rank correlation coefficient $rho=0.89$, $p<0.001$; fig. 1).

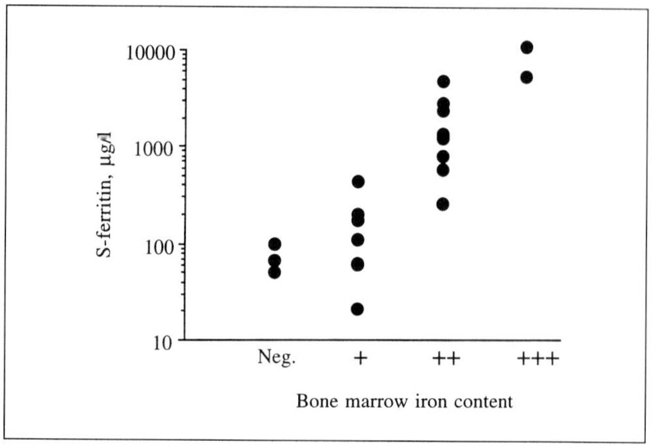

Fig. 1. Bone marrow iron content semiquantitatively assessed in relation to serum ferritin in 28 anemic dialysis patients. The correlation is significant ($rho=0.89$, $p<0.001$). + = Positive iron; ++ = markedly increased iron; +++ = markedly increased iron plus iron positive in endothelial cells.

Muscle and Liver (table 1)

The iron contents in muscle and liver are shown in table 1. The muscle iron concentrations were similar to the control values (0.23 ± 0.10) in 4 of the patients (0.33 ± 0.10 μg Fe/mg protein, mean ± SD) with serum ferritin values ranging from 172 to 620 μg/l. In patient No. 8, who had a serum ferritin concentration of 9,435 μg/l, the muscle (1.30 μg Fe/mg protein) and liver iron (26.7 μg Fe/mg protein) concentrations were high. During 10 years on maintenance hemodialysis treatment, this 24-year-old female had received approximately 90 transfusion units. Splenectomy had been performed 6 years before the study because of hypersplenism. Histological evidence of hemosiderosis was found in the liver and bone marrow. There were no histological signs of fibrosis or cirrhosis in the liver. Computed tomography showed a slightly enlarged liver with high attenuation (80 Hounsfield units). In addition, a bone biopsy revealed hemosiderosis in the skeleton. The three remaining patients (No. 5–7) with serum ferritin levels ranging between 1,150 and 4,550 μg/l had liver iron concentrations (8.7–15.3 μg Fe/mg protein) similar to those found in patients with untreated idiopathic hemochromatosis (10.4 ± 4.8 μg Fe/mg protein, mean ± SD) [9, 10]. Light microscopy examination in patient No. 6 revealed iron excess, but no histological evidence of fibrosis or cirrhosis in the liver.

Serum Iron and Transferrin Saturation

Both the serum iron level (median 13 μmol/l, range 5–44 μmol/l) and the tranferrin saturation (median 27%, range 8–74%) correlated with the bone marrow iron content ($rho=0.61$ and 0.64, respectively, $p<0.005$). The individual values of patients No. 1–8 are shown in table 1.

Table 1. Laboratory data and concentrations of iron in muscle, liver, and bone marrow in 8 anemic hemodialysis patients

Patient No.	age, sex	Transfusions before study units	Serum ferritin µg/l	Serum iron µmol/l	Transferrin saturation %	Muscle iron µg Fe/mg protein	Liver iron µg Fe/mg protein	Bone marrow iron[a]
1	34, F	17	172	18	35	0.25	nd	+
2	34, F	22	222	14	25	0.25	nd	++
3	24, M	30	620	20	38	0.35	nd	nd
4	38, F	62	500	8	15	0.45	nd	++
5	57, M	90	1,150	15	34	nd	10.9	nd
6	68, M	18	1,170	7	16	nd	15.3	++
7	64, F	75	4,550	23	46	nd	8.7	+++
8	24, F	90	9,435	44	66	1.30	26.7	+++
Controls SD			<200	9–38	<65	0.23±0.10 (n=10)	1.4±0.8	
IHC mean SD			1,155±759	40±7	81±10		10.4±4.8	

nd=Not determined; IHC=idiopathic hemochromatosis [see ref. 9].
[a] +=Positive; ++=markedly increased; +++=markedly increased and positive in endothelial cells.

Discussion

Multitransfused patients have been reported to be at risk of complications from hemosiderosis in different organs [12–14]. The findings in the present study demonstrate that previously transfusion-dependent dialysis patients with high serum ferritin levels (>1,000 µg/l) have liver iron concentrations similar to those reported earlier in patients with untreated idiopathic hemochromatosis [9, 10]. Light microscopy examination of the liver tissue was performed in two patients (No. 6 and 8) and did not reveal any signs of fibrosis or cirrhosis. However, one of these patients was young and liver cell damage usually develops slowly after many years of hemosiderosis. Previously, Cecchin et al. [12] reported slightly lower liver iron concentrations in hemodialysis patients with iron overload than in patients with idiopathic hemochromatosis. Only 4/19 of their hemodialysis patients had serum ferritin >1,000 µg/l. Twelve of their patients with iron overload had liver fibrosis, as compared to 2/19 hemodialysis patients without iron overload. In an autopsy study of 18 hemodialysis patients with iron excess, Ali et al. [13] found no cytological evidence of liver cell damage. Finally, Stanbaugh et al. [15] found liver fibrosis in only 1 out of 8 hemodialysis patients with iron overload. The fibrosis was probably a consequence of transfusion-acquired hepatitis. Considering these findings, although the risk of liver cell damage exists, it seems to be lower in transfusional hemosiderosis in dialysis patients than in idiopathic hemochromatosis. Different mechanisms causing liver cell damage may be of importance [10].

The iron concentration in muscle was normal in 4 patients with normal to slightly elevated serum ferritin and was markedly increased in patient No. 8 (table 1) who had a very high serum ferritin (9,435 µg/l). Iron overload in the muscle of hemodialysis patients has been suggested as a cause of myopathy [16].

Serum ferritin is considered to be the best parameter for assessment of iron stores [2–6]. Although it is not very exact in patients with chronic renal failure, serum ferritin was significantly correlated with the bone marrow iron content (fig. 1). However, the relationship between tissue iron and serum ferritin levels in end-stage renal disease seems to be altered, i.e., a relative increase in serum ferritin levels unrelated to iron stores was observed in the dialysis patients. In the patients with negative iron stains, serum ferritin ranged from 44 to 85 µg/l, which is within the normal range for healthy subjects. Patient No. 8, who had the highest iron concentrations in muscle and liver also had the highest serum ferritin value.

The correlation between serum ferritin and iron stores exists only when the iron balance is in equilibrium [1]. When the iron balance is unstable, as during stimulation of erythropoiesis by recombinant human erythropoietin and during treatment with large doses of parenteral iron, serum ferritin is an unreliable tool for monitoring iron metabolism [1, 2]. Accordingly, transferrin saturation is used as an additional routine parameter and values below 20% have been suggested to indicate functional iron deficiency [17, 18] despite adequate serum ferritin concentrations.

Acknowledgments

This study was supported by grants from the Swedish Society of Nephrology, the Swedish Society for Medical Research, the Swedish Medical Research Council (project No. 9127) and the National Association of Kidney Patients in Sweden. We thank Ms. Elsy Digréus and Ms. Ulla Pettersson for expert technical assistance.

References

1 Hörl WH, Cavill I, Macdougall IC, Schaefer RM, Sunder-Plassman G: How to diagnose and correct iron deficiency during r-huEPO therapy – A consensus report. Nephrol Dial Transplant 1996;11: 246–250.

2 Wick M, Pinggera W, Lehmann P: Ferritin in Iron Metabolism. Diagnosis of Anemias, ed 2. Berlin, Springer, 1994.

3 Blumberg AB, Marti HRM, Graber CG: Serum ferritin and bone marrow iron in patients undergoing continuous ambulatory peritoneal dialysis. JAMA 1983;250:3317–3319.

4 Bell JD, Kincaid WR, Morgan RG, Bunce III H, Alperin JB, Sarles HE, Remmers AR Jr: Serum ferritin assay and bone marrow iron stores in patients on maintenance hemodialysis. Kidney Int 1980;17:237–241.

5 Milman N, Bangsbøll S, Strandberg Pedersen N, Visfeldt J: Serum ferritin in non-dialysis patients with chronic renal failure: Relation to bone marrow iron stores. Scand J Haematol 1983;30:337–344.

6 Van de Vyver FL, Vanheule AA, Majelyne WM, D'Haese P, Blockx PP, Bekaert AB, Buyssens N, Keersmaecker W de, Broe ME de: Serum ferritin as a guide for iron stores in chronic hemodialysis patients. Kidney Int 1984;26:451–458.

7 Van Wyck DB, Stivelman JC, Ruiz J, Kirlin LF, Katz MA, Ogden DA: Iron status in patients receiving erythropoietin for dialysis-associated anemia. Kidney Int 1989;35:712–716.

8 Lowry OH, Rosebrough NJ, Farr AL, Randall RJ: Protein measurement with the Folin phenol reagent. J Biol Chem 1951;193:265–275.

9 Hultcrantz R, Angelin B, Björn-Rasmussen E, Ewerth S, Einarsson K: Biliary excretion of iron and ferritin in idiopathic hemochromatosis. Gastroenterology 1989;96:1539–1545.

10 Stål P, Glaumann H, Hultcrantz R: Liver cell damage and lysosomal iron storage in patients with idiopathic hemochromatosis. J Hepatol 1990;11: 172–180.

11 Bergström J: Muscle electrolytes in man: Determination by neutron activation analysis on needle biopsy specimens: A study on normal subjects, kidney patients and patients with chronic diarrhoea. Scand J Clin Lab Invest 1962;14(suppl 68):1–110.

12 Cecchin E, De Marchi S, Querin F, Marin MG, Fiorentino R, Tesio F: Efficacy of hepatic computed tomography to detect iron overload in chronic hemodialysis. Kidney Int 1990;37:943–950.

13 Ali M, Fayemi AO, Rigolosi R, Frascino J, Marsden T, Malcolm D: Hemosiderosis in hemodialysis patients: An autopsy study of 50 cases. J Am Med Assoc 1980;244:343–345.

14 Schafer AI, Cheron RG, Dluhy R, Cooper B, Gleason RE, Soeldner JS, Bunn HF: Clinical consequences of acquired transfusional iron overload in adults. N Engl J Med 1981;304:319–324.

15 Stanbaugh GH, Gillit DM, Holmes AW: Dialysis hemosiderosis mimicking non-A, non-B hepatitis. ASAIO Trans 1984;30:217–221.

16 Bregman H, Gelfand MC, Winchester JF, Manz HJ, Knepshield JH, Schreiner GE: Iron-overload-associated myopathy in patients on maintenance haemodialysis: A histocompatibility-linked disorder. Lancet 1980;ii:882–885.

17 Eschbach JW, Adamson JW: Recombinant human erythropoietin: Implications for nephrology. Am J Kidney Dis 1988;11:203–209.

18 Eschbach JW: The anemia of chronic renal failure: Pathophysiology and the effects of recombinant erythropoietin. Kidney Int 1989;35:134–148.

Mineral and Electrolyte Metabolism

Miner Electrolyte Metab 1997;23:277–282

Irena Pietrzak
Kazimierz Bączyk

Department of Nephrology,
University of Medical Sciences,
Poznań, Poland

Comparison of the Thiamine Level in Blood and Erythrocyte Transketolase Activity in Hemodialyzed and Nondialyzed Patients during Recombinant Human Erythropoietin Therapy

Key Words
End-stage renal disease
Erythrocyte transketolase activity
Erythropoietin
Hemodialysis
Thiamine

Abstract
Thiamine and erythrocyte transketolase activity (ETKA) disturbances in end-stage renal disease are caused mainly by uremia and dialysis treatment. We examined whether recombinant human erythropoietin (rhEPO) can correct these abnormalities in uremic patients.

Thirteen hemodialysis (HD) and 12 nondialyzed (ND) anemic patients showed decreased free and total thiamine levels in plasma and in erythrocytes and decreased ETKA when compared to 20 healthy subjects. Thiamine blood levels (μmol/l) were determined using a fluorimetric technique, and ETKA (μmol/l per minute) was assessed with a photocolorimetric method. Over 20 weeks of study, rhEPO was given intravenously for 8 weeks at 50 UI/kg body weight (BW) three times a week, and subcutaneously for 4 weeks at 25 UI/kg BW, twice a week, and for the last 8 weeks at 25 UI/kg BW once a week. The correction of anemia was associated with an increase in plasma thiamine and erythrocyte total thiamine as well as ETKA in HD patients and with an increase in erythrocyte total thiamine in ND patients only during the period of intravenous infusions.

Introduction

Treatment with recombinant human erythropoietin (rhEPO) enhances erythropoiesis with a subsequent increase in the blood hemoglobin content [1]. Beside these main effects, rhEPO has been useful in the treatment of some metabolic disorders. Moreover, recent studies have shown an impact of EPO on lipid and carbohydrate metabolism, on the immune system and on endocrinological functions [2–8].

Impaired metabolism in end-stage renal disease (ESRD) patients may contribute to thiamine deficiency in blood and to the deterioration of transketolase activity [9–12]. It is worth mentioning that thiamine is utilized as a cofactor for oxidation of pyruvic acid and α-ketoglutaric acid in the TCA cycle and also as a coenzyme of transketolase in the pentose phosphate pathway of nonoxidative glycolysis [13]. It is known that in uremic conditions, chronic thiamine deficiency and/or diminished utilization of thiamine causes a decrease in erythrocyte transketolase activity (ETKA) [14–17]. The thiamine status in hemodialysis (HD) and nondialysis (ND) patients depends on various factors, including residual renal function, vitamin intake including vitamin supplements, uremic toxins, and, in HD patients, losses into dialysate [18]. The influence of rhEPO therapy in predialyzed patients on thiamine in blood and on ETKA has been described in our previous paper [19].

The purpose of this investigation was therefore to determine to what extent rhEPO therapy in HD and ND patients influences the thiamine level in blood and the ETKA. Additionally, in these two groups of ESRD patients, we evaluated possible correlations between hemoglobin (Hb), total protein level ($Prot_{tot}$), thiamine levels in erythrocytes and plasma, and ETKA.

Material and Methods

Thirteen HD and 12 ND patients with ESRD participated in the study. Twenty normals served as controls (table 1). Before starting the rhEPO therapy serum iron levels and total iron-binding capacity were found to be within the limits of normal. There were no signs and symptoms of vitamin B_{12} or folic acid deficiency or aluminum intoxication. During the study all

KARGER
E-Mail karger@karger.ch
Fax +41 61 306 12 34
http://www.karger.ch

0378–0392/97/0236–0277 $12.00/0

Irena Pietrzak, MD
Department of Nephrology, University of Medical Sciences
Przybyszewskiego 49 Str.,
PL–60-355 Poznań (Poland)
Tel. +48 61 671961, fax +48 61 671232

Table 1. General description of studied subjects

Group of studied subjects	Number of subjects	Age, years[a]	Male	Female	Diagnosis			Duration of dialysis, months[a]	Duration of ESRD before enrolling in the study, months[a]
					GN	PyN	PKD		
HD	13	49.30±7.58	7	6	7	5	1	48.9±6.5	58.5±9.3
ND	12	43.00±6.67	2	10	4	6	2	–	42.30±5.6
Normals	20	41.10±7.96	5	15	–	–	–	–	–

GN = Glomerulonephritis; PyN = pyelonephritis; PKD = polycystic kidney disease.
[a] Mean ± SD.

Table 2. rhEPO administration

Period of study	Therapy duration weeks	Dose of rhEPO, UI/kg body weight	Frequency of rhEPO administration per week	Route of rhEPO administration
1	8	50	3	intravenous
2	4	25	2	subcutaneous
3	8	25	1	subcutaneous

Table 3. Biochemical characteristics of studied subjects before starting rhEPO therapy (mean ± SD)

Group of studied subjects	EFTh μmol/l	ETTh μmol/l	PFTh μmol/l	PTTh μmol/l	ETKA μmol/dl per minute	Hb g/d	$Prot_{tot}$ g/dl
HD	0.087 ±0.022	0.124 0.018	0.131 ±0.022	0.235 ±0.050	2.14±0.35**	6.62*** ±0.69	6.09** ±0.60
DN	0.075 ±0.017	0.115 ±0.113	0.145 ±0.023	0.253 ±0.045	1.69±0.44**	7.46** ±0.69	6.66 ±0.58
Normals	0.092 ±0.042	0.139 ±0.056	0.156 ±0.065	0.279 ±0.117	2.59±0.26	12.94 ±0.97	6.88 ±0.46

** $p<0.01$, *** $p<0.001$.

patients received ferrous sulfate supplementation (100 mg up to twice a day) intramuscularly or orally.

The whole study, lasting 20 weeks, comprised three periods: 8, 12 and 20 weeks. rhEPO was given intravenously for 8 weeks and then subcutaneously for 12 weeks (table 2). Venous blood was drawn before starting rhEPO administration and at the end of each period of study. The free and total thiamine levels (μmol/l) in erythrocytes (EFTh, ETTh) and in plasma (PFTh, PTTh) were determined using the fluorimetric method of Blum and Merkel [20]. The TTh comprises FTh and Th bound to phosphates. After separation of the nonsoluble impurities, the released FTh is oxidized in alkaline solution with potassium hexacyanoferrate into thiochrome, which is then extracted from the aqueous solution using isobutanol. Similarly, for the separation of Th from the phosphate, we added takadiastase (10%) to get FTh. The sum of the two fractions of FTh represents TTh.

ETKA (μmol/dl per minute) was determined in red cell hemolysate using the photocolorimetric method according to Bruns et al. [21] and Dische [22]. It is based on the reaction transferring glycolaldehyde from *D*-xylulose-5-phosphate to *D*-ribose-5-phosphate. This reaction generates sedoheptulose-7-phosphate, whose rate of generation is considered to be a measure of transketolase activity. Hb and $Prot_{tot}$ were measured using routine methods. The results were statistically analyzed with a Student's t test and Pearson's test for the coefficient of linear correlation.

Results

Prestudy mean values of EFTh, ETTh, PFTh, PTTh (μmol/l) in HD and ND patients tended to be lower but did not differ significantly from those of normals. The prestudy mean ETKA value (μmol/dl per minute) in HD and ND patients was significantly lower than in normals ($p<0.01$, $p<0.001$, respectively). The mean Hb content (g/dl) at the beginning of the study in both patient groups was significantly lower than in normals ($p<0.001$). The plasma $Prot_{tot}$ levels before rhEPO therapy were significantly lower only in the HD group ($p<0.01$) (table 3).

After 8 weeks of intravenous rhEPO therapy in HD patients, ETTh, PFTh, PTTh levels, and ETKA values increased significantly ($p<0.001$ and $p<0.01$). The EFTh level in HD patients, before starting rhEPO administration, was nearly normal and did not change significantly during the period of study. In ND patients, ETTh levels and ETKA values rose significantly ($p<0.01$ and $p<0.001$, respectively).

The next 4 weeks of rhEPO therapy resulted in a significant increase in ETTh levels in HD patients ($p<0.05$) and in EFTh levels in ND patients ($p<0.05$). The last 8 weeks of this therapy did not present any significant change in thiamine levels in blood or ETKA values in HD patients. In ND patients, ETTh and PTTh levels and ETKA values increased significantly ($p<0.01$ and $p<0.05$, respectively) (tables 4, 5). Hb content rose significantly during each period of study in both groups ($p<0.001$). Plasma total protein increased significantly in HD and ND patients during rhEPO therapy (table 6).

Table 4. Thiamine level in blood and ETKA value in HD patients during rhEPO therapy (mean ± SD and statistical evaluation)

	Before rhEPO	Weeks of rhEPO therapy 8	12	20
EFTh µmol/l	0.087 ± 0.024	0.096 ± 0.015	0.092 ± 0.028	0.093 ± 0.027
(vs previous)		NS	NS	NS
(vs two steps earlier)			NS	NS
(vs before rhEPO)				NS
ETTh µmol/l	0.124 ± 0.018	0.144 ± 0.019	0.148 ± 0.019	0.148 ± 0.018
(vs previous)		$p < 0.001$	$p < 0.05$	NS
(vs two steps earlier)			$p < 0.001$	$p < 0.05$
(vs before rhEPO)				$p < 0.001$
PFTh µmol/l	0.131 ± 0.022	0.153 ± 0.016	0.153 ± 0.016	0.155 ± 0.019
(vs previous)		$p < 0.001$	NS	NS
(vs two steps earlier)			$p < 0.001$	NS
(vs before rhEPO)				$p < 0.001$
PTTh µmol/l	0.235 ± 0.05	0.256 ± 0.036	0.261 ± 0.038	0.269 ± 0.036
(vs previous)		$p < 0.001$	NS	NS
(vs two steps earlier)			$p < 0.001$	NS
(vs before rhEPO)				$p < 0.001$
ETKA µmol/dl per minute	2.14 ± 0.35	2.37 ± 0.27	2.45 ± 0.28	2.46 ± 0.30
(vs previous)		$p < 0.01$	NS	NS
(vs two steps earlier)			$p < 0.01$	NS
(vs before rhEPO)				$p < 0.001$

NS = Nonsignificant.

The differences in changes in the parameters studied in the course of rhEPO therapy are more distinctly visible when the increases are expressed as the percentage of their prestudy values (fig. 1–7).

In HD patients, a correlation was found between EFTh and ETKA ($r = 0.88$), Hb and EFTh ($r = 0.89$) before the rhEPO therapy only, and between Hb and PFTh before ($r = 0.59$) and after 8and 12 weeks of this therapy ($r = 0.58$). On the other hand, a correlation between Hb and ETKA was observed before ($r = 0.87$) administration of rhEPO.

The correlation between plasma $Prot_{tot}$ and PFTh ($r = 0.66$) before administration of rhEPO, and after 8 ($r = 0.58$), 12, and 20 weeks of this therapy between plasma $Prot_{tot}$ and the ETTh level was statistically significant ($r = 59$ and $r = 0.78$ respectively). In ND patients after 8, 12, and 20 weeks of study, a statistically significant correlation was found only between Hb and PFTh ($r = 0.58$, $r = 0.72$, $r = 0.64$ respectively).

Discussion

In our study, the correction of anemia was associated with an increase in thiamine levels in blood and ETKA in HD as well as in ND patients. During the intravenous treatment period of study, we observed a significant increase in ETKA values concomitant with increased ETTh levels in HD and ND patients, in contrast to the EFTh levels in both study groups. The EFTh level in HD patients was almost normal before starting rhEPO and did not change throughout the study. The only significant increase in PFTh and PTTh levels that we observed in the HD patients occurred during the intravenous administration of rhEPO. In ND patients, PFTh and PTTh levels were higher than in HD patients, and only PTTh increased significantly during rhEPO therapy in this group.

The periods of subcutaneous therapy in HD patients significantly influenced the ETTh level only. In ND patients with the subcutaneous route of rhEPO administration, the significant increase in erythrocyte thiamine levels and transketolase activity still occurred. The impact of the intravenous route of rhEPO administration was more apparent than with the subcutaneous administration with regard to the restoration of the decreased thiamine levels and ETKA values to normal in the HD patients.

During the intravenous and subcutaneous periods of rhEPO therapy, Hb levels and plasma $Prot_{tot}$ content rose significantly. This contrasted with the slight influence of subcutaneously administered rhEPO on blood thiamine levels and ETKA in the HD patients. ND patients presented significant increases in Hb levels during

Table 5. Thiamine level in blood and ETKA value in ND patients during rhEPO therapy (mean ± SD and statistical evaluation)

	Before rhEPO	Weeks of rhEPO therapy: 8	12	20	Before vs 8	8 vs 12	12 vs 20	Before vs 12	8 vs 20	Before vs 20
EFTh μmol/l	0.075 ± 0.017	0.078 ± 0.027	0.092 ± 0.014	0.096 ± 0.015	NS	$p < 0.05$	NS	$p < 0.01$	$p < 0.05$	$p < 0.001$
ETTh μmol/l	0.115 ± 0.013	0.121 ± 0.011	0.124 ± 0.012	0.129 ± 0.013	$p < 0.01$	NS	$p < 0.01$	$p < 0.001$	$p < 0.001$	$p < 0.001$
PFTh μmol/l	0.145 ± 0.023	0.146 ± 0.016	0.148 ± 0.013	0.148 ± 0.015	NS	NS	NS	NS	NS	NS
PTTh μmol/l	0.253 ± 0.045	0.258 ± 0.045	0.260 ± 0.045	0.264 ± 0.046	NS	NS	$p < 0.05$	$p < 0.05$		$p < 0.05$
ETKA μmol/dl per minute	1.96 ± 0.44	1.94 ± 0.35	1.98 ± 0.35	2.17 ± 0.29	$p < 0.001$	NS	$p < 0.05$	$p < 0.001$	$p < 0.01$	$p < 0.001$

NS = Nonsignificant.

Table 6. Hb and Rot_{tot} levels in HD and ND patients during rhEPO therapy (mean ± SD and statistical evaluation)

	Before rhEPO	Weeks of rhEPO therapy: 8	12	20	Before vs 8	8 vs 12	12 vs 20	Before vs 12	8 vs 20	Before vs 20
HD Hb g/dl	6.62 ± 0.53	8.76 ± 0.70	10.10 ± 0.91	11.67 ± 0.57	$p < 0.001$	$p < 0.001$	$p < 0.001$	$p < 0.001$	$p < 0.001$	$p < 0.001$
$Prot_{tot}$ g/dl	6.13 ± 0.60	6.53 ± 0.56	6.63 ± 0.52	6.71 ± 0.55	$p < 0.001$	$p < 0.05$	$p < 0.05$	$p < 0.001$	$p < 0.01$	$p < 0.001$
ND Hb g/dl	7.46 ± 0.69	10.27 ± 1.02	11.56 ± 0.86	12.56 ± 0.66	$p < 0.001$	$p < 0.001$	$p < 0.001$	$p < 0.001$	$p < 0.001$	$p < 0.001$
$Prot_{tot}$ g/dl	6.66 ± 0.58	7.10 ± 0.36	7.21 ± 0.36	7.33 ± 0.35	$p < 0.001$	$p < 0.01$	$p < 0.01$	$p < 0.001$	$p < 0.001$	$p < 0.001$

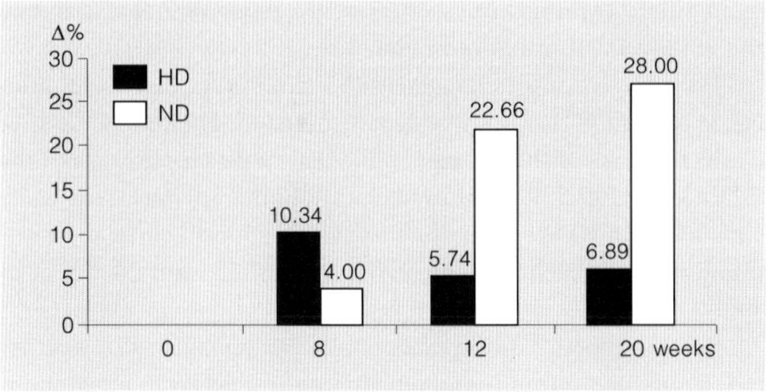

Fig. 1. Differences in mean levels of EFTh during rhEPO therapy expressed as the percentage of the initial value regarded as 100 (Δ%).

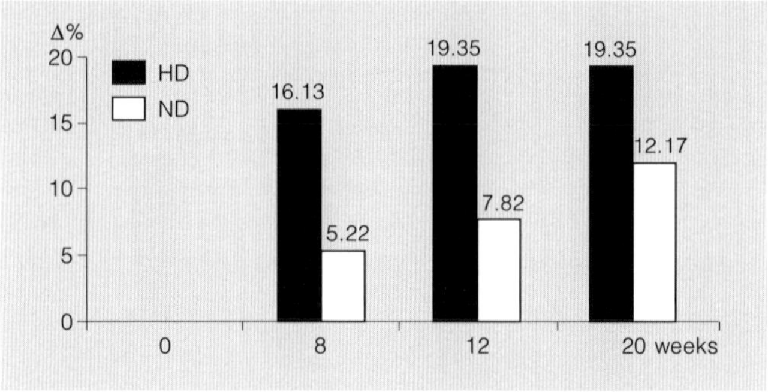

Fig. 2. Differences in mean levels of ETTh during rhEPO therapy expressed as the percentage of the initial value regarded as 100 (Δ%).

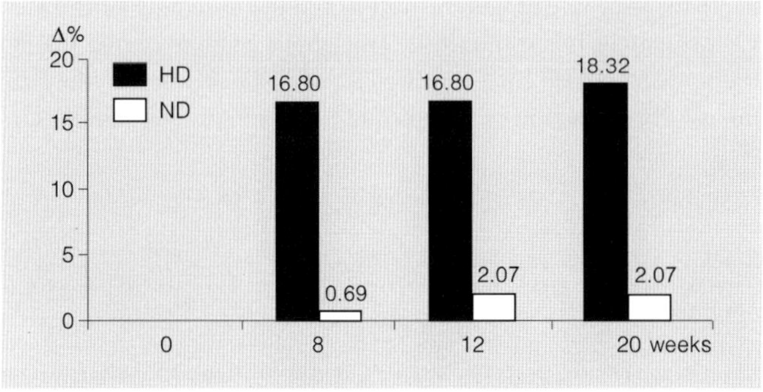

Fig. 3. Differences in the mean levels of PFTh during rhEPO therapy expressed as the percentage of the initial value regarded as 100 (Δ%).

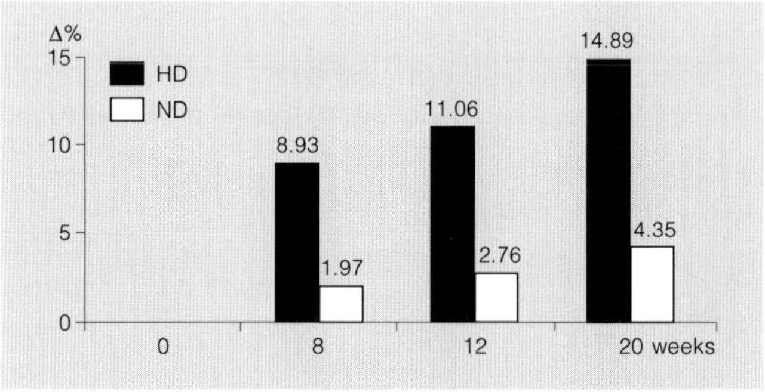

Fig. 4. Differences in the mean levels of PTTh during rhEPO therapy expressed as the percentage of the initial value regarded as 100 (Δ%).

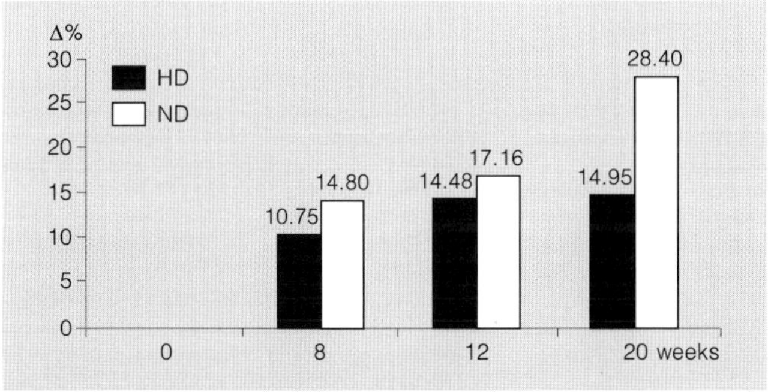

Fig. 5. Differences in the mean levels of ETKA during rhEPO therapy expressed as the percentage of the initial value regarded as 100 (Δ%).

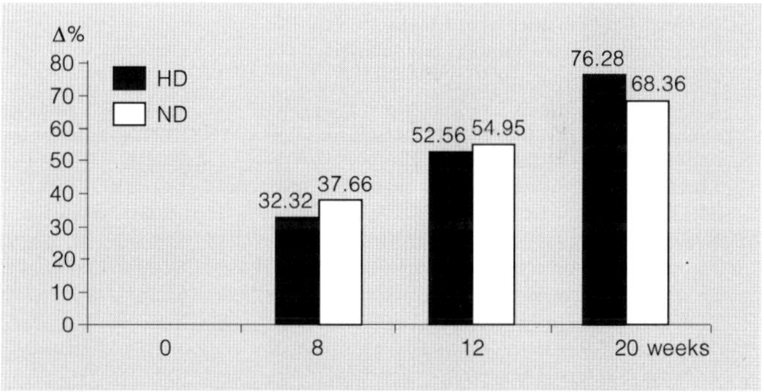

Fig. 6. Differences in the mean levels of Hb during rhEPO therapy expressed as the percentage of the initial value regarded as 100 (Δ%).

the subcutaneous periods of study together with increasing EFTh, ETTh levels and ETKA values.

The increase in ETKA values and erythrocyte thiamine levels can be explained by improved oxygenation during rhEPO treatment, which ameliorates the activity of the pentose phosphate shunt. Such an explanation is supported by an interesting study demonstrating that rhEPO therapy stimulates intracellular energy metabolism. The intraerythrocyte concentrations of 2,3,-diphosphoglycerate and ATP prior to starting rhEPO therapy were found to be increased by approximately 10–20% compared to the control group. After the start of rhEPO therapy, these values first decreased and then clearly increased. The intraerythrocyte pH, slightly lower in the pre-rhEPO patient group, remained unchanged during rhEPO therapy. This study was performed using ^{31}P-NMR spectroscopy [23]. The more pronounced thiamine levels in erythrocytes than in plasma in ND patients during rhEPO therapy makes possible the assumption that improved oxygen supply increases the carbohydrate metabolism in erythrocytes, with subsequent stimulation of thiamine influx into

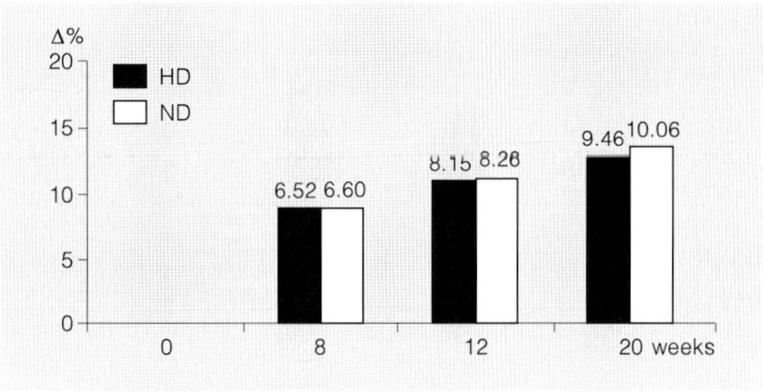

Fig. 7. Differences in the mean levels of $Prot_{tot}$ during rhEPO therapy expressed as the percentage of the initial value regarded as 100 (Δ%).

the cells. The increase in PFTh and PTTh in HD and in PTTh levels in ND patients only during rhEPO therapy might be explained by increased transport of thiamine across the intestinal wall due to the improvement in red cell mass.

The correlation of changes in the parameters that were influenced by rhEPO therapy, was demonstrated for Hb and PFTh after 8 and 12 weeks of this therapy, while that of plasma $Prot_{tot}$ and the ETTh level after 8, 12, and 20 weeks of this therapy was statistically significant. In ND patients, a statistically significant correlation was found only between Hb and PFTh after 8, 12, and 20 weeks of study.

Recent reports indicate that controversy remains as to whether the subcutaneous route of administration is superior to intravenous injection in terms of achieving and maintaining a target Hb concentration at a lower weekly dosage of rhEPO [24]. Our results show that the correction of anemia was associated with an increase in thiamine levels in blood with a concomitant rise in ETKA, independently of the route of administration and the dose of rhEPO given. Unfortunately, it is still unclear if metabolic changes during rhEPO therapy are caused by a direct effect of EPO or whether they are due to the correction of the anemia.

References

1 Eschbach JW, Adamson JW: Anemia in end-stage renal disease. Kidney Int 1985;28:1–5.

2 Horina JH, Schwaberger G, Brussee H, Sauseng-Fellegger G, Holzer H, Krejs GJ: Increased red cell 2,3-diphosphoglycerate levels in haemodialysis patients treated with erythropoietin. Nephrol Dial Transplant 1993;8:1219–1222.

3 Kokot F, Więcek A: Endocrine changes in chronic dialysis patients; in Maher JF (ed): Replacement of Renal Function by Dialysis, ed 3. Dordrecht, Kluwer, 1989, pp 953–971.

4 Krawczyk W, Dmoszyńska A, Ledwozyw A, Soroka-Wojtaszko M: Influence of human recombinant erythropoietin on platelet phospholipids in haemodialysed patients. Acta Haematol Pol 1993; 24:327–331.

5 Krawczyk W, Dmoszyńska A, Ledwozyw A, Marczewski K: Human erythropoietin changes fatty acids composition of blood platelet phospholipids in hemodialyzed patients. Nephron 1995;69:355.

6 Manitius J, Szołkiewicz M, Myśliwska J, Zorena K, Myśliwski A, Jakubowski Z, Łysiak-Szydłowska W, Rutkowski B: Influence of nonhematological doses of erythropoietin on lipid-carbohydrate metabolism and life quality in hemodialysis patients. Nephron 1995;69:363–364.

7 Stafensen G, Aunsholt NA: Does erythropoietin cause hormonal changes in haemodialysis patients? Nephrol Dial Transplant 1993;8:1215–1218.

8 Suzuki Y: Focus on erythropoietin: Effects on red cell enzymatic activity. 25th Congr EDTA-ERA. Madrid, 1988, p 4.

9 Brin M,Ostahever AS: The effect of thiamine deficiency on the activity of erythrocyte hemolysate transketolase. J Nutr 1960;71:273–281.

10 Horecker BL, Smyrniotis PZ: Coenzyme function of thiamine pyrophosphate in pentose phosphate metabolism. J Am Chem Soc 1953;75:1009.

11 Kopple JD, Dirige OV, Jacob M, Wang M, Swendseid ME: Transketolase activity in red blood cells in chronic uremia. Trans Am Soc Artif Intern Organs 1972;18:250–256.

12 Racker E, De La Haba G, Leder IG: Thiamine pyrophosphate, coenzyme of transketolase. J Am Chem Soc 1953;75:1010–1011.

13 Balestri PL, Rindi P, Biagini M, Giovannetti S: Effects of uremic serum urea, creatinine, and methylguanidine on glucose metabolism. Clin Sci 1972;42:395–404.

14 Cohen BD, Stein IM, Bonas JE: Guanidinosuccinic aciduria in uremia. Am J Med 1968;45:63–68.

15 Kurijama M, Mizuma A, Yokomine R, Igata A, Otuji Y: Erythrocyte transketolase in uremia. Clin Chim Acta 1980;108:169–177.

16 Lonergan ET, Semar M, Lange K: Transketolase activity in uremia. Arch Intern Med 1970;126:851–854.

17 Lonergan ET, Semar M, Sterzel RB, Treser G, Needle MA, Voyles BS, Lange K: Erythrocyte transketolase activity in dialyzed patients: A reversible metabolic lesion of uremia. N Engl J Med 1971; 284:1399–1403.

18 Gentile MG, Porrini M, Manna GM, Ciceri R, Cofano F, Simonnetti P, Amico GD: Water and fat soluble vitamins status in chronic renal insufficiency patients; in Guarnieri G, Panzetta G, Toigo G (eds): Metabolic and Nutritional Abnormalities in Kidney Disease. Contrib Nephrol. Basel, Karger, 1992, vol 98, pp 89–97.

19 Pietrzak I, Bączyk K, Kubiak W: Recombinant human erythropoietin administration improves thiamine content in blood and erythrocyte transketolase activity in pre-dialysed patients. Ann Univ Mariae Curie-Skłodowska 1994;48: 29–37.

20 Blum KU, Merkel D: Bemerkungen zum Nachweis des Thiamin in Serum mit der Thiochrom Methode. Z Klin Chem Klin Biochem 1974;10:437–439.

21 Bruns FH, Dünwald E, Nattmann E, Über den Stoffwechsel von Riboso-5-phosphat in Hämolysaten. III: Quantitative Bestimmung von Sedoheptulose-7-phosphat und einige Eigenschaften der Transketolase der Erythrocyten und des Blutserums. Biochem Z 1958;330:497–502.

22 Dische Z: Qualitative and quantitative colorimetric determination of heptose. J Biol Chem 1953;204: 983–997.

23 Nishida A, Kubok K, Ohtsuka H: The changes of energy metabolism in erythrocytes of dialysis patients treated with erythropoietin measured by ^{31}P-NMR spectroscopy (abstract). 27th Congr EDTA-ERA, Vienna, 1990, p 253.

24 Bommer J, Barth HP, Zeier M, Mandelbaum A, Bommer G, Ritz E, Reichel H, Novack R: Efficacy comparison of intravenous and subcutaneous recombinant human erythropoietin administration in hemodialysis patients; in Gurland HJ, Moran J, Samtleben W, Scigalla P, Wieczorek L (eds): Erythropoietin in Renal and Non-Renal Anemias. Contrib Nephrol. Basel Karger, 1991, vol 88, pp 136–143.

Mineral and Electrolyte Metabolism

Miner Electrolyte Metab 1997;23:283–286

Natale G. De Santo
Pietro Anastasio
Lucia Spitali
Domenico Santoro
Dorotea Capodicasa
Eleonora Cirillo
Giovambattista Capasso

Chair of Nephrology,
Second University of Naples, Italy

Renal Reserve Is Normal in Adults Born with Unilateral Renal Agenesis and Is Not Related to Hyperfiltration or Renal Failure

Key Words
Blood pressure
Protein meal
Renal agenesis
Renal disease
Renal reserve

Abstract
This study was carried out to examine the renal hemodynamic response in adult patients with single kidneys born with unilateral renal agenesis. A group of 21 patients with unilateral renal agenesis were divided into three groups according to their glomerular filtration rate (GFR): 112 ± 3 ml/min × 1.73 m^2 in group A, 68 ± 3.2 ml/min × 1.73 m^2 in group B, and 40.7 ± 3.3 ml/min × 1.73 m^2 in group C. Mean arterial blood pressure was significantly higher in the patients of group C who were also proteinuric. The renal hemodynamic response to an oral protein load (2 g/kg of protein as beefsteak) was normal in all groups and unrelated to hyperfiltration or to renal failure and proteinuria. The study indicates that in patients with renal agenesis, the hemodynamic response to a protein challenge is similar to that of kidney donors, renal transplant recipients and uninephrectomized patients. The paper also demonstrates that the renal response to a protein challenge is inadequate to identify patients with renal agenesis who are at risk of developing renal disease. Finally, in renal agenesis with renal disease, creatinine clearance overestimated the GFR by an average of 32.7%.

In patients with single kidneys, the renal hemodynamic response to an acute protein load has been extensively studied in recent years [1–3]. The available information generally points to a normal renal reserve in single kidneys caused by various phenomena. However, such results might have been biased either by poor selection criteria or by the stimuli used to achieve the renal response, or by the methods used to measure renal function. In particular, uninephrectomized patients, kidney donors, and renal transplant recipients along with a minority of patients with unilateral renal agenesis have been lumped together in clinical studies of this question.

This study was therefore devised to study the renal hemodynamic response in adult patients with single kidneys born with unilateral renal agenesis (1 case in every 1,300–1,500 births), which should represent the most physiological, long-lasting and frequent model of adaptation to a 50% lack of renal mass.

Material and Methods

Patients

A total of 21 adult patients born with unilateral renal agenesis participated in this study. They were all living in the city of Naples, and their unilateral renal agenesis was of recent discovery in the course of an abdominal ultrasound examination. Patients were divided into three groups (table 1, fig. 1) according to their glomerular filtration rate (GFR). Group A included all patients with normal GFR (112 ± 3.2 ml/min × 1.73 m^2). Group B included patients with a mean GFR of 68 ± 3.2 ml/min × 1.73 m^2 (range 61–81). Group C included patients with renal failure and proteinuria (600 ± 20 mg/day; their GFR averaged 40.7 ± 3.3 ml/min × 1.73 m^2 (range 28–52). The male/female ratio was 1 in groups A and B, in group C there were 3 males and 4 females. At the time of the study, they ate diets providing 30–40 kcal/kg body weight and 0.8–0.9 g of protein, as ascertained by urea excretion in three consecutive 24-hour urine collections [4, 5]. Mean arterial blood pressure was significantly higher in group C ($p < 0.001$ vs. groups A and B). Plasma urea averaged 22.6 ± 1.4 mg/dl in group A, 43.3 ± 1.2 mg/dl in group B ($p < 0.01$ vs. group A), and 57.4 ± 2.1 mg/dl in group C ($p < 0.01$ vs. groups A and B). Creatinine clearance was always higher than inulin clearance (32.7% in group C).

Measurement of GFR and Oral Protein Load

Patients were admitted individually after an overnight fast at 08:30 hours. They received an oral water load (20 ml/kg). Subsequent water replacement was provided to replace urinary losses. Urine was collected by an indwelling catheter. The GFR was measured as the inulin clearance. Inulin was given as a bolus (40 mg/kg) followed by a constant intravenous infusion to keep the plasma concentration at 20 mg/dl.

The study was started after 90 min of equilibration (as described elsewhere [6–8]) by performing three baseline GFR determinations, each lasting 30 min. At the completion of the baseline study, a meat meal in the form of a beefsteak was ingested within 20–25 min. A total of five GFR measurements were obtained 30, 60, 90, 120, and 180 min after the start of the meal.

KARGER

E-Mail karger@karger.ch
Fax + 41 61 306 12 34
http://www.karger.ch

0378–0392/97/0236–0283 $12.00/0

Natale G. De Santo, MD
Chair of Nephrology, Second University of Naples, Department of Pediatrics
Via Pansini 5, Policlinico Pad. 17, I–80131 Naples (Italy)
Tel. + 39 81 5666650, fax + 39 81 5666655

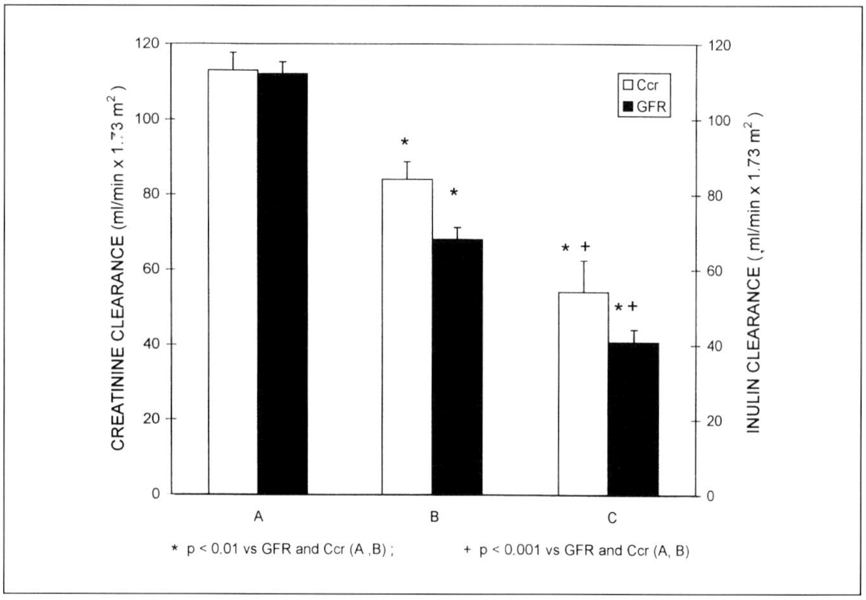

Fig. 1. Creatinine clearance and inulin clearance in experimental groups A, B, and C. Ccr = Creatinine clearance; GFR = glomerular filtration rate.

Table 1. Characteristic of patients enrolled for the study

	Group A (n = 8)	Group B (n = 6)	Group C (n = 7)
Age, years	32.8 ± 5.2	50 ± 3.7*	52.8 ± 4.6*
Mean arterial pressure, mm Hg	102.1 ± 5.06	104.6 ± 4.04	114.3 ± 1.6*, **
Protein intake, g/kg body weight	0.87 ± 0.06	0.86 ± 0.01	0.84 ± 0.06
Plasma creatinine, mg/dl	0.94 ± 0.004	1.32 ± 0.16*	2.30 ± 0.14*, **
Plasma urea, mg/dl	22.6 ± 1.4	43 ± 1.2*	57.4 ± 2.1*, **

Statistically significant versus A (*) and versus B (**).

Laboratory Analysis

Inulin was measured by a colorimetric method, urea and creatinine by autoanalyzer.

Calculations and Statistics

The GFR was adjusted to 1.73 m² of body surface area measured by height and weight. The three baseline GFR measurements were averaged [6–8]. Three different indices of renal hemodynamic response were calculated: renal reserve was calculated as peak postmeal GFR – baseline GFR [9], filtration capacity as the maximal postmeal GFR [9], and percent renal reserve as peak postmeal GFR/baseline GFR × 100.

Data are given as the mean ± SE. Statistical significance was assessed by paired t test. $p < 0.05$ was considered statistically significant.

Results

Renal reserve (fig. 2) averaged 12.7 ± 4.2 ml/min × 1.73 m² in group A, 10.7 ± 4.4 ml/min × 1.73 m² in group B, and 11.6 ± 3.8 ml/min × 1.73 m² in group C (NS).

Percent renal reserve (fig. 3) averaged 11.5 ± 3.8% in group A, 33 ± 4.4% in group B ($p < 0.01$ vs. A) and 30.9 ± 10.4% in group C ($p < 0.01$ vs. A).

Filtration capacity (fig. 4) averaged 125 ± 4.8 ml/min × 1.73 m² in group A, 92 ± 4.4 ml/min × 1.73 m² in group B ($p < 0.01$ vs. A) and 52.3 ± 4.3 ml/min × 1.73 m² in group C ($p < 0.01$).

Discussion

In this large series of adult patients born with unilateral renal agenesis, the renal hemodynamic response to an oral protein load was found to be normal and unrelated to hyperfiltration. The renal hemodynamic response was normal even in patients with moderate renal failure and proteinuria. The novelty of this work consists in the notion that single kidneys from various etiologies follow a common pathway in the renal hemodynamic response to a protein challenge. Therefore, patients with renal agenesis, uninephrectomized patients, kidney donors, and renal transplant recipients should be considered equivalent models for the assessment of renal reserve.

The data also stress the inadequacy of renal reserve to explore the progression to end-stage renal disease in patients with renal agenesis, thus confirming results obtained in other categories of patients with single kidneys [3]. The present findings are therefore

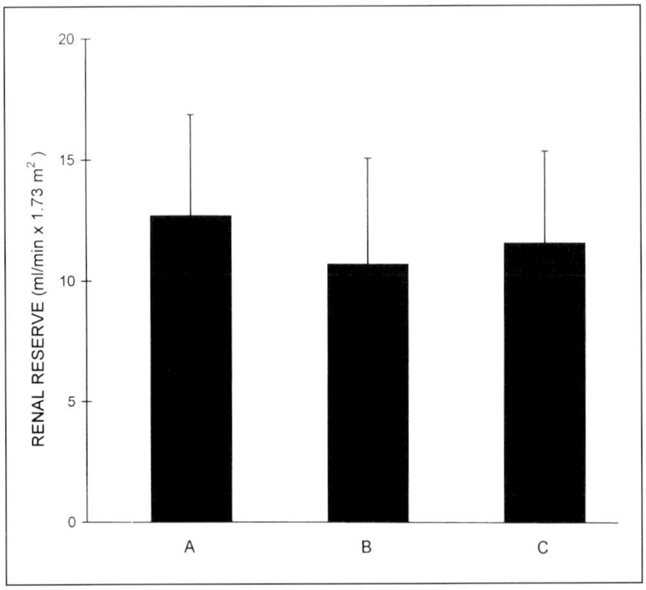

2

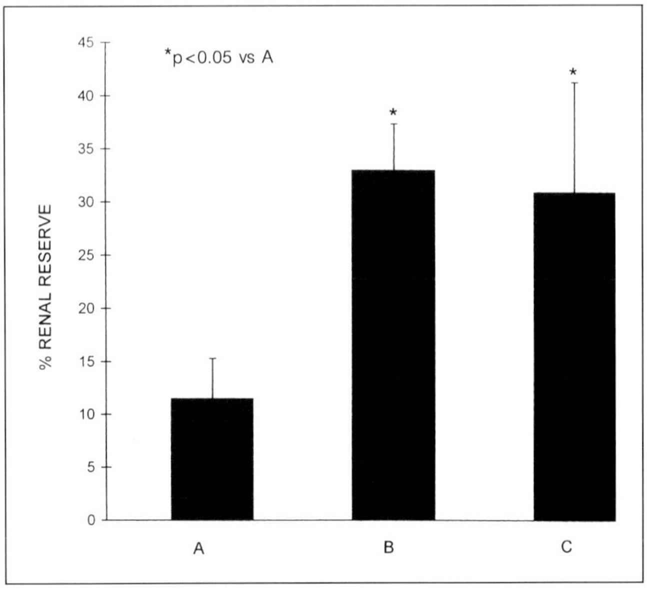

3

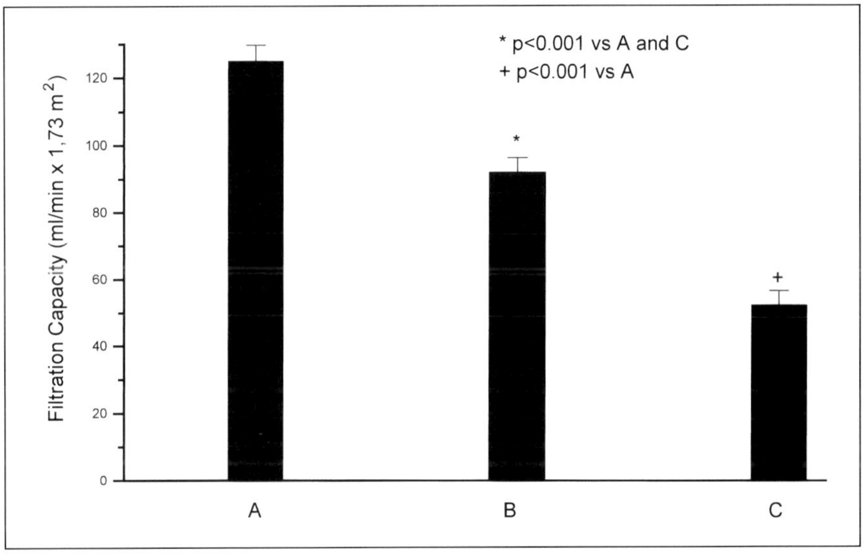

4

Fig. 2. Renal reserve.
Fig. 3. Percent renal reserve.
Fig. 4. Filtration capacity.

consistent with our original suggestion [10] and demonstration [6] related to the normalcy of renal reserve in renal disease. This finding is in agreement with the results of Cassidy and Beck [11], Zuccalà et al. [12], Rugiu et al. [1, 3] and ter Wee et al. [2, 13]. The present results also indicate that the factors and mechanisms responsible for the protein-dependent increase in GFR are different from those mediating the postuninephrectomy increase in GFR [13]. The discrepancy between the present findings and the data reported by Bosch et al. [14] and Rodriguez-Iturbe et al. [15] is difficult to explain but could be due to differences in the previous sodium and protein intake.

Finally, the data indicate that there is an overestimation of GFR when measured by the creatinine clearance, which was greatest (32.7%) in patients with renal disease and proteinuria. Although there is no reason to suspect that renal agenesis with renal disease might represent a special category of patients for whom the use of creatinine clearance may be particularly inaccurate, this possibility warrants further studies.

Acknowledgements

Supported by contracts to the Chair of Nephrology from the Consiglio Nazionale delle Ricerche and by National and Local Funds (40% and 60%) from Ministero dell'Università e della Ricerca scientifica.

References

1 Rugiu C, Oldrizzi L, Maschio G: Effects of an oral protein load on glomerular filtration rate in patients with solitary kidneys. Kidney Int 1987;32:S29–S31.

2 Wee PM ter, Tegzess AM, Donker AJM: Pair-tests renal reserve filtration capacity in kidney recipients and their donors. J Am Soc Nephrol 1994;4:1798–1808.

3 Rugiu C, Oldrizzi L, Maschio G: Renal reserve in patients with solitary kidneys. Semin Nephrol 1995; 15:468–472.

4 Isakson B: Urinary nitrogen output is a valid test in dietary surveys. Am J Clin Nutr 1980;33:4–12.

5 Steffe WP, Goldsmith RS, Penckartz PB, Srinshaw NS, Young VR: Dietary practice uptake and dynamic aspects of whole body nitrogen metabolism in adult humans. Metabolism 1976;25:281–297.

6 DeSanto NG, Capasso G, Anastasio P: The renal hemodynamic response following a meat meal in children with chronic renal failure and in healthy controls. Nephron 1980;56:136–142.

7 DeSanto NG, Coppola S, Anastasio P, Coscarella G, Capasso G, DeMercato R, Bellini L, Strazzullo P, Guadagno A, Pignatelli CP: Pancreatectomy abolishes the renal hemodynamic response to a meat meal in man. Nephron 1990;55:85–86.

8 DeSanto NG, Capasso G, Coppola S: Renal handling of sodium after an oral protein load in adult humans. Renal Physiol Biochem 1992;15:41–52.

9 Bosch JP: Renal reserve: A functional view of glomerular filtration rate. Semin Nephrol 1995;15:381–385.

10 DeSanto NG, Capasso G, Anastasio P: The renal functional reserve is not altered in children with chronic renal failure (abstract). Kidney Int 1988; 33:188.

11 Cassidy MJD, Beck RM: Renal functional reserve in live related kidney donors. Am J Med 1988;11: 468–472.

12 Zuccalà A, Gaggi R, Zucchelli A, Zucchelli P: Renal functional reserve in patients with a reduced number of functioning glomeruli. Clin Nephrol 1989;32:229–234.

13 Wee PM ter, Donker AJM: Renal reserve filtration capacity: Can it predict progression of chronic renal failure. Am J Kidney Dis 1991;27:S71–S75.

14 Bosch JP, Lauer JP, Glabman S: Short-term protein loading in assessment of patients with renal disease. Am J Med 1984;77:873–879.

15 Rodriguez-Iturbe B, Herrera J, Garcia R: Response to acute protein load in kidney donors and apparently normal postacute glomerulonephritis patients: Evidence for glomerular hyperfiltration. Lancet 1985;ii:461–464.

Mineral and Electrolyte Metabolism

Miner Electrolyte Metab 1997;23:287–290

Glomerular Atherosclerosis

John F. Moorhead
Carol Brunton
Zachariah Varghese

Department of Nephrology,
Royal Free Hospital and School of Medicine,
Hampstead, London, UK

Key Words
Glomerular atherosclerosis
Lipoprotein
Low-density lipoprotein
Progressive renal disease

Abstract
This paper suggests a two-hit model for lipoprotein-mediated progressive renal disease, in which postsecretory modification of low-density lipoprotein may favour the transformation of mesangial cells, monocytes and macrophages to glomerular foam cells. Proteinuria and lipiduria would mediate tubulointerstitial damage. Based on this, careful treatment with lipid-lowering agents, lipopheresis and antioxidants may ameliorate the progression of glomerular and tubulointerstitial pathology.

Introduction

Abnormalities of lipid metabolism are present in the nephrotic syndrome, end-stage renal failure, dialysis patients and after renal transplantation. In animals, substantial experimental evidence supports a concept of lipoprotein-mediated exacerbation of renal disease. We have focused attention on the possibility that the nephron might be damaged by lipoproteins, resulting in progressive renal disease (PRD) and end-stage renal disease. Since smooth muscle cells proliferated [1] and became foam cells [2] after exposure to low-density lipoprotein (LDL), we surmised that mesangial cells, which have smooth muscle characteristics, would react similarly, resulting in 'glomerular atherosclerosis'. We also suggested that tubular lipid would mediate tubulointerstitial disease. Thus lipoprotein would cause persistent glomerular injury [3]. In this 'two-hit' model, the original disease could be replaced by lipid-mediated disease. Persistent albuminuria stimulates excess lipoprotein synthesis by the liver so maintaining the vicious cycle.

Nephrotic Syndrome

The nephrotic syndrome – proteinuria, hypoalbuminaemia, oedema, hyperlipidaemia, and lipiduria – dramatically increases plasma lipids in approximately inverse proportion to hypoalbuminaemia. Interstitial and arterial lipidosis was found in diabetics by Kimmelstein and Wilson [4], and glomerular lipidosis by Wilens et al. [5] in intercapillary lesions in diabetic and other glomerular diseases. Hutchison [6] found that serum cholesterol depended only on the renal clearance of albumin in eight nephrotic patients on high- and low-protein diets, consistent with the findings of Wheeler et al. [7] in six nephrotic patients that LDL turnover correlated with plasma albumin. Recently, Yamauchi et al. [8] showed that that oncotic pressure regulates gene transcription of both albumin and apolipoprotein (apo)B in cultured rat hepatoma cells.

Therapeutic reduction of proteinuria is important because disease progression is proportional to both its magnitude and to renal pathology; PRD rarely occurs if proteinuria is reduced or remits [9]. Normal kidneys both excrete and metabolise high-density lipoprotein (HDL) [10]. Saku et al. [11] using a filtering and non-filtering isolated perfused rat kidney model showed that a filtering normal kidney sequestered significantly more rat HDL than controls, and nephrotic kidneys sequestered two- to fourfold more HDL than non-nephrotic kidneys.

Pathogenesis

The pathological effects of lipoproteins in PRD are probably similar to those in atherosclerosis [3, 12], but occur in the microvasculature. The proposed model implies trapping and oxidation of LDL, monocyte recruitment, smooth muscle (mesangial) cell proliferation, and fibrous tissue formation. The glycosaminoglycans of the glomerular basement membrane have long been known to bind human plasma lipoproteins.

There is considerable evidence for the exacerbation of PRD by plasma cholesterol and lipoproteins in experimental models and, recently, a connection has been found in Native Americans between retinopathy, hyperlipidaemia and proteinuria [13]. In another human study, hypertension and hypercholesterolaemia were positively linked with rapid progression of diabetic nephropathy, and hypercholesterolaemia was the major predictor of death [14]. Recently

KARGER
E-Mail karger@karger.ch
Fax +41 61 306 12 34
http://www.karger.ch

0378–0392/97/0236–0287 $12.00/0

John F. Moorhead
Department of Nephrology
Royal Free Hospital and School of Medicine
Hampstead, London, NW3 2QG (UK)

Lee et al. [15] found that 8.4% of 631 patients had ultrastructurally detectable extracellular lipid in non-sclerotic glomeruli, which suggests that there may be an early presclerotic stage of lipoprotein-mediated disease. In humans, the strongest clinical association with PRD is proteinuria [16].

Haemodynamic Influences

Increased glomerular pressure may contribute to the pathogenesis of glomerular sclerosis [17]; unfortunately however, the nephrotic experimental models from which haemodynamic data were derived were not controlled for hyperlipidaemia. Although angiotensin-converting enzyme inhibitors may lower plasma lipoproteins [18], the lack of relevant controls makes these experiments impossible to interpret in the context of lipoprotein-mediated injury.

Oxidised LDL (OxLDL) and Its Cellular Interactions

Lipid-stimulated monocyte influx was first shown by French [19] in cholesterol-fed rabbits and later characterised more fully [20]. OxLDL is taken up by the scavenger receptor [21] to form foam cells, a prominent feature of both atherosclerosis and glomerulosclerosis. This process may depend on LDL undergoing spontaneous oxidation in vivo. Minimally oxidised LDL (mOxLDL) stimulates monocyte endothelial cell interaction [22], and is chemotactic for monocytes which secrete transforming growth factor-β which in turn stimulates fibronectin production by glomerular cells, and collagen and proteoglycan synthesis. The fact that LDL is oxidised by both rat [23] and human mesangial cells, increases the potential for glomerular damage [24]. Such postsecretory modifications of LDL stimulate resident monocytes to produce chemokines, attracting further monocyte influx [20]. The glomerulus may be particularly susceptible, since LDL oxidised in vitro by glomerular macrophages binds preferentially to the glo- merulus when injected intra-arterially in the rat [25]. As in atherosclerosis, fibrosis probably continues independently of the initial insult.

Lipid Deposition

The injured kidney may never be 'safe' from circulating LDL. For example, in experiments with mesangial cell cultures in vitro, a low concentration of OxLDL (200 μg/ml) arrested mesangial cell growth, measured by ^{3}H thymidine incorporation [26]. Both massive obesity and inherited lecithin cholesterol acyltransferase deficiency with abnormal LDL cause renal failure [27], although familial hypercholesterolaemia is not associated with renal disease. Extensive renal lipid deposition may occur in patients with arteriohepatic dysplasia (Alagille's syndrome), in which small kidneys and raised serum creatinine and proteinuria are associated with hypercholesterolaemia and extensive lipid deposition in the mesangium and glomerular and tubular basement membranes [28].

Tubulointerstitial Lesions

Risdon et al. [29] observed that tubulointerstitial disease was a better predictor of PRD than glomerular disease. In nephrotics, fatty acids bound to albumin are metabolised into neutral lipids in the tubules [30] and apoA and apoB have been found in the tubules and interstitium of nephrotic patients [31]; nephrotics of HDL excrete large amounts [32]. Studies in rabbit proximal straight tubule segments microperfused in vitro have shown progressive uptake and cellular accumulation of radiolabelled HDL. We found that native LDL was not toxic to human proximal tubular cells in culture [33], but Ox and mOxLDL caused cell detachment and death. Renal tubular cells may oxidise lipid [23], and thus filtered lipid may damage the renal tubule.

Treatment

PRD in animal models is exacerbated by high-fat diets [34] and ameliorated by lipid-lowering regimes [35], possibly because the latter reduce arachidonic-acid-derived inflammatory mediators. Lipids and immune complexes acted synergistically in lupus nephritis in mice, and fish oil rich in eicosapentaenoic acid prolongs the life of these animals [36]. Total deprivation of lipid mediators through essential-fatty-acid-deficient diets has also shown to prevent progression in glomerulonephritis [37] in rats.

Treatment in Humans

In spite of strong clinical and experimental links between hyperlipidaemia and PRD, it has been tantalizingly difficult to ameliorate disease with lipid-lowering drugs, perhaps because hyperlipidaemia in nephrotics can seldom be normalised, and patient trial numbers are low. Thus lipid-lowering agents reduce total cholesterol by a maximum of 30–35% in most nephrotics, hence are relatively ineffective when total cholesterol is more than 10.0 mmol/l. Angiotensin-converting enzyme inhibitors combined with vastatins and dietary measures may lower plasma cholesterol by reducing proteinuria and increasing cholesterol excretion. Careful management of lipid abnormalities may reduce risk factors associated with chronic haemodialysis, possibly ameliorate the rate of progression of glomerular and tubulointerstitial pathology, and reduce the risk of chronic renal graft failure.

Management of Hyperlipidaemia by LDL Apheresis

Atherosclerosis regresses with reduction of serum cholesterol [38], so why not renal disease? One approach is to use plasma LDL apheresis in nephrotic patients. This rapidly reduces the LDL concentration. Selective elimination of LDL, very-low-density lipoprotein, and lipoprotein(a), achieved by passing hyperlipidaemic plasma over negatively charged dextran sulphate bound to porous cellulose beads, is valuable in the management of familial hypercholesterolaemia; weekly treatments effectively maintain LDL-C within the normal range [39]. Side effects are rare but significant falls occur in postapheresis fibrinogen, due to nonspecific binding by dextran sulphate. Recently, Sakai et al. [40] reported a multicentre trial of LDL apheresis in 16 drug-resistant nephrotics. Treatment was given twice weekly for 3 weeks, 2.5–4.0 litres of plasma being treated per session. Total cholesterol (TC) and triglycerides were reduced by 60–70% after each treatment, HDL-C remaining unchanged. One month later, mean TC was 30–40% of the initial level ($p<0.01$), albumin increased by 10–30% ($p<0.05$), and in 8 patients, urine protein fell to less than 1 g/day, with overall significance of $p<0.05$. No improvement was shown in membranous nephropathy, but 9 patients with focal segmental glomerulosclerosis (FSGS) had improvements ranging from 'very useful' to 'slight'. In a 15-year-old child with drug-resistant FSGS, a combination of LDL apheresis and pravastatin resulted in a fall in plasma TC and urine protein. Urinary excretion of thromboxane B_2 fell strikingly after the initiation of apheresis, suggesting a decrease in the supply of substrate for inflammatory prostaglandins [41]. While this technique offers a new approach, judgement as to its efficacy must await a controlled trial.

Lipoproteins may affect human renal disease progression, but it may be difficult to demonstrate a beneficial effect of lipid lowering in the nephrotic syndrome when the level of cholesterol is greater than 10 mmol/l. Recent advances in understanding the biology and chemistry of lipoproteins, however, suggest that a rational approach to therapy would combine lipid lowering, antioxidants, and lipopheresis.

References

1 Fischer-Dzoga K, Wissler RW: Stimulation of proliferation in stationary primary cultures of monkey aortic smooth muscle cells: Effect of varying concentrations of hyperlipemic serum and low density lipoproteins of varying dietary fat origins. Atherosclerosis 1976;24:515–525.

2 Brown MS, Goldstein JL: Regulation of the activity of the low density lipoprotein receptor in human fibroblasts. Cell 1975;6:307–316.

3 Moorhead JF, Chan MK, El Nahas AM, Varghese Z: Lipid nephrotoxicity in chronic progressive glomerular and tubulo-interstitial disease. Lancet 1982;ii:1309–1312.

4 Kimmelstiel P, Wilson C: Intercapillary lesions in the glomeruli of the kidney. Am J Pathol 1936;12: 83–98.

5 Wilens SL, Elster SK, Baker JP: Glomerular lipidosis in intercapillary glomerulosclerosis. Ann Intern Med 1951;24:592–607.

6 Hutchison FN: Proteinuria, hyperlipidemia, and the kidney. Miner Electrolyte Metab 1993;19:127–136.

7 Wheeler DC, Varghese Z, Moorhead JF: Hyperlipidemia in nephrotic syndrome. Am J Nephrol 1989; 9(suppl 1):78–84.

8 Yamauchi A, Fukuhara Y, Yamamoto S, Yano F, Takenaka M, Imai E, Noguchi T, Tanaka T, Kamada T, Ueda N: Oncotic pressure regulates gene transcription of albumin and apolipoprotein B in cultured rat hepatoma cells. Am J Physiol 1992;263: C397–C404.

9 Zucchelli P, Ponticelli C, Cagnoli L, Passerini P: Long-term outcome of idiopathic membranous nephropathy with nephrotic syndrome. Nephrol Dial Transplant 1987;2:73–78.

10 Peterson DR, Hjelle JT, Carone FA, Moore PA: Renal handling of plasma high density lipoprotein. Kidney Int 1984;26:411–421.

11 Saku K, Reddy GS, Hynd BA, Kashyap ML: Renal handling of high-density lipoproteins by isolated perfused kidneys. Metabolism 1984;33:432–438.

12 Ross R: The pathogenesis of atherosclerosis: A perspective for the 1990s. Nature 1993;362:801–809.

13 West KM, Erdreich LJ, Stober JA: A detailed study of risk factors for retinopathy and nephropathy in diabetes. Diabetes 1980;29:501–508.

14 Krolewski AS, Warram JH, Christlieb AR: Hypercholesterolemia – A determinant of renal function loss and death in IDDM patients with nephropathy. Kidney Int 1994;45:S125–S131.

15 Lee HS, Lee JS, Ko KW: Intraglomerular lipid deposition in routine biopsies. Clin Nephrol 1991; 36:67–75.

16 Cameron SJ: Proteinuria and progression in human glomerular diseases. Am J Nephrol 1990;10(suppl 1): 81–87.

17 Anderson S, Brenner BM: Intraglomerular hypertension: Implications and drug treatment. Annu Rev Med 1988;39:243–253.

18 Pollare T, Lithell H, Berne C: A comparison of the effects of hydrochlorothiazide and captopril on glucose and lipid metabolism in patients with hypertension. N Engl J Med 1989;321:868–873.

19 French SW, Yamanaka W, Ostwald R: Dietary induced glomerulosclerosis in the guinea pig. Arch Pathol 1967;83:204–210.

20 Schreiner GF: The role of the macrophage in glomerular injury. Semin Nephrol 1991;11:268–275.

21 Parthasarathy S, Printz DJ, Boyd D, Joy L, Steinburg D: Macrophage oxidation of low density lipoprotein generates a modified form recognised by the scavenger receptor. Arteriosclerosis 1986;6: 505–510.

22 Berliner JA, Territo MC, Sevanian A, Ramin S, Kim JA, Bamshad B, Esterson M, Fogelman AM: Minimally modified low density lipoprotein stimulates monocyte endothelial interactions. J Clin Invest 1990;85:1260–1266.

23 Fernando RL, Varghese Z, Moorhead JF: Oxidation of low-density lipoproteins by rat mesangial cells and the interaction of oxidized low-density lipoproteins with rat mesangial cells in vitro. Nephrol Dial Transplant 1993;8:512–518.

24 Wheeler DC, Chana RS, Tolpey N, Petersen MM, Davies M, Williams JD: Oxidation of low density lipoprotein by mesangial cells may promote glomerular injury. Kidney Int 1994;45:1628–1636.

25 Coritsidis G, Rifici V, Gupta S, Rie J, Shan ZH, Neugarten J, Schlondorff D: Preferential binding of oxidized LDL to rat glomeruli in vivo and cultured mesangial cells in vitro. Kidney Int 1991;39:858–866.

26 Wheeler DC, Persaud JW, Fernando R, Sweny P, Varghese Z, Moorhead JF: Effects of low-density lipoproteins on mesangial cell growth and viability in vitro. Nephrol Dial Transplant 1990;5:185–191.

27 Gjone E, Blomhoff JP, Skarbovik AJ: Possible association between an abnormal low density lipoprotein and nephropathy in lecithin cholesterol acyltransferase deficiency. Clin Chim Acta 1974;54: 11–18.

28 Chung-Park M, Petrelli M, Tavill AS, Hall PW 3rd, Henoch MS, Dahms BB: Renal lipidosis associated with arteriohepatic dysplasia (Alagille's syndrome). Clin Nephrol 1982;18:314–320.

29 Risdon RA, Sloper JC, De Wardener HE: Relationship between renal function and histological changes found in renal biopsy specimens from patients with persistent glomerular nephritis. Lancet 1968;ii:363–366.

30 Thomas ME, Schreiner GF: Proximal tubule (PT) cells show differential metabolism of endocytosed fatty acids (FA) into neutral lipids. J Am Soc Nephrol 1992;3:752.

31 Kashyap ML, Ooi BS, Hynd BA: Sequestration and excretion of high density and low density lipoproteins by the kidney in human nephrotic syndrome. Artery 1979;6:108–121.

32 Short CD, Durrington PN, Mallick NP, Hunt LP, Tetlow L, Ishola M: Serum and urinary high density lipoproteins in glomerular disease with proteinuria. Kidney Int 1986;29:1224–1228.

33 Ong ACM, Moorhead JF: Tubular lipidosis: Epiphenomenon or pathogenetic lesion in human renal disease? Kidney Int 1994;45:753–762.

34 Pavlovic NM, Malik S, Chan MK, Varghese Z, Persaud JW, Moorhead JF: The influence of high fat diet on the proteinuria in puromycin induced nephrotic syndrome in rats. Clin Sci 1984;66:26P.

35 Wheeler DC, Nair DR, Persaud JW, Jeremy JY, Chappell ME, Varghese Z, Moorhead JF: Effects of dietary fatty acids in an animal model of focal glomerulosclerosis. Kidney Int 1991;39:930–937.

36 Kelley VE, Ferretti A, Izui S, Strom TB: A fish oil diet rich in eicosapentaenoic acid reduces cyclooxygenase metabolites, and suppresses lupus in MRL-1pr mice. J Immunol 1985;134:1914–1919.

37 Lefkowith JB, Rovin B, Schreiner GF: Depletion of resident glomerular macrophages by essential fatty acid deficiency protects against glomerulonephritis. Adv Prostaglandin Thromboxane Leukotriene Res 1989;19:560–563.
38 Wissler RW: Progression and regression of atherosclerotic lesions. Adv Exp Med Biol 1978;104:77–101.
39 Olbricht CJ, Schulzeck P: LDL cholesterol apheresis by dextran sulfate cellulose adsorption: Long-term experience in patients with familial hypercholesterolemia. ASAIO Trans 1991;37:M492–M493.
40 Sakai S, Mune M, Iino Y: Effect of LDL apheresis in patients with drug-resistant nephrotic syndrome. Multi-center cooperative study. Kidney Dial 1992;33:321–328.
41 Hattori M, Kawaguchi H, Tanaka T, Kubota R, Khono M: Treatment with a combination of low-density lipoprotein apheresis and pravastatin of a patient with drug-resistant nephrotic syndrome due to focal segmental glomerulosclerosis. Pediatr Nephrol 1993;7:196–198.

Mineral and Electrolyte Metabolism

Miner Electrolyte Metab 1997;23:291–295

Effects of Protease Therapy in the Remnant Kidney Model of Progressive Renal Failure

K. Sebekova[a]
L. Paczek[b]
J. Dämmrich[c]
H. Ling[d]
V. Spustova[a]
Z. Gaciong[b]
A. Heidland[d]

[a] Institute of Preventive and Clinical Medicine, Bratislava, Slowakia;
[b] The Transplantation Institute Warsaw, Poland;
[c] Institute of Pathology, and
[d] Department of Internal Medicine, University of Würzburg, Germany

Key Words
Interstitial fibrosis
Proteases
Remnant kidney

Abstract
This study investigated whether protease treatment ameliorates the progressive course of chronic failure in the rat model of subtotal nephrectomy. Fourteen male Wistar rats underwent 5/6 nephrectomy, and were randomized into a control group (C, $n = 7$) given 2 ml of 0.9% NaCl intraperitoneally (i.p.) daily, and a study group (P, $n = 7$) treated with 12 mg Phlogenzym® (combination of trypsin, bromelain, and rutosid) in 2 ml saline i.p. daily. After 6 weeks treatment, the Phlogenzym group showed lower proteinuria (C: 19.6 ± 9.1 vs. 10.2 ± 6.2 mg/24 h, $p < 0.05$). Endogenous creatinine clearance was higher (C: 192.3 ± 99.4, P: 300.5 ± 47.9 μl/min per 100 g, $p < 0.05$), while plasma creatinine was decreased (C: 106.7 ± 33.9, P: 76.0 ± 6.3 μmol/l, $p < 0.01$). Blood urea nitrogen levels did not change, although urea clearance tended to a higher level in the protease-treated rats. Decreased renal formation of cytokines was reflected by a lower urinary excretion ratio of transforming growth factor (TGF)-β/creatinine (C: 0.363 ± 0.183, P: 0.232 ± 0.085 ng TGF-β/mg creatinine, $p < 0.05$). Renal morphology revealed less infiltration of mononuclear cells and an amelioration of interstitial fibrosis as expressed by the volume index of the cortical region (C: 17.17 ± 1.43; P: 12.3 ± 0.5%, $p < 0.001$). In addition, the activities of lysosomal proteinases (cathepsin B, L + B, and H), which are decreased in the remnant kidney model of chronic renal failure, were significantly higher in the enzyme-treated group both in isolated glomeruli and proximal tubules. The body and kidney weight tended to be lower, probably due to a catabolic action of the enzymes. In summary, we provide evidence that protease treatment may be beneficial in a nonimmune mediated renal disease. Phlogenzym ameliorated the course of chronic renal failure in the rat model of subtotal nephrectomy and retarded the development of tubulointerstitial fibrosis. Decreased cytokine formation in the remnant kidney is supposed to play a key role.

Introduction

Proteases are of fundamental importance in cellular and extracellular protein turnover in the glomeruli [1, 2]. In various models of chronic renal diseases, proteolytic activity in isolated glomeruli and also in tubules is markedly reduced [3–9]. Such a decrease was demonstrated both by employing the azocasein test and by determining various proteases (cathepsin B, L + B, and H, as well as collagenase and gelatinase). Moreover, inhibition of the plasmin protease system has been found in glomerulosclerosis [10–12]. As a consequence of impaired protease activities, cellular hypertrophy and accumulation of extracellular matrix in glomeruli and tubulointerstitial tissue may occur [12].

It was recently shown that systemic administration of proteases ameliorates various models of immune-mediated glomerulonephritis [13, 14; E.J. McKelvey et al., submitted]. Furthermore, administration of proteolytic enzymes exerted salutary effects on the accelerated arteriosclerosis of aortic allografts [15]. Up to now there are no data available on whether protease treatment improves non-immune-mediated renal diseases. We therefore investigated whether sustained administration of Phlogenzym (a combination of trypsin, bromelain, and rutosid) ameliorates the progressive course of chronic renal insufficiency in the rat remnant kidney model.

KARGER
E-Mail karger@karger.ch
Fax +41 61 306 12 34
http://www.karger.ch

0378–0392/97/0236–0291 $12.00/0

Professor Dr. Dr. h.c. A. Heidland
Department of Internal Medicine
Division of Nephrology, University of Würzburg
Josef-Schneider-Strasse 2
D–97080 Würzburg (Germany)

Material and Methods

The study was approved by the Institutional Ethics Committee for Animals in Bratislava.

Animals

Fourteen male Wistar rats (Velas, Praha, Czech Republic) weighing 180–220 g underwent 5/6 nephrectomy (NX), according to Morrison [16]. They were fed a standard rat chow in the form of pellets with a protein content of 20%; drinking water was provided ad libitum.

Experimental Protocol

Starting from the second day after 5/6 NX, 2 ml of 0.9% NaCl solution were given intraperitoneally to the control group (n = 7) while 12 mg of Phlogenzym (trypsin 2.5 mg, bromelain 4.59 mg, and rutosid 5.1 mg; Mucos Pharma, Geretsried, Germany) in 2 ml of 0.9% NaCl were administered to the study group (n = 7) daily. A day before sacrifice, each animal was placed into a metabolic cage designed for quantitative urine collection.

Biochemical Analysis

Blood. Plasma levels of electrolytes, creatinine, urea, uric acid, glucose, total proteins, cholesterol, and triglycerides were determined employing a Kodak Ektachem 700 analyser (Rochester, N.Y. USA).

Urine. Analysis was performed for proteinuria (Biuret method), creatinine, and urea (Kodak Ektachem 700 analyser), N-acetyl-β-*D*-glucosaminidase (β-NAG) activity (Hitachi 911 analyser, Boehringer Mannheim, Germany), and transforming growth factor-β (TGF-β) (Elisa Kit Prodicta®, Genzyme, Cambridge, Mass., USA).

Kidneys. Glomeruli and proximal tubules were isolated by a differential sieving technique [17] for determination of protein [18], DNA [19] and lysosomal activities of cathepsin B, L + B and H [20] in the cell lysates.

Renal Histology

Renal tissue of the remnant kidney was fixed in formalin and embedded in paraffin. Tissue sections (4 μm thick) were stained with hematoxylin and eosin (HE), periodic acid-Schiff, and PAMS. A double-blind evaluation was performed by a pathologist using light microscopy. The percent volume fraction of renal interstitial tissue (Vi%) in the cortical region was calculated with a point counting method: ten square test fields (100 intersection points, 0.16 mm^2 test field area) engraved on the ocular were evaluated using a magnification of 400 [21].

Statistics

Results are given as the mean and 95% confidence limits (95% CL). The Wilcoxon test for unpaired samples was used to compare the means between the groups ($p < 0.05$ was considered significant).

Table 1. Effects of protease treatment on body and kidney weights in 5/6 NX rats

	5/6 NX-NaCl (6 weeks; n = 6)	5/6 NX-Enzyme (6 weeks; n = 7)
Body weight, g	373.3 ± 54.2	330.7 ± 47.9
Kidney weight, g	1.20 ± 0.25	1.03 ± 0.14
Kidney/body weight, × 10^3	3.24 ± 0.71	3.12 ± 0.27

Data are given as the mean ± 95% CL.

Table 2. Effect of protease therapy on blood chemistry in 5/6 NX animals

	5/6 NX-NaCl (6 weeks; n = 6)	5/6 NX-Enzyme (6 weeks; n = 7)
pH	7.27 ± 0.05	7.33 ± 0.04*
Total protein, g/l	62.2 ± 3.8	56.3 ± 4.44
Cholesterol, mmol/l	1.92 ± 0.32	1.68 ± 0.24
Triglycerides, mmol/l	0.48 ± 0.28	0.65 ± 0.30
Glucose, mmol/l	5.1 ± 2.0	7.9 ± 2.9

Data are given as the mean ± 95% CL; * $p < 0.05$.

Results

Physical Parameters

After 6 weeks, the body and kidney weight of the protease-treated group tended to be lower (table 1). The kidney/body weight ratio was identical in both groups.

Blood Pressure

At sacrifice, the blood pressure of normal control rats averaged 114.2 ± 5.5 mm Hg. In the NX rats, the blood pressure rose to 148.0 ± 8.6 mm Hg in the placebo-treated group and to 146.3 ± 7.5 mm Hg in the enzyme treated rats.

Blood Chemistry

The mean values of plasma electrolytes (Na, K, Ca, Mg) did not differ between the groups (data not given). A trend for lower concentrations of total protein and cholesterol as well as higher levels of triglyceride and glucose was observed in the enzyme-treated group (table 2).

Renal Function

Diuresis tended to be lower in the enzyme-treated rats, while proteinuria was reduced. Plasma creatinine was also significantly less due to a higher creatinine clearance (table 3). Urea clearance tended to be higher in the protease-treated animals, although the mean plasma urea level was unchanged (table 3). The urinary excretion ratio of TGF-β/creatinine was lower in the actively treated group, while lysosomal cathepsin (B, L + B, and H) activities both in tubules and glomeruli were higher during protease treatment (table 4). Urinary activity of β-NAG did not differ between the groups.

Table 3. Effects of protease therapy on parameters of renal function

	5/6 NX-NaCl (6 weeks; n=6)	5/6 NX-Enzyme (6 weeks; n=7)
Creatinine clearance, µl/min per 100 g	192.3±99.4	300.5±47.9*
Plasma creatinine, µmol/l	106.7±33.9	76.0±6.3**
Urea clearance, µl/min per 100 g	58.0±29.8	85.6±29.3
Plasma urea, mmol/l	10.2±2.5	9.9±2.6
Diuresis, ml/24 h	19.3±8.9	11.6±3.5
Proteinuria, mg/24 h	19.6±9.1	10.2±6.2*
Urine TGF-β, µg/mg creatinine	0.363±0.183	0.232±0.095*
Urine β-NAG, µkat/mmol creatinine	17.5±5.5	18.4±4.1

Data are given as the mean ±95% CL; *$p<0.05$; **$p<0.01$.

Table 4. Activities of lysosomal proteases (cathepsin B, L+B, and H; mmol/ml per minute per microgram DNA) in isolated glomeruli and tubules of subtotally nephrectomized rats after 6 weeks treatment

	Glomeruli		Tubules	
	5/6 NX, placebo (n=6)	5/6 NX, enzyme (n=7)	5/6 NX, placebo (n=6)	5/6 NX, enxyme (n=7)
Cathepsin B	108.68±6.53	173.34±19.02**	280.46±32.76	361.52±39.17
Cathepsin L+B	430.05±30.77	919.88±75.48**	1,375.05±241.17	1,799.64±167.44
Cathepsin H	33.29±3.21	48.66±5.46	93.08±10.50	114.44±13.45

*$p<0.05$; **$p<0.01$.

Renal Morphology

In the NaCl-treated 5/6 NX rats, the interstitium of the cortical tissue was increased with focal accentuation around small arterial vessels and arterioles. The amount of collagen fibers was enhanced and the number of infiltrating mononuclear cells, especially lymphocytes and macrophages, was clearly increased (fig. 1a). These alterations were markedly less in the protease-treated animals (fig. 1b). Morphometrically, the Vi% in renal cortex was significantly elevated in untreated 5/6 NX rats (17.2±1.4%) compared to the protease-treated rats (12.3±0.5%, $P<0.01$), in which the mean value was only slightly higher than normal values (9.3±0.3%).

Tolerance and Toxicity

Intraperitoneal administration of Plogenzym to the 5/6 NX rats in a dose of 12 mg/day seemed to be safe: no signs of intolerance were observed. One rat administered saline died after 4 weeks, while all enzyme-treated rats survived.

Discussion

The remnant kidney model induced by 5/6-nephrectomy is characterized by a hypertrophic response of the residual nephrons with a time-dependent development of glomerulosclerosis and tubulointerstitial fibrosis. The glomerular protein/DNA ratio is enhanced, while cysteine and metalloproteinases are impaired both in isolated glomeruli and proximal tubules [9]. These lowered activities may be mediated by the action of TGF-β, which increases the tissue levels of the specific matrix metalloproteinase inhibitors [21] and the plasminogen activator inhibitor-1 [10]; it decreases the synthesis of metalloproteinases [21] and impairs the activity of cathepsins [22]. In various models of glomerulosclerosis and tubulointerstitial fibrosis, including the remnant kidney, overproduction of this cytokine results in cellular hypertrophy and an increase in extracellular matrix [23].

In the present study, daily intraperitoneal administration of proteases to 5/6 NX rats elevated cathepsin activity both in isolated glomeruli and tubules, with an associated marked improvement of the tubulointerstitial fibrosis. The volume fraction of cortical tissue, proteinuria and urinary excretion ratio of TGF-β/creatinine decreased while creatinine clearance increased significantly. If the urinary levels of TGF-β reflect the concentration of the active component of this cytokine in renal tissue, both the enhanced cathepsin levels as well as the amelioration of tubulointerstitial fibrosis could be explained by this alteration.

In contrast to the decline of plasma creatinine concentrations, blood urea levels did not change in the protease-treated 5/6 NX rats, although the clearance of urea tended to be higher. This dissociation may be in part due to a decreased diuresis in the protease-treated group which favors the back-diffusion of urea along the nephron. Furthermore, the unchanged blood urea concentration despite enhanced creatinine clearance may be a consequence of a catabolic action of Phlogenzym. In line with this assumption is the finding that the protease-treated group showed an insignificant decline in body weight as well as a slightly lowered total protein and cholesterol concentration in the plasma while blood glucose levels tended to be higher. A similar tendency to a lower body weight was also observed by our group [Sebekova et al. unpublished

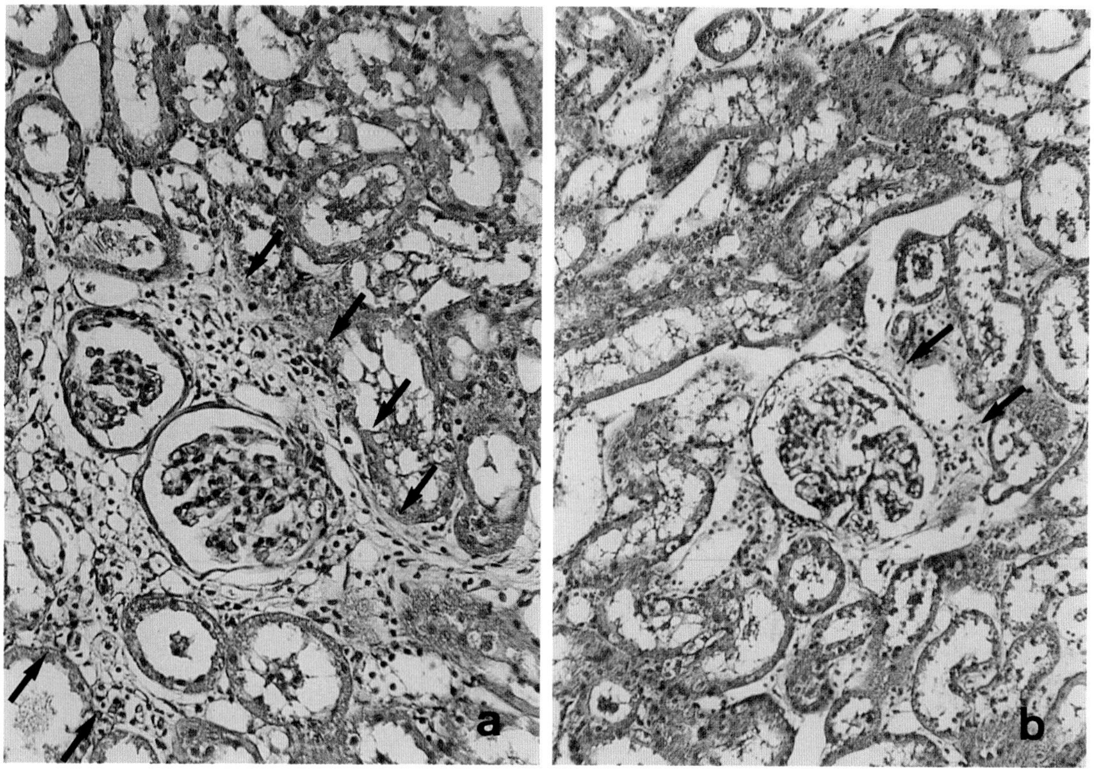

Fig. 1. Histomorphological aspect of renal cortex in the remnant kidney. In untreated 5/6 NX rats (**a**) focal interstitial fibrosis with mononuclear infiltrations was observed, accentuated around arterioles near glomeruli, but in Phlogenzym-treated animals (**b**) there was much less interstitial fibrosis. HE. × 180.

data] under strict pair-feeding conditions in a renewed experiment in NX rats as well as in Goldblatt hypertensive rats, underlining the catabolic properties of intraperitoneal administered hydrolytic enzymes.

How can the beneficial action of protease therapy in the remnant kidney model be explained? It may be argued that systemic administration of proteases cannot be of any value because they will bind to protease inhibitors such as α_2-macroglobulin (α2M) and α_1-antitrypsin (α_1-AT). However, binding of proteases to α2M to the amide bonds differs from that of the classic inhibitor α_1-AT [24]. Thus, the protease active sites are not totally blocked resulting in persistant proteolytic activity at least for reactions with smaller sized proteins. Furthermore, protease binding is followed by an activation of native α2M to its fast form which generates new binding sites at the thiol ester for cytokines and growth factors [25]. Thiol ester activation of α2M enhances binding of TGF-β, platelet-derived growth factor, fibroblast growth factor and interleukin-1β [26]. In addition, the protease-induced structural changes of α2M generate new binding sites for specific α2M receptors, resulting in an enhanced clearance of these cytokines [27]. As a consequence, cytokine and growth factor levels may be lowered in the damaged tissue, allowing an improvement of abnormal cell growth and cell proliferation and reduced extracellular matrix formation in the remnant kidney model [28].

Another potential effect of protease therapy may be a selective effect on cellular adhesion molecules. Thus in vitro and in vivo studies have shown that trypsin cleaves the adhesion molecules CD44, CD4, and B7-1 of activated T cells and macrophages [29]. Bromelain, the other component of the employed enzyme preparation, also cleaves CD44 and may thereby enhance the response to trypsin [30]. As a consequence of these effects, activation of T lymphocytes and macrophages in the interstitium, which mediate the inflammatory response, could be down-regulated.

Finally, the administered proteases trypsin and bromelain may activate various matrix metalloproteinases which are synthesized and secreted as inactive zymogens. Following limited proteolysis of the amino terminus, they become actively proteolytic. This has been demonstrated in in vitro studies [1] and in recent in vivo investigations, in which intravenous administration of trypsin to unilateral NX rats activated the latent collagenases MMP2 and MMP9 in the renal tissue [31]. The salutary effect of Phlogenzym therapy was not caused by a lowering of arterial hypertension since the blood pressure was identical in the treated and nontreated NX rats.

In summary, the data demonstrate that intraperitoneal administration of proteases in the remnant kidney model ameliorates the severity of tubulointerstitial fibrosis. It is assumed that the salutary actions of administered proteolytic enzymes result from the combined effects of inactivation and enhanced removal of growth promoters and cytokines (following their binding to an activated α2M-protease complex), cleavage of various adhesion molecules, and restoration of some impaired proteolytic activities, thereby ameliorating various amplificatory mechanisms involved in the renal damage in 5/6 NX rats.

Acknowledgements

The generous support by Dipl.-Ing. Klaus(†) and Nikolaus Sorg (Lohr am Main) is gratefully appreciated.

References

1 Lovett DH, Sterzel RB, Kashgarian M, Ryan JL: Neutral proteinase activities produced in vitro by cells of the glomerular mesangium. Kidney Int 1983;23:342–349.

2 Davies M, Martin J, Thomas GJ, Lovett DH: Proteinases and glomerular matrix turnover. Kidney Int 1992;41:671–678.

3 Heidland A, Ling H, Vamvakas S, Paczek L: Impaired proteolytic activity as a potential cause of progressive renal disease. Miner Electrolyte Metab 1996;22:157–161.

4 Teschner M, Schaefer RM, Svarnas A, Heidland U, Heidland A: Decreased proteinase activity in isolated glomeruli of streptozotocin diabetic rats. Am J Nephrol 1989;9:464–469.

5 Paczek L, Teschner M, Schaefer RM, Kovar J, Romen W, Heidland A: Intraglomerular proteinase activity in adriamycin-induced nephropathy. Nephron 1992;60:81–86.

6 Olbricht CJ, Geissinger B: Renal hypertrophy in streptozotocin diabetic rats: Role of proteolytic enzymes. Kidney Int 1992;41:966–972.

7 Reckelhoff JF, Tygart VL, Mitias MM, Walcott JL: STZ-induced diabetes results in decreased activity of glomerular cathepsin and metalloproteinase in rats. Diabetes 1993;42:1425–1432.

8 Shechter P, Boner G, Rabkin R: Tubular cell protein degradation in early diabetic renal hypertrophy. J Am Soc Nephrol 1984;4:1582–1587.

9 Schaefer L, Lorenz T, Daemmrich J, Heidland A, Schaefer RM: Role of proteinases in renal hypertrophy and matrix accumulation. Nephrol Dial Transplant 1995;10:801–807.

10 Tomooka S, Border WA, Marshall BC, Noble NA: Glomerular matrix accumulation is linked to inhibition of the plasmin protease system. Kidney Int 1992;42:1462–1469.

11 Feng I, Tang WW, Loskutoff DJ, Wilson CB: Dysfunction of glomerular fibrinolysis in experimental antiglomerular basement membrane antibody glomerulonephritis. J Am Soc Nephrol 1993; 3:1753–1764.

12 Baricos WH: Decreased proteolysis of the glomerular extracellular matrix as a pathogenetic mechanism in progressive renal disease. J Nephrol 1994; 7:322–328.

13 Nakazawa M, Emancipator SN, Lamm ME: Removal of glomerular immune complexes in passive serum thickness nephritis by treatment in vivo with proteolytic enzymes. Lab Invest 1986;55:551–556.

14 White RB, Lowrie L, Stork LE, Iskander SS, Lamm ME, Emancipator SN: Targeted enzyme therapy of experimental glomerulonephritis in rats. J Clin Invest 1991;87:1819–1827.

15 Gaciong Z, Paczek L, Bojakowski K, Socha K, Wisniewski M, Heidland A: Beneficial effect of proteases on allograft arteriosclerosis in a rat aortic model. Nephrol Dial Transplant 1996;11:987–989.

16 Morrison AB: Experimentally induced chronic renal insufficiency in the rat. Lab Invest 1962;11: 321–332.

17 Spiro RG: Studies on the renal glomerular basement membrane: Preparation and chemical composition. J Biol Chem 1984;242:1915–1919.

18 Smith PK, Krohn RJ, Hermanson GT, Mallia AK, Garter FH, Provenzano M, Fujmoto EK, Goeke NM, Olson BJ, Klenk DC: Measurement of protein using bicinochinic acid. Anal Biochem 1985;150: 76–85.

19 Labarca C, Peigen K: A simple, rapid, and sensitive DNA assay procedure. Anal Biochem 1979;102: 344–352.

20 Barrett AJ, Kirschke H: Cathepsin B, cathepsin H and cathepsin L. Methods Enzymol 1981;80:535–540.

21 Edwards DR, Murphy G, Reynolds JJ, Whitman SE, Docherty AJP, Angel P, Heath JK: TGF-β modulates the expression of collagenase and metalloproteinase inhibitor. EMBO J 1987;6:1899–1904.

22 Ling H, Vamvakas S, Busch G, Daemmrich J, Schramm L, Lang F, Heidland A: Suppressing role of transforming growth factor β_1 on cathepsin activity in cultured tubule cells. Am J Physiol 1995;269: F911–F917.

23 Border WA, Ruoslahti E: Transforming growth factor β_1 in disease: The dark side of tissue repair. J Clin Invest 1992;90:1–7.

24 Feinman RD: The protease-binding reaction of α_2M. Ann NY Acad Sci 1994;737:245–262.

25 Borth W: α_2-Macroglobulin, a multifunctional binding protein with targeting characteristic. FASEB J 1992;6:3345–3353.

26 James K: Interaction between cytokines and α_2-macroglobulin. Immunol Today 1990;11:163–166.

27 LaMarre J, Hayes MO, Wollenberg GK, Hussaini J, Hall SW, Gonias StL: An α_2-macroglobulin receptor dependent growth factor-β_1 in mice. J Clin Invest 1991;87:39–44.

28 Mannhalter JW, Barth W, Eibl MM: Modulation of antigen-induced T cell proliferation by α_2M-trypsin complexes. J Immunol 1986;136:2792–2799.

29 Lehmann PV: Immunomodulation by proteolytic enzymes. Nephrol Dial Transplant 1996;11:953–955.

30 Harrach T, Gebauer F, Eckert K, Kunze R, Maurer HR: Bromelain proteinases modulate the CD44 expression on Molt 4/8 leukemia and SK-Mel 28 melanoma cells in vitro. Int J Oncol 1994;5:485–488.

31 Jones CI, Forbes JM, Walker RG, Becker GJ: Activation of renal type IV collagenases. J Am Soc Nephrol 1996;7:2549A.

Mineral and Electrolyte Metabolism

Miner Electrolyte Metab 1997;23:296–300

Bertram L. Kasiske
J.D. Anthony Lakatua

The Division of Nephrology,
Department of Medicine,
University of Minnesota College of Medicine,
Hennepin County Medical Center,
Minneapolis, Minn., USA

The Effects of Dietary Protein Restriction on Chronic Progressive Renal Disease

Key Words
Meta-analysis
Treatment
Low protein diet
Kidney disease

Abstract
There have been a large number of clinical trials and two meta-analyses attempting to determine if dietary protein restriction retards the rate of renal disease progression. Indeed, in a systematic search of prospective, controlled trials examining the effects of protein restriction over at least 6 months, we located 23 published between 1980 and 1996. Only 12 of these trials were randomized and controlled. The results of the 23 studies were heterogeneous. Two meta-analyses combined the results of randomized, controlled trials. Although both meta-analyses concluded that dietary protein restriction significantly reduced the number of patients who died or required treatment for end-stage renal disease, the well-known effects of protein restriction on the signs and symptoms of uremia leave open the question of whether protein restriction had a substantial effect on renal disease progression per se. Adding to the uncertainty is the inconclusive result of the largest, best-designed clinical trial, the Modification of Diet in Renal Disease Study. In any case, the effect of protein restriction on the rate of decline in renal function is arguably modest, making it difficult to demonstrate statistical significance, even in large, well-designed clinical trials. All of these trials suggest that we need therapies that are more effective than dietary protein restriction to halt the progression of renal disease.

Introduction

Dietary protein restriction has been used in patients with chronic renal insufficiency since the beginning of this century. Volhard [1] wrote of using a low-protein diet to ameliorate uremic symptoms in 1918. Smith [2] described prolonged renal survival in a patient maintained on a low-protein diet in 1927. Thomas Addis [3] also suggested using low-protein diets to retard the progression of renal disease.

More recently, a number of clinical trials have attempted to determine whether dietary protein restriction retards the rate of chronic renal disease progression. In 1992, Fouque et al. [4] published a meta-analysis of randomized, controlled clinical trials examining the effects of dietary protein restriction in 890 patients with nondiabetic, chronic, progressive renal disease. They concluded that the data strongly supported the effectiveness of low-protein diets in delaying the onset of end-stage renal disease (ESRD). To confirm the results of these smaller trials, the Modification of Diet in Renal Disease (MDRD) study was carried out in 840 patients with nondiabetic renal disease [5]. This was a well-designed, multicenter trial of adequate statistical power to test the hypothesis that diet would slow the rate of renal disease progression. Nevertheless, the results of the MDRD trial were inconclusive. Subsequently, another meta-analysis was carried out whose authors concluded that dietary protein restriction effectively slows the progression of both diabetic and nondiabetic renal diseases [6]. Thus, clinicians are left with the dilemma that, while the largest and arguably best-designed clinical trial was inconclusive, two meta-analyses suggest low-protein diets are beneficial. In this review, we will examine the clinical trials and the meta-analyses in an attempt to better understand possible reasons for their seemingly conflicting results.

Clinical Trials

There have been a large number of controlled and uncontrolled clinical trials examining the effects of dietary restriction on chronic, progressive renal disease. We carried out a systematic search for controlled clinical trials published since 1980. We used MEDLINE and bibliographies from recent articles and reviews. We included both randomized, controlled trials, as well as trials where controls and treated patients were not randomly allocated, and sequential trials where patients served as their own controls. In these latter studies, a control diet period was usually followed by protein restriction, and in many cases data collected in the control period were collected retrospectively. All studies measured changes in renal function over time, either (1) serial changes in glomerular filtration rate (GFR), (2) serial changes in creatinine clearance, or (3) the slope

KARGER
E-Mail karger@karger.ch
Fax +41 61 306 12 34
http://www.karger.ch

0378–0392/97/0236–0296 $12.00/0

Bertram L. Kasiske, MD
Department of Medicine, Hennepin County Medical Center
701 Park Avenue, Minneapolis, MN 55415 (USA)
Tel. +1 612 347 6088, fax +1 612 347 2003

Table 1. Controlled, clinical trials examining the effects of dietary protein restriction on renal disease progression

Study	Year	RCT	Diabetic	Number	Effective	Ineffective	Inconclusive
Barsotti et al. [16]	1981			56	×		
Bennett et al. [17]	1983			31	×		
Barsotti et al. [18]	1984			55	×		
Mitch et al. [19]	1984			17	×		
El Nahas et al. [20]	1984			39	×		
Oldrizzi et al. [21]	1985			100	×		
Lucas et al. [22]	1986			12		×	
Acchiardo et al. [23]	1986			30	×		
Jungers et al. [7]	1987	×		19	×		
Barsotti et al. [24]	1988		×	8	×		
Evanoff et al. [25]	1989		×	11	×		
Walker et al. [26]	1989		×	19	×		
Ihle et al. [10]	1989	×		64	×		
Rosman et al. [8]	1989	×		248			×
Bergström et al. [27]	1989	×		18		×	
Brouhard and LaGrone [28]	1990	×	×	16	×		
Zeller et al. [29]	1991	×	×	47	×		
Locatelli et al. [9]	1991	×		456		×	
Williams et al. [11]	1991	×		95		×	
Dullaart et al. [30]	1993	×	×	31	×		
Raal et al. [31]	1994	×	×	22	×		
Klahr et al. [5]	1994	×		840			×
D'Amico et al. [32]	1994	×		128	×		

Year is the year of publication. RCT indicates if a randomized, controlled trial design was used. Diabetic indicates that only diabetic patients were studied. Number is the number of treated and control patients. Effective, ineffective or inconclusive is the conclusion drawn by the authors.

of the inverse serum creatinine. In addition, some studies examined the number of patients who died or developed ESRD. We excluded retrospective studies that were not true clinical trials, and trials that did not follow patients for a mean or median of at least 6 months. We also excluded trials that did not provide data on the rate of change in renal function.

We located 23 clinical trials fulfilling the above criteria (table 1). There were a total of 2,362 patients in these 23 trials, but 14 included less than 50 patients, and 9 included less than 25. There were 12 randomized, controlled trials. Interestingly, only 1 of the 10 trials published before 1989 was randomized and controlled, in contrast to 11 of the subsequent 13. The authors of 17 studies concluded that a low-protein diet was effective (7 of these studies were randomized and controlled). However, the authors of 4 studies (3 of which were randomized and controlled) concluded that diet was ineffective, and in 2 (both were randomized and controlled) the results were felt to be inconclusive. Of the 3 randomized trials with over 200 patients each, 1 was judged to be negative and 2 were felt to be inconclusive. Thus, it is very difficult to conclude from these individual trials whether dietary protein restriction is effective in retarding the rate of renal disease progression.

A Meta-Analysis of Randomized Controlled Trials in Nondiabetic Renal Disease

Fouque et al. [4] published a meta-analysis of randomized, controlled trials in 1992, and concluded that these trials unequivocally demonstrated a benefit of low-protein diet. They searched the medical literature for published and unpublished trials, and included in their analysis four investigations published between 1982 and 1991 [7–10]. In addition, they included one study by Bagros and coworkers that has never been published. They also included a study by Williams et al. [11] that had only appeared in abstract form at the time they carried out the meta-analysis, but has subsequently been published. They excluded trials that studied only patients with diabetes. Despite the fact that most of the trials that they combined were designed to measure changes in renal function as an endpoint, the authors used renal death as an endpoint. Renal death was defined as the initiation of dialysis or death during the study. In addition, patients transplanted before starting dialysis were counted as having had a renal death [4]. In some cases, the data on renal deaths were not included in the published results of the trials and were, therefore, provided to the reviewers without ever having been subjected to peer review.

The authors carried out an intention-to-treat analysis of the logarithm of the odds ratios (the ratio of the number of renal deaths in the treatment group to the number of patients in the treatment group who did not have a renal death, divided by the ratio of the number of renal deaths in the control group to the number of control patients without renal death). In five of the six trials, the odds ratio tended to favor the low-protein diet group, but in only one of the trials was this statistically significant. However, the combined odds ratio was 0.54 (95% confidence interval 0.37–0.79, $p<0.002$), indicating a statistically significant difference between the diet and control groups. The authors pointed out that this indicated there was a longer period before the start of dialysis or transplantation in the low-protein diets compared to controls, and did not necessarily prove that there was a reduction in the rate of progression of renal disease. In other words, the diets may have delayed signs and symptoms of uremia without altering the natural history of the disease. The authors also pointed out three other limitations of their analysis including (1) the heterogeneity in the populations studied and in the degrees of protein restriction used, (2) differences in criteria used to initiate dialysis, and (3) the tendency of patients in the study to eat more or less than the prescribed amounts of protein. They concluded that a large randomized controlled trial should be carried out to confirm the results of their meta-analysis, a trial that indeed was already in progress.

The MDRD Trial

The MDRD trial was mandated by an act of the United States Congress [12]. This mandate led the National Institute of Diabetes and Digestive and Kidney Diseases to form a multicenter cooperative trial to define the influence of dietary protein and phosphorus restriction on the rate of progression of chronic renal disease. The effort included a planning phase, a feasibility trial, and a full-scale clinical trial. The design of the full-scale clinical trial was based on data from the feasibility trial, and the full-scale trial was designed to enroll enough patients to adequately test the hypothesis.

The full-scale trial was divided into two studies. Study 1 randomly allocated 585 patients with a GFR of 25–55 ml/min per 1.73 m^2 to a usual protein diet or a low-protein diet (1.3 or 0.58 g/kg per day) and to a usual or a low blood pressure group [5]. The primary endpoint for study 1 was the rate of change in the GFR (the slope) calculated on the basis of the final baseline and follow-up GFRs. There was no difference between the two diet groups in the primary endpoint: the decline in GFR (ml/min per 3 years) in the low-protein group was 10.9 (95% confidence interval 9.2–12.5) compared to 12.1 (10.5–13.8) in the usual-protein group (NS). However, the rate of decline between baseline and 4 months was greater in the low-protein group ($p = 0.004$), so that the rate of decline between 4 months and the last follow-up was significantly less in the low-protein group compared to the usual-protein diet group ($p = 0.009$). The authors concluded, therefore, that, 'Among patients with moderate renal insufficiency, the slower decline in renal function that started four months after the introduction of a low-protein diet suggests a small benefit of this dietary intervention' [5].

In study 2, 255 patients with a GFR 13–24 ml/min per 1.73 m^2 were randomly assigned to either a low-protein (0.58 g/kg per day) or a very-low-protein diet (0.28 g/kg per day) with keto amino acid supplement. Among these patients with more advanced renal disease, there was a trend towards a slower rate of decline in patients treated with dietary protein restriction, but this difference did not achieve statistical significance ($p = 0.07$). There was no difference in the incidence of death or in the occurrence of ESRD ($p = 0.62$), leading to the conclusion that '... a very-low protein diet, as compared with a low-protein diet, did not significantly slow the progression of renal disease' [5].

In retrospect, it is possible that the results of the MDRD trial may have suggested a more favorable effect of a low-protein diet if the follow-up period had been longer. If the differences in the 4 month to last follow-up GFR slope persisted, it is possible that a longer follow-up period would have led to a statistically significant difference in the primary endpoint. It is also possible that the inclusion of patients with different underlying cause-specific rates of renal disease progression made it difficult to see the benefits of a low-protein diet in the whole population. The rate of decline in patients with polycystic kidney disease, for example, was significantly greater compared to patients with other renal diseases [13]. The inclusion of a disproportionally large number of patients with polycystic kidney disease may have made it more difficult to detect the benefits of a low-protein diet in patients with other renal diseases. In study 2, the outcome may have been different if the control group had consumed a usual-protein diet instead of a low-protein diet. These and other reasons may have obscured what otherwise would have been a more beneficial effect of dietary protein restricition. However, this speculation does not alter the fact that the MDRD study, despite the extraordinary effort that went into its planning and execution, failed to demonstrate anything but a very marginal benefit from dietary protein restriction.

A Meta-Analysis of Diabetic and Nondiabetic Renal Disease Progression

Recently, Pedrini et al. [6] carried out a meta-analysis of randomized controlled trials. This was actually two separate meta-analyses, one combining trials in patients with nondiabetic renal disease, and the other combining trials in patients with diabetic nephropathy. The analysis of nondiabetic renal disease excluded two studies previously included by Fouque et al. [4] in their meta-analysis: one had not been published (Bagros et al.), and the other was felt not to have included an adequate control group [7]. However, this meta-analysis did include the results of the MDRD trial that were not available to Fouque et al. [4]. Altogether, five randomized controlled trials, including a total of 1,413 patients with nondiabetic renal disease, were combined [5, 8–11]. The endpoint for this analysis was renal failure or death. In the case of the MDRD trial, the unpublished data for this endpoint were obtained directly from the investigators. The authors used a random-effects model to combine the data in these trials. According to this analysis, only one of the individual studies found a significant reduction in the risk for renal failure or death among patients with nondiabetic renal disease [8]. However, compared to the control diet, the relative risk of renal failure or death in the combined protein restriction groups was 0.67 (95% confidence interval 0.50–0.89, $p = 0.007$).

Pedrini et al. [6] also examined the effects of protein restriction in diabetic renal disease. However, there were too few trials and too few patients in these trials to examine renal failure and death as an endpoint. Therefore, the authors combined the results of three randomized and two cross-over design trials to examine the effects of a low-protein diet in preventing either a decline in GFR greater than 1.2 ml/min per year or a 10% increase in urine albumin excretion. There were only 108 patients in these combined trials. In only one individual trial was this endpoint reduced by the low-protein diet. However, the risk ratio of achieving this endpoint for the low-protein compared to control diet was 0.56 (95% confidence interval 0.40–0.77, p = 0.00036) [6].

The authors of this meta-analysis concluded that a low-protein diet effectively slows the rate of progression of both diabetic and nondiabetic renal disease. Indeed, they suggested that there is now sufficient justification to recommend dietary protein restriction in patients with nondiabetic, chronic renal disease. However, they did admit that the number of patients in the combined trials of patients with diabetes was too small to make firm conclusions.

Where Do We Go from Here?

Two meta-analyses have been carried out combining the results of randomized controlled trials, and both concluded that there was a small, but statistically significant, reduction in the incidence of ESRD from dietary protein restriction. Does this mean that we should routinely prescribe a low-protein diet to patients with progressive renal disease? Does this mean that additional clinical trials are no longer necessary? Although meta-analysis is a useful tool for looking at the results of multiple clinical trials, the limitations of meta-analyses in general, and these meta-analyses in particular, should be kept in mind.

The rate of renal disease progression was the primary endpoint in each of the clinical trials that were combined in the two meta-analyses. However, the meta-analyses did not examine this endpoint, but rather examined renal death. It is quite plausible that a low-protein diet may reduce the signs and symptoms of uremia without altering the rate of decline in renal function. Thus, a low-protein diet may result in small differences in the rate of renal death without influencing the rate of functional decline. Moreover, it should be kept in mind that these trials were not blinded. It is possible that investigators may have been influenced in their decision to begin renal replacement therapy by the knowledge that patients were less likely to become symptomatic if treated with dietary protein restriction.

A second problem with the meta-analyses is the assumption that the treatment and the groups of patients treated were homogeneous. The approach taken in both meta-analyses is one that is commonly taken: the authors chose to combine only randomized, controlled trials, and examined whether the overall treatment effect achieved statistical significance. This approach assumes that the combined study groups are homogeneous. Although the authors' statistical tests for a lack of homogeneity were negative, we know from clinical studies that it is unlikely that the effects of dietary protein restriction are homogeneous in different patient populations.

Age, gender, the degree of proteinuria, blood pressure, and the cause of renal disease may each influence the rate of renal disease progression and could modulate the effects of dietary protein restriction. The patient populations in the different studies indeed varied in ways that could affect the results. For example, the proportion of patients with autosomal dominant polycystic kidney disease varied between the different studies. Even more problematic is the fact that patients in the different studies were treated with very different levels of protein restriction. Indeed, the level of dietary protein restriction in the intervention groups of some studies was equal to the level of protein intake in the control groups of others. It is unlikely that the effects of these different levels of dietary protein on renal function were the same, thereby making the interpretation of the meta-analyses combining them difficult. Finally, even if we assumed that the meta-analysis assumption of treatment effect homogeneity were correct and the result was valid, applying this result to an individual patient who does not resemble the collective mean is problematic.

A problem shared by all clinical trials examining the rate of progression of renal disease is the fact that there are no perfect measures of renal function. Moreover, the cruder measures of renal disease progression, such as inverse serum creatinine and creatinine clearance may be influenced by dietary protein restriction [14]. We would like to infer that dietary protein restriction is preventing renal histologic damage. Unfortunately, no clinical trials have examined the effects of dietary protein restriction on renal histology.

The individual trials and the meta-analyses that have been conducted to date suggest that additional clinical trials are needed to understand the role of dietary protein restriction in treating patients with chronic progressive renal disease. Future clinical trials should focus on the effects of diet in homogeneous subsets of patients with renal disease. How age, gender, the level of baseline renal function, the degree of proteinuria, and differences in the cause of renal disease modulate the effects of dietary protein restriction need to be clarified. The number of patients required for such trials may be large. However, the number of patients needed may be reduced by designing trials that have very long follow-ups.

Finally, it should be remembered that the results of clinical trials may not always be directly applicable to clinical practice. The degree of adherence to therapy in clinical trials may be difficult to achieve in standard clinical practice. Indeed, the level of dietary instruction and reenforcement used in well-funded clinical trials are not necessarily routinely available in clinical practice. Similarly, the fact that a low-protein diet did not appear to have adverse effects on nutrition in the MDRD study [15] does not automatically ensure that nutritional adequacy will be achieved in clinical practice. Indeed, the attention paid to nutrition and the resources available to investigators may make it easier to ensure nutritional adequacy in an investigational setting than in the standard clinical situation.

References

1 Volhard F: Die doppelseitigen hämatogenen Nierenerkrankungen (Bright'sche Krankheit); in: Mohr L, Staehelin R (Hrsg): Handbuch der Inneren Medizin. Berlin, Springer, 1918, pp S1149–S1722.

2 Smith M: A case of chronic nephritis maintained for six months on an average daily protein intake of 0.26 grams per kilogram of body weight. Boston Med Surg J 1927;196:941–946.

3 Addis T: Glomerular Nephritis: Diagnosis and Treatment. New York, Macmillan, 1948.

4 Fouque D, Laville M, Boissel JP, Chifflet R, Labeeuw M, Zech PY: Controlled low protein diets in chronic renal insufficiency: Meta-analysis. BMJ 1992;304:216–220.

5 Klahr S, Levey AS, Beck GJ, Caggiula AW, Hunsicker L, Kusek JW, Striker G: The effects of dietary protein restriction and blood-pressure control on the progression of chronic renal disease. N Engl J Med 1994;330:877–884.

6 Pedrini MT, Levey AS, Lau J, Chalmers TC, Wang PH: The effect of dietary protein restriction on the progression of diabetic and nondiabetic renal diseases: A meta-analysis. Ann Intern Med 1996;124:627–632.

7 Jungers P, Chauveau P, Ployard F, Lebkiri B, Ciancioni C, Man NK: Comparison of ketoacids and low protein diet on advanced chronic renal failure progression. Kidney Int 1987;32(suppl 22):S67–S71.

8 Rosman JB, Langer K, Brandl M, Piers-Becht TP, Hem GK van der, Wee PM ter, Donker AJ: Protein-restricted diets in chronic renal failure: A four year follow-up shows limited indications. Kidney Int 1989;36(suppl 27):S96–S102.

9 Locatelli F, Alberti D, Graziani G, Buccianti G, Redaelli B, Giangrande A: Porospective, randomised, multicentre trial of effect of protein restriction on progression of chronic renal insufficiency. Northern Italian Cooperative Study Group. Lancet 1991;337:1299–1304.

10 Ihle BU, Becker GJ, Whitworth JA, Charlwood RA, Kincaid-Smith PS: The effect of protein restriction on the progression of renal insufficiency. N Engl J Med 1989;321:1773–1777.

11 Williams PS, Stevens ME, Fass G, Irons L, Bone JM: Failure of dietary protein and phosphate restriction to retard the rate of progression of chronic renal failure: A prospective, randomized, controlled trial. Q J Med 1991;81:837–855.

12 Klahr S: The modification of diet in renal disease study. N Engl J Med 1989;320:864–866.

13 Klahr S, Breyer JA, Beck GJ, Dennis VW, Hartman JA, Roth D, Steinman TI, Wang SR, Yamamoto ME (Modification of Diet in Renal Disease Study Group): Dietary protein restriction, blood pressure control, and the progression of polycystic kidney disease. J Am Soc Nephrol 1995;5:2037–2047.

14 Levey AS, Bosch JP, Coggins CH, Greene T, Mitch WE, Schluchter MD, Schwab SJ (Modification of Diet in Renal Disease Study Group): Effects of diet and antihypertensive therapy on creatinine clearance and serum creatinine concentration in the modification of diet in renal disease study. J Am Soc Nephrol 1996;7:556–565.

15 Kopple JD, Berg R, Houser H, Steinman TI, Teschan P: Nutritional status of patients with different levels of chronic renal insufficiency. Kidney Int 1989;36(suppl 27):S184–S194.

16 Barsotti G, Guiducci A, Ciardella F, Giovannetti S: Effects on renal function of a low-nitrogen diet supplemented with essential amino acids and ketoanalogues and of hemodialysis and free protein supply in patients with chronic renal failure. Nephron 1981;27:113–117.

17 Bennett SE, Russell GI, Walls J: Low protein diets in uraemia. BMJ 1983;287:1344–1345.

18 Barsotti G, Giannoni A, Morelli E, Lazzeri M, Vlamis I, Baldi R, Giovannetti S: The decline of renal function slowed by very low phosphorus intake in chronic renal patients following a low nitrogen diet. Clin Nephrol 1984;21:54–59.

19 Mitch WE, Walser M, Steinman TI, Hill S, Zeger S, Tungsanga K: The effect of a keto acid-amino acid supplement to a restricted diet on the progression of chronic renal failure. N Engl J Med 1984;311:623–629.

20 El Nahas AM, Masters-Thomas A, Brady SA, Farrington K, Wilkinson V, Hilson AJ, Varghese Z, Moorhead JF: Selective effect of low protein diets in chronic renal diseases. BMJ 1984;289:1337–1341.

21 Oldrizzi L, Rugiu C, Valvo E, Lupo A, Loschiavo C, Gammaro L, Tessitore N, Fabris A, Panzetta G, Maschio G: Progression of renal failure in patients with renal disease of diverse etiology on protein-restricted diet. Kidney Int 1985;27:553–557.

22 Lucas PA, Meadows JH, Roberts DE, Coles GA: The risks and benefits of a low protein-essential amino acid-keto acid diet. Kidney Int 1986;29:995–1003.

23 Acchiardo SR, Moore LW, Cockrell S: Does low protein diet halt the progression of renal insufficiency? Clin Nephrol 1986;25:289–294.

24 Barsotti G, Ciardella F, Morelli E, Cupisti A, Mantovanelli A, Giovannetti S: Nutritional treatment of renal failure in type 1 diabetic nephropathy. Clin Nephrol 1988;29:280–287.

25 Evanoff G, Thompson C, Brown J, Weinman E: Prolonged dietary protein restriction in diabetic nephropathy. Arch Intern Med 1989;149:1129–1133.

26 Walker JD, Bending JJ, Dodds RA, Mattock MB, Murrells TJ, Keen H, Viberti GC: Restriction of dietary protein and progression of renal failure in diabetic nephropathy [published erratum appears in Lancet 1989;ii:1540]. Lancet 1989;ii:1411–1415.

27 Bergström J, Alvestrand A, Bucht H, Guiterrez A: Stockholm clinical study on progression of chronic renal failure – An interim report. Kidney Int 1989;36(suppl 27):S110–S114.

28 Brouhard BH, LaGrone L: Effect of dietary protein restriction on functional renal reserve in diabetic nephropathy. Am J Med 1990;89:427–431.

29 Zeller K, Whittaker E, Sullivan L, Raskin P, Jacobson HR: Effect of restricting dietary protein on the progression of renal failure in patients with insulin-dependent diabetes mellitus. N Engl J Med 1991;324:78–84.

30 Dullaart RPF, Beusekamp BJ, Meijer S, Doormaal JJ van, Sluiter WJ: Long-term effects of protein-restricted diet on albuminuria and renal function in IDDM patients without clinical nephropathy and hypertension. Diabetes Care 1993;16:483–492.

31 Raal FJ, Kalk WJ, Lawson M, Esser JD, Buys R, Fourie L, Panz VR: Effect of moderate dietary protein restriction on the progression of overt diabetic nephropathy: A 6-mo prospective study. Am J Clin Nutr 1994;60:579–585.

32 D'Amico G, Gentile MG, Fellin G, Manna G, Cofano F: Effect of dietary protein restriction on the progression of renal failure: A prospective randomized trial. Nephrol Dial Transplant 1994;9:1590–1594.

Mineral and Electrolyte Metabolism

Miner Electrolyte Metab 1997;23:301–305

Francesco Locatelli
Celestina Manzoni
Daniele Marcelli

Department of Nephrology and Dialysis,
Hospital of Lecco, Italy

Factors Affecting Progression of Renal Insufficiency

Key Words
Chronic renal insufficiency
Factors of progression
Glomerular hypertension
Proteinuria
Renin-angiotensin system
Systemic hypertension

Abstract
The aim of this study was to analyze the factors affecting chronic renal insufficiency (CRI) progression at diagnosis (markers of progression), their spontaneous or therapy-induced behavior, and their relationship to CRI progression during follow-up. The underlying disease is the 'determinant factor' of progression and although clinical trials usually report crude cumulative renal survival without taking into account concomitant risk factors, it is known that diabetic nephropathy, polycystic kidney disease, and glomerulonephritis are more progressive than nephroangiosclerosis and interstitial nephropathy. Among the 'effect modifiers,' the baseline level of renal function, hypertension, and proteinuria are the most important. The adverse synergistic effects of proteinuria and high blood pressure have been confirmed, and the importance of correcting hypertension (systemic and glomerular) and proteinuria for slowing disease progression has also been demonstrated. The potential adverse role of a high-protein intake, strongly suggested by experimental studies and the clinical data of uncontrolled trials, has been challenged by the data coming from large controlled trials. The role of lipids needs to be clarified by prospective randomized trials, but the effects of therapeutic interventions aimed at correcting lipid abnormalities seem very promising. The association between the DD genotype of the gene encoding the angiotensin-converting enzyme (ACE) and an increased risk of renal function loss is under evaluation with the aim of identifying the patients who may most benefit from ACE inhibition.

Introduction

It is well known that if there has been sufficient initial damage due to any cause, chronic insufficiency (CRI) usually progresses even when the original disease is no longer active [1]. This implies that factors other than intrinsic disease activity contribute towards the progression of CRI by causing glomerular and tubulointerstitial lesions. Various clinical trials have been performed to identify these factors and ascertain whether at least some of them can be modified by treatment, thus making it possible to halt or at least slow CRI progression, and postpone the need for dialysis.

It is methodologically important to separate the 'determinant' factor affecting CRI progression (underlying renal disease) from 'effect modifiers' (table 1). Furthermore, it is also useful to analyze separately the effect on CRI progression of the factors identified at the time of diagnosis (markers of progression) from those of spontaneous or treatment-induced behavior during follow-up. In this review, we will analyze some of the factors that have been further clarified by recent large-scale clinical trials.

Role of Underlying Renal Disease

Although authors usually report crude cumulative renal survival on the basis of the different nephropathies, without taking into account concomitant risk factors (Cox's analysis), a number of studies [2–7] have stressed the importance of the underlying disease as a factor determining the rate of CRI progression: diabetic nephropathy (DN), polycystic kidney disease (PKD), and glomerulonephritis (GN) are more progressive than interstitial nephritis (IN) and nephrosclerosis. However, the rate of decline in glomerular filtration rate (GFR) is not only different between groups of patients with CRI due to different causes, but also among patients with the same disease. By measuring the increases in plasma creatinine levels, Locatelli et al. [4] found that the progression of CRI in patients with PKD, GN, and IN is sevenfold, threefold, and twice as high, respectively, as in patients with nephrosclerosis, regardless of baseline values of renal function. The Modification of Diet in Renal Disease (MDRD) Study [5] found that the GFR declined more rapidly in patients with PKD than in those with glomerular disease (and no proteinuria), or in those with other diagnoses.

KARGER
E-Mail karger@karger.ch
Fax +41 61 306 12 34
http://www.karger.ch

0378–0392/97/0236–0301 $12.00/0

Francesco Locatelli, MD
Department of Nephrology and Dialysis Hospital of Lecco
Via Ghislanzoni 22, I–22053 Lecco (Italy)
Tel. +39 341 489470, fax +39 341 489474

Table 1. Factors involved in CRI progression

Determinant	
Underlying renal disease	
Effect modifiers	
Initial plasma creatinine levels	
Hypertension	
Systemic	
Glomerular	
Proteinuria	
Dyslipidemia	
Calcium-phosphate	
Hyperuricemia	
Hyperglycemia	
Genetic factors	

On the basis of a univariate analysis according to underlying renal diseases, the Angiotensin-Converting Enzyme Inhibition in Progressive Renal Insufficiency (AIPRI) Study [6] showed that the proportion of patients with nephrosclerosis reaching an endpoint was low (2% in the benazepril group and 4% in the placebo group), as was that of those with IN (5% and 8%); but it was high in those with PKD (27% and 26%) or GN (12% and 27%), and even higher in those with DN (17% and 47%).

Locatelli et al. [7], using renal death as the endpoint, and after taking into account all of the known factors affecting CRI progression, showed that PKD and nephroangiosclerosis had a relative renal death risk of 2.47 and 3.55, respectively, in comparison with IN (fig. 1). However, after taking into account the different baseline levels of renal function, multivariate analysis showed that the patients with PKD had a relatively lower risk of renal death, and PKD was as progressive as IN; similarly, after taking into account baseline 24-hour proteinuria, it was found that the relative risk in GN patients was not significantly higher than in those with IN. However, it is important to stress that hypertension is frequently present in PKD, and hypertension and proteinuria in glomerulonephrites. Multivariate analysis is therefore a very important tool for clarifying the importance of the various individual factors affecting CRI progression, but is somehow misleading in defining the importance of underlying disease.

Initial Plasma Creatinine Levels

High plasma creatinine levels have long been recognized as a factor associated with faster CRI progression [4], and a number of trials [2, 6–7] have stratified patients on the basis of the degree of renal function deterioration. The progression of CRI accelerates as renal function deteriorates, probably because there is a greater chance of encountering patients with a more progressive underlying disease at higher levels of CRI, even among those with the same underlying nephropathy. However, the exponential increase in plasma creatinine with the progressive deterioration in creatinine clearance must be taken into account when analyzing this phenomenon.

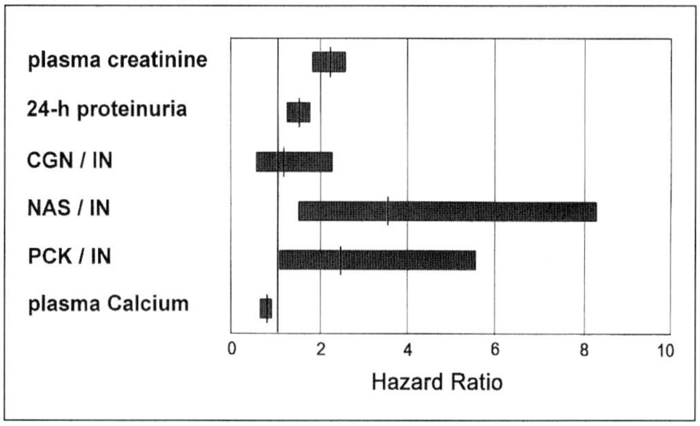

Fig. 1. Relative risk of renal death in different kidney diseases, adjusted for the known factors affecting CRI progression [7]. CGN = Chronic glomerulonephritis; NAS = nephroangiosclerosis; PCK = polycystic kidney; IN = interstitial nephritis.

Systemic and/or Glomerular Hypertension and Proteinuria

When considering hypertension as a factor directly determining CRI, the magnitude of the phenomenon is not very large. Furthermore, it is very difficult to be absolutely sure of the diagnosis, since it is always possible that patients with hypertension and even very low values of proteinuria may have undiagnosed GN; thus, misclassification could misleadingly increase the incidence of CRI in hypertensive patients. However, on the basis of the results of large-scale trials, arterial hypertension is acknowledged to be one of the main factors determining CRI progression. Even more importantly, it was reported several years ago that the lowering of blood pressure per se reduces the rate of decline in renal function in diabetic nephropathy [8] and, more recently, the same effect has also been reported in nondiabetic nephropathy [9]. On the other hand, it has been found [4, 10] that only PKD patients show a relationship between the progression of CRI and blood pressure values; in the other nephropathies, systolic and diastolic blood pressure levels were not statistically related to CRI progression, although the patients who reached an endpoint had significantly higher baseline randomization values [4]. The MDRD study [5] has recently shown a slower rate of renal function deterioration in patients whose blood pressure values were reduced to below the recommended normal values during follow-up, and the effect of controlling hypertension on progression of CRI was especially evident in those with higher baseline proteinuria. It has long been suggested that proteinuria is the other highly important risk factor of CRI progression, and the question as to whether it is only a marker of a more active underlying disease or a direct cause of functional deterioration has been clarified by recent trials [5–7, 11, 12]. It appears that both possibilities could be operative.

It is very difficult to analyze the effect of hypertension (systemic and/or glomerular) and proteinuria separately because of their interrelationships. A number of experimental studies have shown that the relationship between hypertension and CRI progression is at

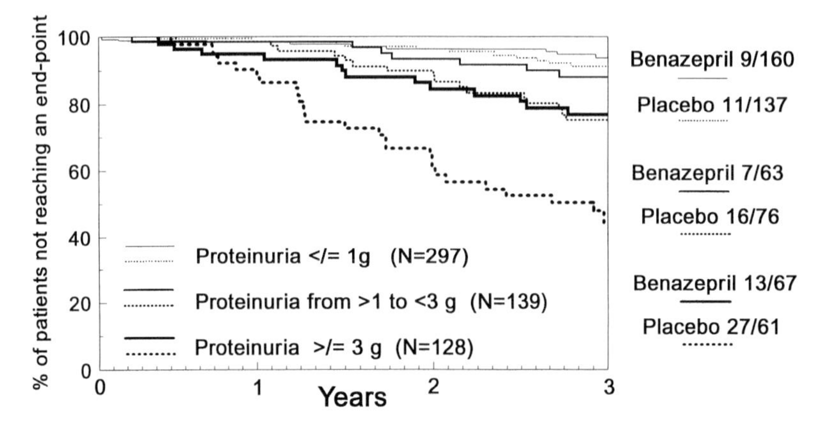

Fig. 2. CRI progression rate in relation to baseline proteinuria according to the AIPRI study [6].

least partly the consequence of impaired renal hemodynamics. Systemic hypertension may cause glomerular hypertension which is the critical determinant of injury involving all of the glomerular constituents [13]. However, glomerular hypertension may also be present in the absence of systemic hypertension, as has been shown in experimental models of diabetes mellitus. It has been suggested that the hemodynamic and cellular effects of the renin-angiotensin system on glomeruli and tubules are significantly involved in the mechanisms of renal injury [14], particularly as the cause of glomerular hypertension and proteinuria. Therefore, in addition to its antihypertensive effect, the blockade of angiotensin II, could also have an antiproteinuric and renoprotective effect. Various studies have shown that angiotensin-converting enzyme (ACE) inhibitors are more capable of decreasing proteinuria and preserving glomerular filtration rate than other antihypertensive agents. Lewis et al. [12] reported the nephroprotective action of captopril versus placebo in a large randomized study of 409 patients with type I diabetes mellitus and overt nephropathy, using the doubling of serum creatinine, renal replacement therapy, or death as endpoints. The incidence of all of the endpoints was significantly lower in the patients receiving captopril therapy; but confounding factors, such as baseline proteinuria and blood pressure control during the trial, were not completely matched between the treated and placebo groups. A recent meta-analysis [15] assessing 84 studies (including 12 randomized and controlled trials involving diabetic patients) has also concluded that ACE inhibitors are more effective than alternative antihypertensive agents in reducing the decline in GFR in patients with incipient and overt diabetic nephropathy, and in patients with other renal diseases. In the AIPRI study [6], 31 of the 300 patients with progressive renal insufficiency treated with an ACE inhibitor, and 57 of the 283 patients treated with placebo reached the endpoint after 3 years of follow-up ($p=0.0006$). The associated reduction in the relative risk of reaching the endpoint was 53% in the ACE-inhibitor group as a whole: 71% in the patients with mild and 46% in those with moderate renal insufficiency. CRI progression was faster (in both the benazepril and the placebo group) going from patients with proteinuria of less than 1 g/24 h to those with proteinuria of more than 3 g/24 h (fig. 2). Proteinuria was reduced to a clinically relevant degree in all of the ACE-inhibitor-treated patients, and the protective effect on renal function was greatest in the patients with substantial proteinuria, even after adjustments were made for diastolic pressure or changes in urinary protein excretion over time. This suggests that the antiproteinuric effect of ACE inhibitors is at least partly related to mechanisms other than their antihypertensive effect per se, and that it may be mediated by hemodynamic changes or by a direct effect of these drugs on the glomerular basement membrane, mesangium, and interstitium.

Experimental studies and clinical data from uncontrolled trials strongly suggest that a low-protein diet may help to delay CRI progression, but the results of largescale controlled studies [2, 5, 10, 11] call for a more cautious approach. In the MDRD study [5], among nondiabetic patients, a nonsignificant trend toward a slower decline in GFR was found in the patients on the very low protein diet (0.4 g/kg per day with amino acid and ketoanalogue supplementation), but there was a nonsignificant difference between the diet groups in terms of the time to the occurrence of end-stage renal disease or death. Nevertheless, secondary analyses of the MDRD study [16] have shown a close correlation between achieved protein intake and the rate of decline in GFR. A recent meta-analysis [17] only considering randomized and controlled studies separately analyzed five trials in diabetic and five in nondiabetic patients. Although the conclusions of the individual studies were inconsistent, the pooled results showed that low dietary protein prescription (mainly but possibly not only via reduced glomerular hyperfiltration) significantly reduced the risk of renal failure or death in comparison with the usual-protein diet ($p=0.007$), thus confirming the results of a previous meta-analysis [18]. The prescription of a low-protein diet (or more importantly, a low-protein intake) therefore does seem to be able to significantly delay the progression of nondiabetic and possibly diabetic renal disease.

Other Effect Modifiers

Williams et al. [11] found only a weak correlation between the calcium phosphate product and the rate of renal function deterioration, whereas Locatelli et al. [4] found that the calcium-phosphate product did correlate with CRI progression although only with regard to the reciprocal of plasma creatinine levels. However, it seems more likely that the behaviour of the calcium-phosphate product is a consequence of the progressive deterioration in renal function (with the accumulation of phosphate being insufficiently controlled by diet or phosphate binders) than an important factor determining CRI progression.

It has been suggested that lipids might also play an important role in CRI progression [19]. However, to our knowledge, no correlation has been found between lipids and the deterioration of chronic renal failure in prospective, randomized large-scale clinical trials, even if the relationship between proteinuria and high plasma cholesterol levels is well known. However, although the role of lipids needs to be clarified by prospective randomized clinical trials, the effects of therapeutic interventions aimed at correcting lipid alterations seem to be very promising.

Considering the important role of the renin-angiotensin system in renal damage progression, as a result of the hemodynamic and cellular effects of angiotensin II, attention has recently been focussed on the structure and polymorphism of the ACE and angiotensin II type 1 receptor genes. A frequent insertion/deletion (I/D) polymorphism of the ACE gene, characterized by the presence/absence of the 287-base-pair Alu sequence situated in intron 16, has been identified and studied as a possible genetic marker of various forms of cardiovascular disease. It is now known that the DD genotype of the gene encoding ACE is accompanied by higher plasma ACE levels. Small studies have suggested an association between the DD genotype and the development of nephropathy in diabetic patients, as well as an association with the better follow-up outcome in IgA nephropathy. It has also been found that the I genotype is a potential determinant of the antiproteinuric effect of ACE inhibitors in patients with proteinuria. It has also been suggested that angiotensin II type 1 receptor gene polymorphism may be associated with proliferative diabetic retinopathy. However, further studies are required to assess the mechanisms underlying these observed associations, as well as the real role of various other polymorphic markers now available, and the possibility of using them in selecting patients who could benefit more from specific treatments (ACE inhibitors and angiotensin II receptor antagonists) [20–21].

Conclusions

If well-established functional and morphological damage is present in the early stages of CRI, prompt treatment is vital in order to break the vicious circle that culminates in end-stage renal disease. Recent trials have definitively demonstrated the importance of correcting hypertension (systemic and above all glomerular) and proteinuria in slowing CRI progresion. A deleterious intrarenal process, mainly mediated by the renin-angiotensin system and its hemodynamic and cellular effects on glomeruli and tubuli, seems to be the main cause of CRI progression, and so the blockade of angiotensin II production by ACE inhibitors is likely to underlie their antiproteinuric and renoprotective effect, in addition to their antihypertensive action. The importance of early intervention is further increased by the advent of new forms of curative treatment at the level of matrix synthesis or matrix induction. The use of drugs that block T cell activation, interfere with T cell recognition, prevent collagen synthesis, and modulate cell adhesion and cytokine formation is also of considerable interest. Although their clinical advantages and long-term side effects are still unknown, these drugs seem to be very promising. In any case, medical care of patients with CRI is of paramount importance in renal outcome.

References

1 Klahr S, Schreiner G, Ichikawa I: The progression of renal disease. N Engl J Med 1988;318:1657–1666.
2 Locatelli F, Alberti D, Graziani G, Buccianti G, Redaelli B, Giangrande A: Prospective, randomised, multicentre trial of effect of protein restriction on progression of chronic renal insufficiency. Lancet 1991;337:1299–1304.
3 Maschio G, Oldrizzi L, Rugiu C, Valvo E, Lupo A, Loschiavo C, Tessitore N, Fabris A, Gammaro L, Panzetta G: Factors affecting progression of renal failure in patients on long-term dietary protein restriction. Kidney Int 1987;32(suppl 22):49–52.
4 Locatelli F, Alberti D, Graziani G, Buccianti G, Redaelli B, Giangrande A, Marcelli D, Francucci BM, and Northern Italian Cooperative Study Group: Factors affecting chronic renal failure progression: Results from a multicentre trial. Miner Electrolyte Metab 1992;18:295–302.
5 Klahr S, Levey AS, Beck GJ, Caggiula AW, Hunsicker L, Kusek JW, Striker G, Modification of Diet in Renal Disease Study Group: The effects of dietary protein restriction and blood pressure control on the progression of chronic renal disease. N Engl J Med 1994;330:877–884.
6 Maschio G, Alberti D, Janin G, Locatelli F, Mann J, Motolese M, Ponticelli C, Ritz E, Zucchelli P, and the Angiotensin-Converting-Enzyme Inhibition in Progressive Renal Insufficiency Study Group: Effect of the angiotensin-converting-enzyme inhibitor benazepril on the progression of chronic renal insufficiency. N Engl J Med 1996;334:939–945.
7 Locatelli F, Marcelli D, Comelli M, Alberti D, Graziani G, Buccianti G, Redaelli B, Giangrande A, and the Northern Italian Cooperative Study Group: Proteinuria and blood pressure as causal components of progression to end-stage renal failure. Nephrol Dial Transplant 1996;11:461–467.
8 Mogensen CE: Long-term antihypertensive treatment inhibiting progression of diabetic nephropathy. Br. Med J 1982;285:685–688.
9 Zucchelli P, Zuccalà A, Borghi M, Fusaroli M, Sasdelli M, Stallone C, Sanna G, Gaggi R: Long-term comparison between captopril and nifedipine in the progression of renal insufficiency. Kidney Int 1992;42:452–458.
10 Rosman JB, Langer K, Brandl M, PiersBecht TM, Van Der Hem GK, Wee PM ter, Donker JM: Protein-restricted diets in chronic renal failure: A four year follow up shows limited indications. Kidney Int 1989;36(suppl 27):96–102.
11 Williams PS, Fass G, Bone JM: Renal pathology and proteinuria determine progression in untreated mild/moderate chronic renal failure. Q J Med 1988; 67:343–354.
12 Lewis EJ, Hunsicker LG, Bain RP, Rohde RD: The effect of angiotensin-converting enzyme inhibition on diabetic nephropathy. N Engl J Med 1993;329: 1456–1462.
13 Anderson S: Progression of chronic renal disease: Role of systemic and glomerular hypertension. Am J Kidney Dis 1989;13(suppl 1):8–12.
14 Klahr S, Ishidoya S, Morrissey J: Role of angiotensin II in the tubulointerstitial fibrosis of obstructive nephropathy. Am J Kidney Dis 1995;26: 141–146.
15 Maki DD, Ma JZ, Louis TA, Kasiske BL: Long-term effects of antihypertensive agents on proteinuria and renal function. Arch Intern Med 1995; 155:1073–1080.
16 Levey AS, Adler S, Caggiula AW, England BK, Greene T, Hunsicker LG, Kusek JW, Rogers NL, Teschan PE: Effects of dietary protein restriction on the progression of advanced renal disease in the Modification of Diet in Renal Disease Study. Am J Kidney Dis 1996;27:652–663.
17 Pedrini MT, Levey AS, Lau J, Chalmers TC, Wang PH: The effect of dietary protein restriction on the progression of diabetic and nondiabetic renal diseases: A meta-analysis. Ann Intern Med 1996;124: 627–632.
18 Fouque D, Laville M, Boissel JP, Chifflet R, Labeeuw M, Zech PY: Controlled low protein diets in chronic renal insufficiency: Meta-analysis. BMJ 1992;304:216–220.
19 Keane WF, Kasiske BL, O'Donnell MP: Lipids and progressive glomerulosclerosis. Am J Nephrol 1988; 8:261–271.
20 Yoshida H, Kon V, Ischikawa I: Polymorphisms of the renin-angiotensin system genes in progressive renal diseases. Kidney Int 1996;50:732–744.
21 Van Essen GG, Rensma PL, De Zeeuw D, Sluiter WJ, Scheffer H, Apperloo AJ, De Jong PE: Association between angiotensin-converting-enzyme gene polymorphism and failure of renoprotective therapy. Lancet 1996;347:94–95.

Mineral and Electrolyte Metabolism

Miner Electrolyte Metab 1997;23:306–310

Beatriz D. Kuizon
Pauline A. Nelson
Isidro B. Salusky

Department of Pediatrics, UCLA School of Medicine, Los Angeles, Calif., USA

Tube Feeding in Children with End-Stage Renal Disease

Key Words
Growth
Infants
Nutrition
Peritoneal dialysis
Tube Feeding

Abstract
Aggressive nutritional support and treatment with peritoneal dialysis (PD) have been advocated in the management of children with end-stage renal disease. In recent years, supplemental enteral feeding has been recommended by the majority of pediatric centers in the US and Europe since it has been shown to improve growth and neurologic development. Tube feeding, most commonly using a nasogastric (NG) or gastrostomy tube (G-tube) must be instituted when voluntary intake does not consistently meet the caloric and protein requirements or when there is poor growth. Recurrent emesis is the most common complication with NG feeds, while PD leak and exit site infection have been described with G-tube feeds. The early initiation of aggressive nutritional therapy to maximize growth potential must be emphasized.

Introduction

While awaiting transplantation, vigorous nutritional support and treatment with peritoneal dialysis (PD) have been the cornerstone in the management of young children who develop end-stage renal disease (ESRD) during infancy [1–6]. Recognizing the profound effects of renal insufficiency on growth and development, these interventions must be initiated during the first 2 years of life, when the potential for growth is highest. During this period, children with intact renal function demonstrate a 50% increase in linear growth, quadrupling of weight, and a 40% increase in head circumference [4]. By contrast, loss of renal function during this period results in severe growth retardation and failure of catch-up growth despite treatment with dialysis or renal transplantation once significant deficits have been incurred [7, 8]. Since approximately 30% of normal adult height is gained during this critical period, chronic renal failure in infancy may have a long-term impact, resulting in permanent short stature.

Neurologic abnormalities have also been described in children who develop chronic renal failure early in life, and the main factors implicated are malnutrition, uremia, aluminum toxicity, and psychosocial deprivation. Some of these abnormalities are not unique to children with chronic renal failure but also develop in other conditions with coexistent malnutrition. Evidence to support the impact of any insult to the brain during the first 2 years of glial proliferation and rapid myelination is seen in animal and human studies. Elimination of aluminum-containing phosphate binders, nutritional therapy, and treatment with dialysis have improved outcome when these interventions were initiated early [9].

Although other factors such as uremia, renal osteodystrophy, vitamin D deficiency, acidosis, and hormonal abnormalities contribute to growth retardation and developmental delay, malnutrition undoubtedly plays a major role. Anorexia is a frequent finding in children with renal failure and leads to inadequate oral intake. Supplemental enteral feeding, therefore, becomes necessary if the recommended caloric requirement is not consistently achieved with voluntary intake. Aggressive nutritional support combined with dialysis treatments must be maintained until the children attain the optimal size for renal transplantation. Although this is center dependent, this corresponds to a body weight that ranges between 7–12 kg.

This paper will address the indications and the different types of tube feeding in children undergoing PD. The advantages and disadvantages associated with each method of tube feeding will also be discussed.

Indications for Supplemental Tube Feeding

Inadequate caloric intake for growth is commonly observed in children with ESRD and it is multifactorial in origin. Anorexia may be secondary to uremia, altered taste from zinc deficiency, or from unpalatable low-phosphorus diets. The presence of PD fluid contributes to abdominal distention and may lead to feeding intolerance. Vomiting is not unusual and may possibly be from an unrecognized gastroesophageal reflux or delayed gastric emptying time because of uremia. Consequently, enteral tube feedings should be initiated when the voluntary oral intake does not consistently meet

KARGER
E-Mail karger@karger.ch
Fax +41 61 306 12 34
http://www.karger.ch

0378–0392/97/0236–0306 $12.00/0

Isidro B. Salusky, MD
UCLA Children's Hospital, Division of Pediatric Nephrology
10833 Le Conte Avenue, A2-383 MDCC, Los Angeles, CA 90095 (USA)
Tel. +1 310 206 6987, fax +1 310 825 0442
E-Mail salusky@crc.medsch.ucla.edu

the recommended daily allowances (RDAs) for calories and protein despite supplementation, or when there is poor weight gain and linear growth [1–5]. Parental anxiety and frustration associated with unsuccessful attempts at feeding may also prompt introduction of tube feedings before an aversive feeding interaction develops between the infant and the parents [3, 10, 11].

Tube feedings may be used either to provide the total nutritional requirement or as supplement to oral intake. Ideally, these should be instituted before the patient has markedly fallen off the growth curve and has developed significant nutritional deficits. Tube feedings may be introduced early in the course of renal failure, prior to the initiation of renal replacement therapy. Other uses for the enteral tube include delivery of medications and fluid supplements in infants with high obligatory urine output.

For infants undergoing peritoneal dialysis, energy intake should exceed the RDA of 110–120 kcal/kg per day. Total caloric intake should be calculated using oral intake and dialysate glucose absorption, which may account for approximately 12% of the total energy consumption. The recommended protein intake in infants should range between 2.5–3 g/kg per day, with the majority consisting of protein of high biologic value. These recommendations are based on the RDA, the total amount of daily protein losses through the dialysate effluent, and the nitrogen balance studies performed in adult patients undergoing peritoneal dialysis [12]. Protein intake should be adjusted to achieve normal levels of serum albumin and prealbumin while maintaining blood urea nitrogen below 80 mg/dl. In addition, patients should receive the following vitamins: folic acid 1 mg/day, pyridoxine (B_6) 5–10 mg/day, and ascorbic acid 75–100 mg/day. Vitamin D sterols (calcitriol) and phosphate-binding agents, primarily calcium carbonate, are recommended to prevent and control renal bone disease. The requirements for sodium, potassium, and water intake must be managed individually.

Commercially available formulas as well as carbohydrate, protein, and fat supplements may be used for tube feeding (table 1). Because of its low potassium and phosphorus content, Similac PM 60/40 (Ross Laboratories) may be appropriate in children with renal failure who are less than 1 year of age.

Methods of Tube Feeding

Supplemental enteral feedings may be administered via nasogastric (NG), transpyloric, gastrostomy (G-tube/button) or gastrojejunostomy (GJ) tube. While each method of enteral feeding has its own advantages and complications, the preferred tube feeding modality is largely center dependent. The most commonly chosen modalities are the NG- or G-tube feedings.

A variety of options are available for administration of feedings, such as daytime intermittent bolus feeds, continuous drip for 24 h or for 10–12 h at night while the patient is undergoing dialysis. The method of delivery is dependent on the age of the patient, the quantity to be delivered, and the composition of the formula. In continuous infusion, the patients are disturbed less often and have fewer problems with gastric distention. Scheduling the feeding time at night and during dialysis offers the child opportunities for spontaneous oral intake and freedom from being attached to equipment during the day.

Continuous drip feedings may be initiated using isotonic formulas at a rate of approximately 1–2 ml/kg per hour [3], and the rate gradually increased as tolerated. The infant should be observed frequently during feeding for signs of intolerance such as increased gastric residual volume, diarrhea, vomiting, and abdominal distention.

NG-Tube Feeding

Supplemental feeding via the NG tube has been the most common method of nutritional support in children undergoing PD. It has the major advantage of not requiring a surgical procedure. The parents can be taught the technique of passing a NG tube and to confirm placement by auscultation in less than 1 day. The length of the feeding tube is measured from the tip of the nose, around the ear, to the xiphoid. Warady et al. [3] recommend using either polyurethane or silicone NG tubes because of their flexibility, decreased local irritation, and capacity for long-term use. These NG tubes are changed every 4 weeks. In contrast, PVC NG tubes may stiffen and require replacement every 2–4 days because of the risk for gastric and intestinal perforation. Since infants are obligatory nose breathers, the tube with the smallest possible external diameter should be used to prevent airway obstruction.

Considerable clinical experience has been gained with NG tubes. Conley [2] described the outcome of ten infants treated with PD and aggressive nutritional support after 164 patient-months (range 3–30 months) of follow-up. NG feedings, which supplied the majority of intake, were maintained in nine out of ten patients. The patients received greater than 100% of the RDA for the same height age, with mean intake of 120% RDA for energy and 183% for protein. Biochemical parameters were monitored and the mean serum urea nitrogen was maintained at 64 mg/dl, creatinine at 4.6 mg/dl, and albumin at 3.1 g/dl. Patients exhibited a good weight gain although obesity was noted in some patients. All except one patient had an initial standard deviation score (SDS) for height of less than –2 and at the end of follow-up, catch-up growth was seen in three patients and the remainder grew within 1 SDS of normal. Intellectual development also improved in all patients except for two who had significant neurologic deficits at the start of therapy.

Brewer [13] reported the 203-month experience with NG tube feeding in 14 children (age 3 days – 37 months; weight 6.5 ± 2.8 kg at start of dialysis) undergoing PD. NG-tube feedings were initiated in 4 patients prior to dialysis. A high protein intake (3–4 g/kg per day) was required in most patients to maintain serum albumin above 3 g/dl. Height SDSs were maintained within the normal range (> -2.00) in 5 patients. Three out of 4 patients who had a height SDS between –2.00 and –4.00 at the start of PD showed improvement in height SDS to the normal range. Of the 5 infants who had a height SDS < -4.00 at the start of PD, 3 showed an increase of ≥ 1.00 height SDS during treatment. Weight SDSs were maintained within the normal range (> -2.00) or improved in 12 patients. Nine out

Table 1. Infant formulas and nutritional supplements

	Energy kcal/ml	Osmolality mOsm/kg H_2O	Protein g/l	Fat g/l	Carbohydrate g/l	Na/K mEq/l
Infant formulas						
Similac PM 60/40 (Ross)	0.68	260	16	38	69	7/15
Similac (Ross)	0.68	300	15	36	72	8/18
Milk, cow	0.65	288	35	35	48	22/37
Milk, human	0.68	290	11	39	72	8/13
Carbohydrate modules						
Corn syrup	3.9	NA	0	0	1,027	40/2
Polycose (Ross)	2 (4/g)	900	0	0	94/100 g	5/0.3 per 100 g
Sumacal (Sherwood)	(4/g)	NA	0	0	95/100 g	4/≤1 per 100 g
Moducal (Mead-Johnson)	(4/g)	NA	0	0	95/100g	3/≤1 per 100 g
Protein modules						
Casec (Mead-Johnson)	(4/g)	NA	88/100 g	2/100 g	0	7/3 per 100 g
Propac (Sherwood)	(4/g)	30	78/100 g	8/100 g	5/100 g	10/13 per 100 g
ProMod (Ross)	(4/g)	NA	100/100 g	12/100 g	13/100 g	13/33 per 100 g
Fat modules						
MCT Oil (Mead-Johnson)	7.7	0	0	858	0	0
Vegetable oil, corn	8.0	0	0	898	0	0
Essential amino acid supplements						
Aminaid (McGaw)	1.9	700	19	46	366	14/5
Travasorb Renal (Clinitec)	1.35	590	23	18	271	0/0
Calories/protein supplements (over 1yr of age)						
Ensure (Ross)	1.0	470	37	37	145	37/40
Sustacal (Mead-Johnson)	1.0	650	61	23	140	40/53
Pediasure (Ross)	1.0	310	30	50	110	16/33
Suplena (Ross)	2.0	600	30	96	255	34/29
Nepro (Ross)	2.0	635	70	96	215	36/27

NA = Information not available.

of 10 patients were successfully transplanted. It is noteworthy that poor outcome was observed despite dialysis and aggressive nutritional therapy in the 4 patients who started PD at an early age (<4 months of age).

Further evidence of improved outcome with supplemental NG feeding is supported by the report of Warady [14] on 28 infants who started PD before 3 months of age. Supplemental NG feeds were administered to 27 infants. At 1 year of age, all patients had weight >90% of the ideal body weight calculated as the weight for height at the 50th percentile. The mean head circumference SDS was -1.03 ± 1.04. Developmental delay was noted only in 2 patients.

Recurrent emesis is the most common complication associated with NG feeding. These episodes may or may not be associated with aspiration. Persistent vomiting may be so severe as to cause PD exit site leak requiring surgical repair [15]. The concurrent presence of significant gastroesophageal reflux may necessitate Nissen fundoplication and placement of a G-tube. Other potential complications of long-term NG tube use include sinusitis, otitis, nasoseptal erosion, pulmonay placement with aspiration, and problems with swallowing and speech. NG tubes are also cosmetically unattractive and repeated passage of the catheter may result in considerable distress to the child and parents.

Transpyloric Tube Feeding

Transpyloric catheters, which were passed directly to the duodenum (nasoduodenal) or jejunum (nasojejunal), were subsequently introduced to prevent vomiting, reflux, and aspiration associated with NG-tube feeding. Tube migration, however, was frequently observed. Furthermore, the proper placement of the catheters was difficult and often required radiologic confirmation. Transpyloric feeding was also associated with intestinal perforation, diarrhea, alteration in the gastrointestinal flora, and impaired fat absorption. Because of lack of gastric reservoir capacity, continuous drip feeding is required in this modality.

G-Tube/Button

The G-tube or button has facilitated the delivery of nutritional support in an increasing number of patients. The G-tube is placed by percutaneous endoscopic insertion or by surgical implantation after excluding the presence of gastroesophageal reflux. It may be necessary to with hold PD for 24–48 h after G-tube placement and to start with low-volume dialysis exchanges [11]. G-tube feeding is started once intestinal function has returned.

G-tube feeding is preferred at UCLA because the episodes of vomiting with its resultant complications are less frequently encountered. Possible dislodgment of the NG tube and consequent aspiration may also be prevented using the G-tube. In addition, this modality prevents considerable stress and trauma, both to the patient and the parents, associated with repeated placement of the NG tube. Continuous infusions for 10–12 h at night are given to encourage oral intake during waking hours.

Balfe et al. [11] reported the outcome of tube feeding in 20 children receiving PD. Two patients received NG tube feeding while 18 received G-tube feeding. The G-tube was placed endoscopically in 5 children and surgically in 13 children. Nissen fundoplication was necessary in 4 patients. Most of the children were less than 2 years of age (mean age 3.9 ± 3.8 years) and half received tube feeding before the initiation of dialysis. Mean follow-up time while receiving tube feeding and peritoneal dialysis was 14.1 ± 11.4 months. After initiation of tube feeding, 17 patients had improvement in SDSs for weight and half had improvement in SDS for height. Although there was improvement in mean SDS for weight after initiation of PD, there was no difference in either SDS for weight or height in patients who received both tube feeding and dialysis treatments. Relatively few complications were noted. Gastrocutaneous fistula developed in 3 children and 2 required surgical correction. G-tube exit site infection developed in 4 children, subsequently leading to peritonitis with the same organism in one child. Vomiting and diarrhea also occurred although it was not serious.

Wood et al. [15] compared the complications associated with NG tube and G-tube feedings in children with ESRD. In six patients (age 1 week – 16 months) who received NG feedings for 5.1 ± 5.1 months, half were converted to G-tube feedings for the following complications: (1) persistent vomiting with aspiration and coexisting gastroesophageal reflux requiring Nissen fundoplication, (2) persistent vomiting with PD exit site leak requiring surgical repair, and (3) patient and parental intolerance. Of the three patients who were maintained on NG feeding, one had recurrent emesis with aspiration that was controlled with conversion to nasoduodenal feeding. The other two had intermittent vomiting. Sinusitis was frequently encountered in one patient. The complication rate with NG feeding was 1 per 6.4 patient months. This group of patients was compared with nine children (age 4 days–11 years) who received G-tube feedings. The complication rate with G-tube feeding was similar at 1 per 7.1 patient months. There were fewer episodes of vomiting in this group although mechanical problems were more frequently encountered. The following complications were observed: (1) PD fluid leak around the G-tube exit site which led to peritonitis, (2) G-tube exit site infection associated with peritonitis in one patient, (3) G-tube obstruction, (4) tube migration to the duodenum causing partial bowel obstruction, and (5) gastrocutaneous fistula requiring surgical closure in two patients.

GJ-Tube Feedings

Percutaneous GJ-tube feedings have been recommended by O'Regan et al. [16] because of the increased morbidity associated with NG- and G-tube feedings. In a study of 13 children (age 2 months–11 years) receiving PD and GJ feedings, no peritoneal fluid leak was noted from the GJ exit site. Of the three episodes of exit site infections described, all responded to oral antibiotic therapy and were not complicated by peritonitis. Other complications included intestinal pseudo-obstruction because the tube used was too large, catheter retraction, catheter obstruction, and peritonitis secondary to peritoneal leak. Except for exit site infection, morbidity has been neglibible in the last 8 patients who have undergone this procedure. Thus, the authors recommended this method for caloric supplementation because of minimal morbidity and hospitalization as well as esthetic acceptability.

Summary

Supplemental feeding has markedly improved the outcome of young children with ESRD. The importance of instituting early aggressive nutritional therapy to maximize growth potential must be emphasized. Its significance in older children, however, has not been determined. The risk for impaired oral-motor development and coordination during tube feeding in infancy has been raised [3, 17]. Extensive retraining on how to eat should be an integral part of the overall management of these children. Nonnutritive sucking as well as oral stimulation with a variety of textures should also be encouraged. A multi-disciplinary team approach involving the physicians, nurses, dietitians, occupational therapists, and behavioral psychologists may often be required [3].

Some of the observed complications are unique to this patient population. Peritoneal fluid leak either at the G-tube or PD catheter exit site has been described and may be attributable to increased intraabdominal pressure from the dialysate fluid. G-tube obstruction may be encountered frequently, especially with the the use of highly concentrated formulas and viscous medications. Increased risk for fungal peritonitis has also been reported in patients receiving G-tube feeds and PD [18].

The availability of recombinant human growth hormone (rhGH) and the potential for marked acceleration in growth velocity has made a tremendous impact in the management of these patients [19–21]. The significant increase in muscle mass as evidenced by increased midarm muscle circumference and decreased triceps skinfold provide evidence for the positive anabolic effect of rhGH therapy. Thus, aggresive nutritional support in combination with rhGH therapy may provide the best regimen for optimal growth in these patients. However, the guidelines for use of rhGH with vigorous nutritional support need to be established.

References

1 Kohaut EC, Whelchel J, Waldo FB, Diethelm AG: Aggressive therapy of infants with renal failure. Pediatr Nephrol 1987;1:150–153.

2 Conley SB: Supplemental (NG) feedings of infants undergoing continuous peritoneal dialysis; in Andreucci VE (ed): Chronic Ambulatory Peritoneal Dialysis (CAPD) and Chronic Cycling Peritoneal Dialysis (CCPD) in Children. Boston, Nijhoff, 1987, pp 263–269.

3 Warady BA, Weis L, Johnson L: Nasogastric tube feeding in infants on peritoneal dialysis. Periton Dial Int 1996;16(suppl 1):S521–S525.

4 Wassner SJ, Albitol C, Alexander S, Conley S, Grupe W, Holliday M, Rigden S, Salusky IB: Nutritional requirements for infants with renal failure. Am J Kidney Dis 1986;7:300–305.

5 Warady BA, Kriley M, Lovell H, Farrell SE, Hellerstein S: Growth and development of infants with end-stage renal disease receiving long-term peritoneal dialysis. J Pediatr 1988;112:714–719.

6 Loirat C, Ehrich JHH, Geerlings W, Jones EH, Landais P, Mallick NP, Margreiter R, Raine AE, Salmela K, Selwood NH, Tufveson G, Valderrabano F: Report on the management of renal failure in children in Europe, XXIII, 1992. Nephrol Dial Transplant 1994;1:26–40.

7 Kleinknecht C, Broyer M, Gagnadoux M, Marthihenneberg C, Dartois AM, Kermanach C, Pouliquen M, Degoulet P, Usberti M, Roy MP: Growth in children treated with long term dialysis. A study of 76 ptients. Adv Nephrol 1980;9:133–163.

8 Chantler C, Carter JE, Bewick M, Counahan R, Cameron JS, Ogg CS, Williams DG, Winder E: Ten years experience with hemodialysis and renal transplant. Arch Dis Child 1980;55:435–445.

9 Elzouki A, Carroll J, Butinar D, Moosa A: Improved neurologic outcome in children with chronic renal disease from infancy. Pediatr Nephrol 1994; 8:205–210.

10 Warady BA, Kriley M, Belden B, Hellerstein S, Alon U: Nutritional and behavioral aspects of nasogastric tube feeding in infants receiving chronic peritoneal dialysis; in Khanna R, Nolph KD, Prowant BF, Twadorski ZJ, Oreopoulos DG (eds): Advances in Peritoneal Dialysis. Toronto, Peritoneal Dialysis Bulletin, 1990, vol 6, pp 265–268.

11 Balfe JW, Secker DJ, Coulter PE, Balfe JA, Geary DF: Tube feeding in children on chronic peritoneal dialysis; in Khanna R, Nolph KD, Prowant BF, Twadorski ZJ, Oreopoulos DG (eds): Advances in Peritoneal Dialysis. Toronto, Peritoneal Dialysis Bulletin, 1990, vol 6, pp 257–261.

12 Blumenkrantz MJ, Kopple JD, Moran JK, Coburn JW: Metabolic balance studies and dietary protein requirements in patients undergoing continuous ambulatory peritoneal dialysis. Kidney Int 1982;21: 849–861.

13 Brewer ED: Growth of small children managed with chronic peritoneal dialysis and nasogastric tube feedings: 203-Month experience in 14 patients; in Khanna R, Nolph KD, Prowant BF, Twadorski ZJ, Oreopoulos DG (eds): Advances in Peritoneal Dialysis. Toronto, Peritoneal Dialysis Bulletin, 1990, vol 6, pp 269–272.

14 Warady BA: Neuro-developmental issues in infants on peritoneal dialysis. Presented at the Sixth Annual Pediatric PD Symposium, Baltimore, Feb 1995.

15 Wood EG, Bunchman TE, Khurana R, Fleming SS, Lynch RE: Complications of nasogastric and gastrostomy tube feedings in children with end stage renal disease; in Khanna R, Nolph KD, Prowant BF, Twadorski ZJ, Oreopoulos DG (eds): Advances in Peritoneal Dialysis. Toronto, Peritoneal Dialysis Bulletin, 1990, vol 6, pp 262–264.

16 O'Regan S, Garel L: Percutaneous gastrojejunostomy for caloric supplementation in children on peritoneal dialysis; in Khanna R, Nolph KD, Prowant BF, Twadorski ZJ, Oreopoulos DG (eds): Advances in Peritoneal Dialysis. Toronto, Peritoneal Dialysis Bulletin, 1990, vol 6, pp 273–275.

17 Kamen RS: Impaired development of oral-motor functions required for normal oral feeding as a consequence of tube feeding during infancy; in Khanna R, Nolph KD, Prowant BF, Twadorski ZJ, Oreopoulos DG (eds): Advances in Peritoneal Dialysis. Toronto, Peritoneal Dialysis Bulletin, 1990, vol 6, pp 276–278.

18 Murugasu B, Conley SB, Lemire JM, Portman RJ: Fungal peritonitis in children treated with peritoneal dialysis and gastrostomy feeding. Pediatr Nephrol 1991;5:620–621.

19 Koch VH, Lippe B, Nelson P, Boechat MI, Sherman B, Fine RN: Accelerated growth after recombinant human growth hormone treatment of children with chronic renal failure. J Pediatr 1989;115:365–371.

20 Mehls O, Tonshoff B, Tonshoff C, Haffner D, Blum WF: Therapeutic value or recombinant human growth hormone in children with chronic renal failure. Miner Electrolyte Metab 1992;18:320–324.

21 Van Renen MJ, Hogg RJ, Sweeney A, Henning PH, Penfold JL, Jureidini KF: Accelerated growth in short children with chronic renal failure treated with both strict dietary therapy and recombinant growth hormone. Pediatr Nephrol 1992;6:451–458.

Mineral and Electrolyte Metabolism

Miner Electrolyte Metab 1997;23:311–313

Luigi Iorio
Giuseppe Saltarelli
Remo G. Nacca
Roberto Simonelli
Fabio Violi

G. DeBosis Hospital,
Cassino, Italy

Hyperlipidemia in Dialysis Patients: What Treatment?

Key Words
Hyperlipidemia
Maintenance dialysis
Dialytic techniques
High-flux membranes
Biocompatibility
Omega-3 fatty acids

Abstract
The incidence of hyperlipidemia in the dialysis population in our hospital reflects the incidence reported by other authors but we ascertained a high incidence of hypercholesterolemic patients. The importance of cardiovascular illnesses as the cause of death in patients on dialysis means taking into consideration risk factors. In our study, omega-3 fatty acids proved to be effective in about half the patients treated and did not have side effects. In a comparison of patients who underwent alternative techniques with high-flux membranes and those who underwent standard dialysis, the first group showed a virtually normal lipid profile. High-flux membranes could be a therapeutical possibility for hypertriglyceridemic patients under dialysis.

Introduction

Patients in maintenance dialysis frequently demonstrate high levels of lipoproteins [1–4]. It is well known that hyperlipidemia is a major factor in cardiovascular illnesses, its degree of severity being correlated with the patient's life expectancy [5–8]. Evaluation of the lipid profile is a common practice for dialysis patients but opting for the therapeutical treatment is complicated because drugs frequently produce serious side effects [9, 10]. Therefore, in an attempt to resolve this problem, a different pharmacological approach and/or alternative treatment needs to be found. We compared the lipid profile of these patients undergoing standard dialysis with that of those undergoing alternative dialysis treatment. We also examined the effects of omega-3 fatty acids in hypertriglyceridemic patients.

Materials and Methods

We studied 86 patients excluding those with diseases or those treated with drugs which could have altered lipid metabolism. Of these 86, 57 patients used cuprophan or hemophan filters and acetate buffer (standard dialysis), while 29 were under alternative treatments [hemofiltration, biofiltration, hemodiafiltration, and acetate-free biofiltration (AFB)] using polyamide AN69HF filters (acrylonitrile and sodium methyl sulfonate copolymer) or high-permeability polysulfone filters. Patients treated with bicarbonate buffer suffering from metabolic alterations (e.g., diabetes mellitus or liver illnesses) were excluded from the study. The dialysis regimen was between 3.5 and 4 h three times a week. The patients followed a 35–40 kcal/kg and 1.2 g protein/kg of ideal weight diet. The blood samples were drawn from vascular access before administering anticoagulants, the patients having fasted for at least 12 h. We analyzed the average values obtained after the determination of total cholesterol, high-density lipoprotein (HDL) cholesterol, low-density lipoprotein (LDL) cholesterol, and triglycerides. The results of patients undergoing standard dialysis were compared with those of patients under alternative therapies.

The latter were subsequently divided into subgroups, comparing those who underwent AFB (11 patients) with those medicated by means of acetate buffer during biofiltration (12 patients), and the group under AN69HF filter treatment (13 patients) with those using polysulfone filters (10 patients). No further subgroups were taken into consideration due to the exiguous number of patients (6) undergoing treatment using the polyamide filter.

In the second stage, 31 hypertriglyceridemic patients (>200 mg/dl), who were undergoing dialysis with acetate buffer, were treated with omega-3 fatty acids: docosahexaenoic acid (DHA) and eicosapentaenoic acid (EPA) in 3.4 g/day doses. Total cholesterol, HDL and LDL cholesterol, and triglyceride values were calculated at the beginning of therapy and after 4 and 12 weeks.

Results

The average value of cholesterolemia was high in 43.8% (36 patients), while that of HDL cholesterol was low. About 50% of our patients (n = 43) had triglyceridemia above the normal level (table 1). The lipid profile of patients treated with alternative techniques was generally better than those managed with standard dialysis using acetate buffer (table 2). Statistical analysis (t test), usually difficult when applied to triglycerides due to the high numeric variation and standard deviations, revealed the importance of comparing both triglycerides and HDL cholesterol average values. The

KARGER
E-Mail karger@karger.ch
Fax +41 61 306 12 34
http://www.karger.ch

0378–0392/97/0236–0311 $12.00/0

Luigi Iorio, MD
G. DeBosis Hospital, via Casilina Nord
I–03043 Cassino FR (Italy)
Tel. +39 776 303243, fax +39 776 302232

Table 1. Mean (±SE) values of triglycerides, total cholesterol, HDL cholesterol, and LDL cholesterol for the entire patient population (n=86)

Triglycerides	Total cholesterol	HDL cholesterol	LDL cholesterol
235.8±141.2	194.6±51.4	38.4±16.0	127.9±37.4

Table 2. Mean (±SE) values of triglycerides, total cholesterol, HDL cholesterol and LDL cholesterol: standard dialysis compared to alternative techniques

	Triglycerides	Total cholesterol	HDL cholesterol	LDL cholesterol
Alternative techniques (n=29)	173.2±113.3	192.0±41.1	41.6±14.5	121.6±36.5
Standard dialysis (n=57)	247.1±142.4*	196.9±52.2	36.4±6.1*	131.7±38.3

*p<0.05.

Table 3. Mean (±SE) values of triglycerides, total cholesterol, HDL cholesterol, and LDL cholesterol: AFB versus alternative techniques with acetate buffer

	Triglycerides	Total cholesterol	HDL cholesterol	LDL cholesterol
AFB (n=11)	151.1±97.5	181.4±36.3	41.7±13.2	134.3±32.5
Alternative techniques with acetate buffer (n=12)	198.6±145.8	200.1±45.6	44.1±15.7	115.0±23.7

Table 4. Mean (±SE) values of triglycerides, total cholesterol, HDL cholesterol, and LDL cholesterol: AN69HF versus high-flux polysulfone

	Triglycerides	Total cholesterol	HDL cholesterol	LDL cholesterol
AN69HF (n=11)	152.5±152.5	182.2±37.0	47.0±15.7	119.2±12.2
Polysulfone (n=12)	210.4±147.1	201.5±47.2	37.4±10.3	121.8±35.1

subgroup of patients that underwent alternative treatment without acetate in the dialysate (AFB) had a better lipid profile than those treated with acetate dialysate (table 3). Those patients treated with the AN69HF filter had better values than those subjects treated with high-permeability polysulfone (table 4). Of the 31 patients who took the omega-3 fatty acids, 5 dropped out: 2 interrupted the therapy due to gastrointestinal disturbances (nausea and diarrhea); the other 3 interrupted the treatment spontaneously. Total cholesterol, HDL cholesterol, and LDL cholesterol did not undergo any significant change. Twelve patients, 46.1%, with hypertriglyceridemia responded to the treatment with omega-3 fatty acids. The high lipid values returned to normal levels or decreased by over 100 mg/dl; the other indices of the lipid profile in this group did not show any significant change.

Discussion

During the first part of the study, we evaluated the incidence of dyslipidemia in the maintenance dialysis patients in our hospital. The results confirmed a high percentage of hyperlipidemia and 43.8% of our patients were hypercholesterolemic. In most of these patients, the hyperlipoproteinemia was revealed in the increase in triglycerides which may participate in the development of cardiovascular illnesses. Triglyceride and cholesterol metabolism are strictly linked via the cholesterol ester transfer protein. This reduces HDL cholesterol when the level of lipoproteins rich in triglycerides is high [11]. Currently, there is no drug for the treatment of dyslipidemia which guarantees therapeutical effectiveness without serious side effects. We decided, therefore, to treat the hypertriglyceridemic patients with omega-3 fatty acids in minimum doses sufficient to modify the triglyceride levels; our results showed that these drugs are tolerated and do not present any dangerous side effects. Of the treated patients, 46.1% responded to the therapy. The effectiveness of the drugs was evident as early as the first weeks of treatment and the nonresponders remained so during the successive weeks. We evaluated other indices of the lipid profile in the patients who responded to the therapy. The variations noted were not significant, but there was a slight increase in HDL cholesterol. An interesting aspect is the correlation between the high-flux membrane and a better lipid profile. Studies evaluating the relationship between hyperlipidemia and filters used in dialysis have been published but there are no definite explanations for the effect [12, 13]. Many factors are involved in hyperlipidemia (hormonal, metabolic, nutritional, dialysis buffer, recombinant human erythropoietin, reduction of lipoprotein lipase activity). These can overlap and act simultaneously and are, furthermore, linked to glucose and protein metabolism. The high-flux membranes are more biocompatible, hence the reduced activation of the complement system and monocytes, factors responsible for the production of cytokines which play a role in the alteration of metabolism in dialysis patients.

References

1 Grundy SM: Management of hyperlipidemia of kidney disease. Kidney Int 1990;37:847–853.

2 Deferrari G, Fiorini F: Disordini del metabolismo e della nutrizione nell'uremia cronica; in Cambi V (ed): Trattato Italiano Dialisi, vol 2. Milano, Wichtig, 1990, pp 5–7.

3 Sakhrani L: Lipoprotein abnormalities in uraemia; in Nissenson AR, Fine RN, Gentile DE (eds): Clinical Dialysis. Appleton Contury-Crofts, Norwalk, Conn, 1984, pp 439–450.

4 Attman P, Alaupovic P: Lipid abnormalities in chronic renal insufficiency. Kidney Int 1991; 39(suppl 31):16–23.

5 Ma KW, Greene EI, Raij L: Cardiovascular risk factors in chronic renal failure and hemodialysis populations. Am J Kidney Dis 1992;6:505–513.

6 Rostand SG, Brunzell JD, Cannon RO, Victor RG: Cardiovascular complications in renal failure. J Am Soc Nephrol 1991;6:1053–1062.

7 Kaysen GA: Hyperlipidemia of chronic renal failure. Blood Purif 1994;12:60–67.

8 Burr ML, Fehily AM, Gilbert JF, Rogers S, Holliday RM, Sweetnam PM, Elwood PC, Deadman NM: Effects of changes in fat, fish, and fibre intakes on death and myocardial reinfarction: Diet and Reinfarction Trial (DART). Lancet 1989;ii:757–761.

9 Pierides AM, Alvarez-Ude F, Kerr DNS: Clofibrate-induced muscle damage in patients with chronic renal failure. Lancet 1975;ii:1279–1282.

10 Pierce LR, Wysowski DK, Gross TP: Myopathy and rhabdomyolysis associated with lovastatin-gemfibrozil combination therapy. JAMA 1990;264: 71–75.

11 Jones PH, Patsch J, Gotto AM Jr: Biochimica della regolazione lipidica e valutazione delle dislipidemie; in Hurst JW (ed): Il Cuore, vol 2, Milano, McGraw-Hill, 1995, pp 1055–1070.

12 Docci D, Capponcini C, Mengozzi S, Baldrati L, Neri L, Feletti C: Effects of different dialysis membranes on lipid and lipoprotein serum profiles in hemodialysis patients. Nephron 1995;69:323–326.

13 Blankestijn PJ, Vos PF, Rabelink TJ, Van Rijn HJ, Jansen H, Koomans HA: High-flux dialysis membranes improve lipid profile in chronic hemodialysis patients. J Am Soc Nephrol 1995;5:1073–1078.

Manuscript Consultants

Mineral and Electrolyte Metabolism

The following consultants have provided valuable assistance in the reviewing of manuscripts during 1997:

Brautbar, Nachman
Coburn, Jack
Fichman, Marshal
Heidland, August
Kaysen, George
Keane, William
Kleeman, Charles
Kurtz, Ira
Langman, Craig
Malluche, Hartmut H.
Puachette, Jules, B.
Rude, Robert K.
Singer, Frederick R.
Suki, Wadi
Vaziri, Nosratola

Mineral and Electrolyte Metabolism

Author Index Vol. 23, 1997

KARGER
E-Mail karger@karger.ch
Fax + 41 61 306 12 34
http://www.karger.ch

Subject Index Vol. 23, 1997

Mineral and Electrolyte Metabolism

KARGER
E-Mail karger@karger.ch
Fax + 41 61 306 12 34
http://www.karger.ch

Vol. 23, 1997

Mineral and Electrolyte Metabolism

KARGER

Basel · Freiburg · Paris · London · New York · New Delhi · Bangkok · Singapore · Tokyo · Sydney

KARGER

S. Karger
Medical and Scientific Publishers
Basel · Freiburg · Paris · London
New York · New Delhi · Bangkok
Singapore · Tokyo · Sydney

Drug Dosage
The authors and the publisher have exerted every effort to ensure that drug selection and dosage set forth in this text are in accord with current recommendations and practice at the time of publication. However, in view of ongoing research, changes in government regulations, and the constant flow of information relating to drug therapy and drug reactions, the reader is urged to check the package insert for each drug for any change in indications and dosage and for added warnings and precautions. This is particularly important when the recommended agent is a new and/or infrequently employed drug.

Printed in Switzerland on acid-free paper by
Boehm-Hutter AG, Reinach, BL

Contents Vol. 23, 1997

Mineral and Electrolyte Metabolism

No. 1

Original Papers

Case Report

Letter to the Editor

No. 2

Original Papers

KARGER

E-Mail karger@karger.ch
Fax+ 41 61 306 12 34
http://www.karger.ch

The list of contents is available at:
http://www.karger.ch/journals/mem/memcont.htm

No. 3–6

Nutrition and Metabolism in Renal Disease
Proceeding of the 8th International Congress on Renal Nutrition and Metabolism, Naples, October 9–12, 1996
Editors: Joel D. Kopple (Torrance, Calif.); Natale G. De Santo; Giovambattista Capasso (Naples); Shaul G. Massry (Los Angeles, Calif.)